Handbook of Energy Data and Calculations

including Directory of Products and Services

Handbook of Energy Data and Calculations

including Directory of Products and Services

Peter D Osborn, BScEng (Hons), C Eng, FIEE *Engineering Consultant*

Butterworths
London Boston Durban Singapore Sydney Toronto Wellington

First published 1985

© **Butterworth & Co (Publishers) Ltd 1985**

British Library Cataloguing in Publication Data

Osborn, Peter D.
 Handbook of energy data and calculations.
 1. Buildings–Power supply 2. Buildings–
 Energy consumption
 I. Title
 696 TH6025

 ISBN 0-408-01327-3

Library of Congress Cataloging in Publication Data

Osborn, Peter D. (Peter Digby)
 Handbook of energy data and calculations including
directory of products and service

 Includes index.
 1. Energy conservation–Handbooks, manuals, etc.
2. Power (Mechanics)–Handbooks, manuals, etc.
3. Energy industries–Directories. I. Title.
TJ163.3.086 1985 621.042 84-21437
ISBN 0-408-01327-3

Photoset by Butterworths Litho Preparation Department
Printed and bound by Robert Hartnoll Ltd, Bodmin, Cornwall

Preface

The origins of this manual stem from the author's involvement in energy survey work, particularly in factories from the smallest to around 10 000 m² floor area. This work led to a wide range of problems and the consequent need to assemble data, information, calculation procedures and product specifications from every available source.

The largest single problem in assembling the data was the confusion resulting from the different systems of units and for this reason the text which follows allows no compromise with the SI system. All data and calculations procedures are presented on the assumption that the change to SI has already been made using, if need be, the conversion table at **A3**; how much simpler it then becomes!

This small contribution to energy education therefore has a limited but quite definite objective; it seeks to put into the hands of the 'energy executive' whether managing director, works engineer or energy manager, a comprehensive review of practical energy problems. It does require a scientific or engineering educational background and may well point the way for those seeking to broaden their grasp of the all-embracing nature of energy studies, but it avoids getting lost in advanced mathematics and seeks always the practical answer. Lord Kelvin's celebrated quotation is to the point: 'when you can measure what you are speaking about and express it in numbers you know something about it'.

Renewable energy sources (solar, wind and wave) combined heat and power and the role of nuclear energy in future energy resources are outside the scope of this manual. There is ample literature and there are many research and development projects on all of these. Concentration on conserving what we have and using it more effectively results from the author's belief that the World is only on the threshold of realizing the reductions in energy usage to levels which will become the norm for the 21st century.

Peter D. Osborn

Contents

Section B. Calculation and analysis procedures

Section C. Directory of products and services

Section D Bibliography and sources

Notes on presentation

(1) Throughout the manual the emphasis is on the practical approach; definitions are, as far as possible rigorous, but avoid scientific 'fussiness'; similarly values are often compromise figures based on scrutiny of information from many different sources.

(2) Although it is scientifically correct to use negative powers in definition of SI quantities, the author believes that, for the relatively straightforward examples and calculations, the use of the solidus is less likely to confuse, thus:

> for watts per square metre Kelvin
> we use W/m^2K
> rather than $W\,m^{-2}K^{-1}$

(3) In general, calculations are made based on three significant figures in the values used, but invariably the final line is rounded to a practical figure within the limits of the accuracy of the data provided. All calculations are based on the manual approach, but presuppose the use of a scientific calculator.

(4) The drudgery has been taken out of calculation procedures by the computer and the microprocessor with many programs to assist the specialist and much detailed published data derived by computer methods; examples are building heat loss calculations, the design of heat exchangers, balancing pipe systems and the layout of lighting fittings. The best known microprocessor application to energy work is the direct display of boiler losses and efficiency on instruments sensing flue gas content and temperature. Understanding the principles comes first, the aids to ease labour follow! Only by following the manual calculations can the engineer equip himself to challenge the figures which he senses may be wrong.

(5) The directory of manufacturers (**Section C**) inevitably reflects a constantly changing scene, but its objective is to enable the user to make some initial contacts whenever he has a specific supply problem to deal with. To those companies who are not included but who have a service to offer the author apologizes.

How to use this book to the best advantage

(1) Look closely at the definitions in **A1** and **A2** and make full use of the conversion table at **A3**. Search for data in the remainder of **Section A** using the table of contents and the index.

(2) The layout of **Section B** is intended for progressive reading but because of the inter-relationship of subjects a totally logical progression is not possible. The 34 sub-sections cover theoretical background, product technology, case histories and calculation procedures with the emphasis on these four aspects varying according to the subject matter.

(3) **B5**, **B15**, **B24** and **B32** are based on an abridged theoretical approach leading to practical conclusions; the reader may prefer to leave detailed study of these until the second or third reading.

(4) **B12**, **B13**, **B20**, **B21** and **B33** have their emphasis on product technology whilst the remaining sub-sections are concerned primarily with case histories and calculation procedures.

(5) The charts and diagrams in **Section B** are those directly relevant to the subject matter, those in **Section A** cover data for general use.

Section A

Data charts and tables

A1 SI units

The c.g.s. system followed by the formal adoption by the IEC (International Electrotechnical Commission) of MKS in 1935 has been the basis of scientific education for well over a century. The SI system was formally established in 1960. It has a more logical basis, can be more readily applied to practical problems and is simpler to use. The cost to industry of changing from the Imperial system to SI must not be underestimated: machine tools, gauges, drawing office records and all the apparatus of metrology cannot lightly be brushed aside, neither can partial metrication be accepted as a long term solution; nevertheless the eventual World change to SI is inevitable and right. Once the effort has been made to express all quantities in SI and to carry out all calculations in the new units, the whole process becomes simpler in every way.

Length, Mass, Time, Temperature and Electric Current are fundamental to measurement and if we add to these Luminous Intensity and Amount of Substance we have the seven basic units of SI; there are just three supplementary units, namely Plane Angle, Solid Angle and Radioactivity. From these ten the complete system of units is derived.

Energy engineering is now a discipline in its own right and necessitates the use of common units for a proper understanding of the many facets of energy supply and use. For calculation work there are four concepts which must be grasped and understood:

(1) All forms of energy can be expressed in the practical unit of kilojoule (kJ).
(2) Energy flow rate, whether electrical, mechanical, heating or cooling, can be expressed in the practical unit of kilowatt (kW).
(3) The unit of pressure is the pascal (Pa) or kilopascal (kPa), alternatively stated as newton or kilonewton per metre squared (N/m^2 or kN/m^2): $100\,kPa = 1$ bar and 1 bar is very close in value to 1 atmosphere.
(4) The kelvin (K) is the measure of temperature and, as a temperature interval, is identical to 1 Centigrade or Celsius degree; however, the zero of the kelvin scale is absolute zero (approximately $-273\,°C$), so that $20\,°C = 293\,K$.

In applying SI it is wise to use the second as the unit of time whenever possible and to remember to divide by 60 as, for example, when low volume flows are measured in m^3/min. Rates per hour are best avoided! Nevertheless they frequently occur, and in this book the symbol 'hr' is used.

The **A1** Tables list all the basic and supplementary units, together with the derived units which are relevant to energy work in general and to the contents of this manual in particular. **A2.1** and **A2.2** give the basic definitions and international values which are behind the whole system and are probably only of academic interest to the engineer. **A2.3** to **A2.8** inclusive give the derivation and relationships of all the units used in **Section B** of

1

the manual and the reader should gain from referring to these relationships when working with **Section B**, where the necessary cross-references are shown.

Table A3 is of special interest as it is designed to enable a painless transfer to be made into SI units when confronted with the many traditional units which are still quoted.

The advantages of switching to SI units for energy calculation work cannot be too strongly emphasized. It does require a positive mental effort, but this is amply justified by the simplification which then follows.

A1.1 Basic units

Quantity	Dimensional definition	Title
Length (ℓ)	m	metre
Mass (m)	kg	kilogram
Time (t)	s	second
Electric current (I)	A	ampere
Temperature (T, θ)	K	kelvin
Luminous intensity (I)	cd	candela
Amount of substance (n)	mol	mol

A1.2 Supplementary units

Quantity	Dimensional definition	Title
Plane angle (α)	rad	radian
Solid angle (Ω)	sr	steradian
Radioactivity (A)	Ci	curie
Frequency (ν)	1/s $=$ Hz	hertz

A1.3 Prefixes

10^{18}	exa	E	10^{-18}	atto	a
10^{15}	peta	P	10^{-15}	femto	f
10^{12}	tera	T	10^{-12}	pico	p
10^{9}	giga	G	10^{-9}	nano	n
10^{6}	mega	M	10^{-6}	micro	μ
10^{3}	kilo	k	10^{-3}	milli	m

A1.4 Derived units

Quantity	Definition	Title
Area (A)	m^2	square metre
Volume (V)	m^3	cubic metre
Velocity (u)	m/s	metre per second
Angular velocity (ω)	rad/s	radian per second
Acceleration (a)	m/s^2	metre per second squared
Density (ρ)	kg/m^3	kilogram per metre cubed
Momentum (p)	kg m/s	kilogram metre per second = newton second
Rotational frequency	1/s $=$ Hz	reciprocal second (hertz)
Force (F), weight (W)	N	newton ($kg\,m/s^2$)
Work (A), heat (E), energy (H)	J	joule (N m $= kg\,m^2/s^2$)
Power (P)	W	watt ($kg\,m^2/s^3$)

continued

A1.4 Derived units (Continued)

Quantity	Definition	Title
Temperature interval (θ)	K or °C	kelvin or degree Celsius
Heat quantity, (Q, H), latent heat (L), enthalpy (H)	J	joule
Heat flow rate (ϕ)	W	watt
Density of heat flow (q), heat flux (σ)	W/m^2	watt per metre squared
Thermal conductivity (κ)	W/m K	watt per metre kelvin
Thermal transmittance (α)	W/m^2 K	watt per metre squared kelvin
Heat capacity, entropy (S)	J/K	joule per kelvin
Specific heat capacity, specific entropy (s)	J/kg K	joule per kilogram kelvin
Specific energy (v)	J/kg	joule per kilogram
Luminance (L)	cd/m^2	candela per metre squared
Luminous flux (ϕ)	lm	lumen (cd $\times$ sr)
Illuminance (E)	lx	lux (lm/m^2)
Pressure (stress) (ρ)	N/m^2	pascal
Viscosity (dynamic) (μ)	N s/m^2	newton second per metre squared
Viscosity (kinematic) (ψ)	m^2/s	metre squared per second
Mass flux (ϕ)	kg/s	kilogram per second
Surface tension (γ)	N/m	Newton per metre
Specific volume (v)	m^3/kg	metre cube per kilogram
Quantity of electricity, electric charge (Q)	C	coulomb (ampere second)
Electromotive force (E)	V	volt (watt per ampere, joule per coulomb)
Capacitance (C)	F	farad
Current density (J)	A/m^2	ampere per metre squared
Resistance (R)	Ω	ohm (volt/ampere)
Resistivity (ρ)	Ω m	ohm metre
Conductance (G)	S	siemens (reciprocal ohm)
Conductivity (κ)	S/m	siemens per metre
Inductance (L, M)	H	henry (volt second per ampere)
Active power (P)	W	watt
Reactive power, apparent power (Q)	VA	volt ampere
Permittivity (ε)	F/m	farad per metre
Magnetic field strength (H)	A/m	ampere per metre
Permeability (μ)	H/m	henry per metre
Magnetic flux (Φ)	Wb	weber (volt second per ampere)
Magnetic flux density (B)	T	tesla (weber per metre squared)

A2 Definitions, relationships and international values of units and quantities

A2.1 Basic definitions

The concept of INERTIA states that all matter unless influenced by external forces will maintain constant velocity; this is why a satellite when free from the earth's gravity will travel into outer space until it comes under the influence of the gravity of other bodies in space or until an external force acts on it (as will occur when a retrorocket is fired).

MASS in the SI system is measured in kg and is defined as the quantity of matter measured by weighing against standards.

The unit of FORCE is the NEWTON which, when applied to a mass of one kg will cause it to accelerate by one m/s^2; force thus has the units kg m/s^2.

WEIGHT is also measured in NEWTONS; the Earth's gravity will exert a force of 9.8 newton on a body of mass one kg (g, the acceleration due to gravity at the Earth's surface, is about 9.8 m/s^2).

The unit of WORK, HEAT or ENERGY is the JOULE which is the work done when a force of one NEWTON acts for a distance of one METRE; its units are thus N m or kg m^2/s^2.

KINETIC ENERGY is energy by virtue of motion and if a stationary body is accelerated uniformly to a velocity V its kinetic energy will be

$$\tfrac{1}{2} \times \text{mass} \times V^2$$

MATTER IN MOTION exhibits similar characteristics to RADIATION.

A2.2 *International values*

GRAVITY ACCELERATION (g) = 9.80665 m/s^2

LITRE   Volume of one kg of pure water at 4°C (temperature of maximum density).

STANDARD ATMOSPHERE   Pressure exerted by exactly 760 mm of mercury at 0°C and at standard gravity (9.807 m/s^2). Alternatively 101 325 N/m^2. (Density of mercury at 0°C = 13 590 kg/m^3.)

TEMPERATURE   Triple point of water 273.16 K = 0.01°C. Ice point 0°C. Steam point 100°C. Boiling oxygen −182.96°C. Freezing mercury −38.87°C.

MOLAR GAS VOLUME   22.4136 litre/mol (or 22.4136 m^3/kmol). Volume occupied by 32 g (or 32 kg) of oxygen at 0°C and standard atmosphere.

AVOGADRO NUMBER   6.02 × 10^{23} molecules per mole.

$$\text{GAS CONSTANT } (R) = \frac{1 \text{ atmosphere} \times 22.4136 \text{ litres}}{1 \text{ mole} \times 273.15 \text{ K}}$$

$$= 0.082\,056 \text{ litre atmosphere/K mol}$$

Alternatively

$$\frac{101\,325 \times 0.082\,056}{1000}$$

$$= 8.314 \text{ N m/K mol (J/K mol)}$$

A2.3 *Derivation and relationship of units and quantities*

WATT   The unit of power defined by a rate of working one joule per second. The units are kg m^2/s^3. HEAT FLOW RATE and ENERGY FLOW may also be expressed in watts.

PASCAL   The unit of pressure is measured in N/m^2 and thus has units kg/m s^2.

DENSITY   The mass per unit volume at stipulated temperature normally stated in kg/m^3.

SPECIFIC VOLUME   The volume per unit mass stated in m^3/kg.

SPECIFIC LATENT HEAT   Heat required to effect a phase change in a substance at constant temperature, usually expressed in kilojoule per kilogram. The latent heat of vapourization decreases with increase in presssure up to a critical point when it becomes zero (374°C, 221 bar for saturated steam).

SPECIFIC HEAT CAPACITY   This is normally expressed in kilojoule per kilogram per kelvin for gases. Two values are defined, i.e. at constant pressure (C_p) and at constant volume (C_v). For gases the ratio of these $C_p/C_v = \gamma$ varies from 1.66 for monatomic gases to just over 1.00 for polyatomic gases. The specific heat capacity of liquids and solids is essentially independent of pressure but varies with temperature.

MOLAR ENTHALPY   Heat content of fuels expressed in kJ/mol (= MJ/kmol).

$$\text{GROSS CALORIFIC VALUE (MJ/kg)} = \frac{\text{Gross C.V. (MJ/m}^3)}{\text{Ideal gas density (kg/m}^3)}$$

$$= \frac{\text{molar enthalpy (MJ/kmol)}}{\text{Relative molecular mass}}$$

THERMAL CONDUCTIVITY   The ability of a material to transmit heat expressed in watt per square metre of surface area per kelvin of temperature gradient for a unit thickness of one metre, i.e. $(W/m^2 K) \times m = W/m\,K$. THERMAL RESISTIVITY is reciprocal = $m\,K/W$.

THERMAL RESISTANCE $= m^2 K/W$

THERMAL TRANSMITTANCE $= W/m^2 K$.

VISCOSITY (or DYNAMIC VISCOSITY) is defined as the shear stress required to move one layer of fluid past another layer of fluid unit distance away with unit velocity; it is expressed in

$$N s/m^2 = (kg\,m/s^2 \times (s/m^2) = kg/m\,s$$

KINEMATIC VISCOSITY is the ratio of dynamic viscosity to density and its units are therefore $(kg/m\,s) \times (m^3/kg) = m^2/s$.

A2.4 The gas laws

BOYLE'S   The volume occupied by a given mass of any gas kept at constant temperature is inversely proportional to the pressure exerted upon it.

CHARLES'   The volume of a given mass of any gas kept at constant pressure is directly proportional to the absolute temperature.

THE GAS CONSTANT   Boyle's law and Charles' law can be combined into a general ideal gas equation $R = PV/T$ where R may be called the gas constant. The gas laws are based on the 'ideal gas' concept; all real gases deviate from ideal behaviour depending on the balance of molecular forces with increasing pressure.

$$\text{IDEAL GAS DENSITY} = \frac{\text{Relative molecular mass}}{22.4136 \text{ (at } 0\,°C) \text{ or } 23.64 \text{ (at } 15\,°C)}$$

Density (ρ) of *any* gas at any condition of temperature and pressure (ignoring deviation from ideal gas behaviour)

$$= \frac{P \times M}{8314 \times T}$$

where

P is absolute pressure
T is absolute temperature
M is relative molecular mass of the gas

A2.5 Definitions relevant to psychometry

ABSOLUTE HUMIDITY (ω) is the mass ratio water to dry air in an atmosphere carrying water vapour and is defined

$$\omega = \frac{18}{29} \times \frac{p}{P - p}$$

where

p is the partial pressure of water vapour
P is the total pressure of mixture
18 and 29 are taken as the relative molecular masses of water vapour and air
   respectively

RELATIVE HUMIDITY is expressed as a percentage based on the mass of water vapour present in the mixture to the mass required to cause saturation at the same temperature; it may be defined as

$$\text{R.H.} = \frac{p\,(\text{vap})}{P\,(\text{liq})} \times 100$$

where

  p (vap) is the partial pressure of water vapour in the mixture
  P (liq) is the vapour pressure of liquid phase water at the same temperature

SPECIFIC HUMID HEAT CAPACITY (S) of an air–water vapour mixture is the amount of heat (kJ/kg K) required to raise one kg of the mixture by 1 °C and is defined

$$S = (C_pA) + [\omega \times (C_pWV)]$$

where

  C_pA is specific heat capacity of air at constant pressure
  ω is absolute humidity
  C_pWV is specific heat capacity of water vapour

TOTAL SPECIFIC ENTHALPY is the sum of the enthalpies (kJ/kg K) of the dry air and its associated water vapour based on the reference temperature of 0 °C and is in four components:

(1) The air component = specific heat capacity (C_pA) × temperature (°C).
(2) The water component to dew point = specific heat capacity of water (C_pW) × dewpoint temperature (°C).
(3) The latent heat component of vapour at dew point.
(4) The superheat component = specific heat capacity of water vapour (C_pWV) × temperature rise above dewpoint (°C).

HUMID VOLUME (V_G) is the volume of dry air plus its associated water vapour and is measured in terms of kilomole and referred to the absolute temperature, i.e.

$$V_G = \left(\frac{1}{29} + \frac{\omega}{18}\right) \times \frac{22.41 \times \theta}{273}$$

where

  θ is absolute temperature

A2.6 Definitions relevant to heat transfer

RATE OF HEAT TRANSFER OR HEAT FLOW RATE (Q) is defined as overall heat transfer coefficient or thermal transmittance (W/m²K) × area of surface (m²) × temperature difference (K) and is measured in watt. Alternatively as mass flow rate (kg/s) × specific heat capacity (J/kg K) × temperature change (K).

HEAT FLUX = Heat transfer rate per unit surface area (W/m²).

HEAT CONTENT OF A BODY = Mass (kg) × specific heat capacity (J/kg K) × temperature rise above reference level (normally 0 °C).

OVERALL HEAT TRANSFER COEFFICIENT is assessed by means of the reciprocals of the elements of resistance to heat flow and with particular relevance to convective heat flow requires values of inside and outside film coefficients of the fluids flowing on either side of a separating pipe or plate, such that

$$\frac{1}{\text{overall heat transfer coefficient}} = \frac{1}{\text{inside film coefficient}} + \frac{\text{thickness of plate}}{\text{conductivity of plate}} + \frac{1}{\text{outside film coefficient}}$$

A PERFECT BLACK BODY will absorb all incident radiation but will not reflect or transmit; its absorptivity is thus unity. The maximum practical absorptivity is approximately 0.97.

EMISSIVITY is the rate at which a radiating body will lose heat to its surroundings and has a maximum value of unity. The emissivity of a surface is the ratio of the emissive power of the surface to the emissive power of a perfect black surface.

ABSORPTIVITY, REFLECTIVITY AND TRANSMISSIVITY are the rates at which a body receiving radiated energy will absorb, reflect or transmit that energy and the sum of all three is unity.

A2.7 Electrical unit relationships

The AMPERE is that electric current which, when flowing in two parallel conductors of infinite length and negligible cross section 1 m apart in a vacuum, produces a force between them of 2×10^{-7} N per metre of their length.

The COLOUMB is the charge carried by 1 A in 1 s.

The VOLT is the potential difference that exists between two points if it requires 1 J to move 1 C from one point to the other.

    joule = volt × coulomb
    watt = volt × ampere

The OHM is the resistance between two points if 1 V drives a current of 1 A between the points

$$\text{ohm} = \frac{\text{volt}}{\text{ampere}}$$

There is an older system called the ELECTROSTATIC SYSTEM. It is based on the ELECTROSTATIC UNIT OF CHARGE (esu) which is the charge which, when placed 1 cm from a similar charge in a vacuum will exert on it a force of 1 dyne (1 dyne = 10^{-5} N).

    1 coloumb = 2.998×10^9 esu

The FARAD is the unit of capacitance.

$$\text{farad} = \frac{\text{coloumb}}{\text{volt}}$$

The HENRY is the unit of self inductance; it is the self inductance of a circuit in which an emf of 1 V is produced when the current changes at a rate of one ampere per second

$$\text{henry} = \frac{\text{volt}}{\text{ampere per second}}$$

The henry is also used for mutual inductance.

ACTIVE, REACTIVE AND APPARENT POWER are best illustrated by vectors. E is the alternating RMS voltage, I is the alternating RMS current in ampere, ϕ is the angle of lag, and $\cos\phi$ is the POWER FACTOR. Active power is expressed in watt, and reactive power and apparent power are expressed in volt-ampere.

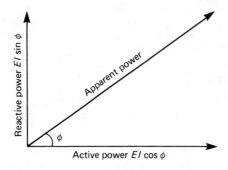

Figure A2.7 *Vector diagram for AC power*

INDUCTIVE REACTANCE $= \omega L$

where

ω is angular frequency of AC supply (rad/s)
L is inductance in henry

CAPACITATIVE REACTANCE $= \dfrac{1}{\text{angular frequency} \times \text{capacitance (farad)}}$

RESONANCE occurs when inductive and capacitative reactance are brought into balance by varying circuit frequency.

IMPEDANCE of a circuit with resistance, inductance and/or capacitance = volt/ampere, and is measured in ohm.

RESISTIVITY (ρ) is the resistance offered to a current by a one metre length of material of cross section area one metre squared and is measured in ohm/m/m^2, i.e. ohm m. It can alternatively be defined as

$$\frac{\text{electric field strength (V/m)}}{\text{current density (A/m}^2\text{)}}$$

MAGNETIC PERMEABILITY (μ) is a measure of the increase in the self inductance of a coil or the mutual inductance of a circuit when a magnetic material is substituted for air within the coils of the circuit. Above a temperature referred to as the CURIE point of the material permeability of magnetic materials becomes unity and magnetic properties vanish. It may also be defined as the ratio of magnetic flux density (B) in a material to the magnetic field strength (H) in which it is placed. B is measured in tesla and H is measured in ampere per metre.

$\mu = \mu_0 \times \mu_r$

where

μ_0 is permeability of a vacuum (1.25664×10^{-6} H/m)
μ_r is relative permeability of a medium

CAPACITORS were formerly called condensers. When a material is placed between the plates of a capacitor, the capacitance of the capacitor changes. The relative permittivity of the material (formerly DIELECTRIC CONSTANT and SPECIFIC INDUCTIVE CAPACITY), ε_r, is given by

$$\varepsilon_r = \frac{\text{capacitance with medium}}{\text{capacitance without medium}}$$

Also

$$\varepsilon = \varepsilon_0 \times \varepsilon_r$$

where

ε_0 is permittivity of a vacuum $(8.85 \times 10^{-12}\,\text{F/m})$
ε_r is relative permittivity of a medium

HIGH FREQUENCY RESISTANCE results from the 'skin' effect in a conductor and increases with frequency according to the approximate relationship

$$\frac{R_f}{R_0} \simeq \sqrt{\frac{\mu f}{\rho}}$$

where

R_f is resistance at frequency f
ρ is resistivity
μ is magnetic permeability
$R_0 = \rho/d^2$
d is diameter of conductor

It can be shown that, if ρ is written in ohm cm $\times 10^9$ or ohm m $\times 10^{11}$, with μ_r as *relative* permeability, this leads to a penetration depth

$$p\ (\text{cm}) = \frac{1}{2\pi} \sqrt{\frac{\rho}{\mu_r f}}$$

(*See* **Section B32.2.2**)

A2.8 Units and definitions used in lighting

The basic SI unit of luminous intensity is the candela.

CANDELA is the illuminating power of a light source in a given direction. The unit of *brightness of a light source (luminance)* is the APOSTILB and is measured in lumens emitted per square metre (ℓm/m^2). The *brightness of an illuminated surface* (also referred to as *luminance*) is measured in candela per square metre (cd/m^2). Note particularly that *illuminance* is the measure of light falling on a surface whereas *luminance* is the light reflected from it or in some cases emitted by it. The unit of illuminance is the lux equal to one lumen/m^2.

LUMINOSITY is the apparent brightness of a surface which to the observer varies with the general level of lighting, i.e. a source which appears bright at night may be barely noticeable in full daylight.

EFFICACY or luminous efficiency of a light source (lamp) is measured in lumens per watt (lm/W).

LUMINOUS FLUX (lumen)   Quantity derived by evaluating the radiation emitted in accordance with the spectral sensitivity of the standard eye.

LUMEN   A small source of uniform intensity of one candela emits a total of 4π lumens in all directions or one lumen within unit solid angle (one steradian).

LUMINOUS INTENSITY (candela) is the luminous flux emitted in a very narrow cone containing the given direction divided by the solid angle of the cone.

A3 Conversions into SI units

Length

1 inch (in)	= 25.4 mm
1 foot (ft)	= 0.3048 m
1 yard	= 0.9144 m
1 mile	= 1.609 km

Area

1 square inch (in^2)	= 645 mm^2
1 square foot (ft^2)	= 0.0929 m^2
1 square yard	= 0.836 m^2
1 hectare	= 10^4 m^2
1 acre	= 4047 m^2

Volume

1 cubic inch (in^3)	= 16 390 mm^3
1 cubic foot (ft^3)	= 0.0283 m^3
1 imperial gallon	= 4.546 l
	= 0.004 546 m^3
1 US gallon	= 0.005 458 m^3

Mass

1 pound (lb)	= 0.4536 kg
1 ton	= 1016 kg
1 ton (US)	= 907 kg
1 tonne	= 1000 kg
1 grain	= 0.065 g

Force

1 lb ft	= 4.448 N
1 ton ft	= 9.964 kN
1 ton (US) ft	= 8.895 kN

Pressure

1 atmosphere (atm)	= 101.3 kPa
1 bar	= 100 kPa
1 inch mercury	= 3.386 kPa
1 inch water gauge	= 249.1 Pa
1 mm water gauge	= 9.807 Pa
1 kg/mm^2	= 9807 Pa
1 lb/in^2	= 6.895 kPa
1 ton/in^2	= 15.440 MPa
1 ton (US)/in^2	= 13.780 MPa

Velocity

100 miles per hour	= 161 km/hr
	= 44.71 m/s
100 ft/s	= 30.48 m/s

Density

1 lb/ft^3	= 16.02 kg/m^3
1 lb/in^3	= 27 680 kg/m^3

Specific volume

1 ft^3/lb	= 0.0624 m^3/kg

Temperature

1 °F	= 0.5556 °C
	= 0.5556 K

32 °F equates with 0 °C

Power

1 horse power (H.P.)	= 0.7457 kW
1 BThU/s	= 1.055 kW
100 000 BThU/hr	= 29.31 kW
1000 lb steam/hr	= 284 kW
1 centigrade thermal unit per second	= 1.899 kW
1 ton (cooling)	= 3.52 kW
= 37.97 kg ice melted per hour	

Energy, work, heat

1 British thermal unit (BThU)	= 1055 J
1 kg cal = 1 kcal	= 4187 J
1 therm	= 100 000 BThU
	= 29.3 kW hr
	= 105.5 MJ
1 kW hr	= 3.6 MJ
1000 lb steam evaporated from and at 100 °C	= 1.0237 GJ
1 H.P. hr	= 2.685 MJ
10^6 tons coal equivalent (mtce)	= 26.4 PJ
10^6 tons oil equivalent (mtoe)	= 44 PJ

Heat content

1 BThU/lb	= 2.326 kJ/kg
1 BThU/ft^3	= 37.26 kJ/m^3
1 BThU/gallon	= 232.1 kJ/m^3
1 BThU/gallon (US)	= 278.7 kJ/m^3

Thermal conductivity

1 BThU in/ft^2 hr °F	= 0.1445 W/m K

Specific heat capacity

1 BThU/lb °F	= 4.87 kJ/kg K

Heat transfer coefficient

1 BThU/ft^2 hr °F	= 5.678 W m^2 K

Fluid movement

1000 ft^3/min	= 0.472 m^3/s
1000 gallons/min	= 4.546 m^3/min
	= 4546 l/min
	= 0.0758 m^3/s
1000 gall (US)/min	= 0.0910 m^3/s
1 ft^3/hr	= 0.028 m^3/hr
	= 7.778 × 10^{-6} m^3/s

Kinematic viscosity

1 stokes	= 1 cm^2/s
	= 0.0001 m^2/s
1 ft^2/s	= 0.093 m^2/s
1 ft^2/hr	= 25.8 × 10^{-6} m^2/s

Dynamic voscosity

1 poise	= 0.1 N s/m^2
	= 0.1 Pa s
1 lb/ft s	= 1.49 Pa s
1 lb/ft hr	= 0.000 413 Pa s
Redwood 100–5000	≈ 24–1200 mm^2/s

A4 Everyday empirical values

For fuller information, refer to the references given

Heat transfer coefficients (**A20, A21**)
(*'U' values in* W/m^2 K)

Walls	Single sheet asbestos	7
	Solid brick	2–3
	Unfilled cavity	1.5
	Double skin asbestos insulated	1.0
	Insulated cavity	0.6
Roofs	Single sheet asbestos	8
	Single glazed light	7
	Double skin asbestos	3
	Ditto, insulated	1
Glazing	Single	6
	Double	3–4
	Triple	2

Ventilation loss (**B23**)

Loss (W/m^3 K) $\simeq$ one third multiplied by the number of air changes per hour

Gross calorific values (**A5, A6, A7, A8**)

Oil	Class C2	36.7 MJ/l
	Class D	38.0 MJ/l
	Class E	40.6 MJ/l
	Class F	40.8 MJ/l
	Class G	41.2 MJ/l
Gas	Propane	50.4 GJ/tonne
	Butane	49.6 GJ/tonne
	N. Sea	38.6 MJ/m^3
Coal		29–35 GJ/tonne
Wood		10–15 GJ/tonne

Heating season (**B5**)

Normal 12–18 hr for 5–6 days for 30–34 weeks	= 1800–3600 hr/year
Maximum 24 hr for 7 days for 38 weeks	= 6400 hr/year

Winter temperatures (**B10, B14**)

Average UK external	6–8 °C
Normal basis for plant rating	0 °C
Statutory internal	19 °C

Efficacy of light sources (**A39, B37**)

	lm/W
Son	90–100
Fluorescent	78–80
Mercury blended	40–50
Tungsten filament	13–15

Metabolic heat (**B12**)
(Body surface area = 0.6–1.5 m^2)

	W/m^2 of body surface area
Sleeping	40
Relaxed	70
Housework	100–180
Heavy manual	up to 350

Outputs of steel radiators (**A37**)

	W/nominal m^2
Single	1270–1460
Double	2030–2300
Single with fins	1700–1750
Double with fins	3100–3260

Specific heat capacities (**A15, A16, A18**)

	kJ/kg
Water (0 °C–100 °C)	4.2
Dry air (0 °C–100 °C)	1.0
Average flue gases (350 °C–750 °C)	1.1–1.2

The *latent heat of evaporation of water* at atmospheric pressure is 2.25 MJ/kg (**A32**)

The *wavelength of electromagnetic radiation* in metres can be found from (**A40, B32.2**)

$$\text{Wavelength} = \frac{300\ 000\ 000}{\text{Frequency/Hz}}$$

A5 Typical composition and properties of some gaseous fuels

	% by volume[a]							Density[b] (kg/m³)	Gross calorific value[c]		C:H ratio[d]	Dewpoint of waste gas (°C)
	N_2	CO_2	CH_4	C_2H_6	C_3H_8	C_4H_{10}	Other		(MJ/m³)	(MJ/kg)		
Substitute natural gas	–	–	95.2	–	2.0	–	2.8	0.762	37.9	49.7	3.05	59
North Sea	1.5	–	94.4	3.0	0.5	0.2	0.4	0.762	38.6	50.7	3.07	59
Lacq	0.3	–	95.5	3.4	0.6	–	0.2	0.749	39.0	52.1	3.07	59
Slochteren	14.0	0.8	81.9	2.7	0.4	0.1	0.1	0.825	33.3	40.4	3.09	59
Gröningen	14.0	–	81.8	2.7	0.4	0.1	1.0	0.823	33.3	40.5	3.07	59
Arzew	0.3	–	86.5	9.4	2.6	1.1	0.1	0.831	42.8	51.5	3.25	57
Ekofisk	0.3	1.9	84.7	8.7	3.1	–	1.1	0.844	42.0	50.0	3.27	57

[a] Composition figures are typical.
[b] Densities are calculated from ideal gas densities of constituents at 0 °C and 101.3 kPa.
[c] Gross calorific values are averages from several sources.
[d] C:H ratios are calculated from the composition.

A6 Typical properties of liquefied petroleum gas[a].

Values stated are at 15 °C and 101.3 kPa and dry

	Property	Units	Commercial	
			Butane[b]	Propane[c]
A	Freezing point	°C	−140	−186
B	Boiling point	°C	−2	−42
C	'Ideal gas' density	kg/m³	2.45	1.84
D	Density of liquid	kg/m³	575	512
E	Calorific value (vaporized)	MJ/m³	122	95
F	Latent heat of vaporization at b.p.	MJ/kg	0.39	0.43
G	Specific heat capacity (gas)	kJ/kg K	1.61	1.55
H	Specific heat capacity (liquid)	kJ/kg K	2.34	2.43
J	Volume of gas per mass of liquid (K/D)	m³/kg	0.43	0.54
K	Volume of gas per volume of liquid (D/C)	m³/m³	233	274
L	Litres per tonne (liquid)	l/tonne	1743	1957
M	m³ per tonne (gas) (K × L/1000)	m³/tonne	405	544
N	Combustion requirement (oxygen)	m³/m³	6.25	4.80
P	Combustion requirement (air)	m³/m³	30.0	23.0
Q	Ignition temperature	°C	480–540	480–540
R	Maximum flame temperature	°C	1996	1980
S	% gas in air for max flame temp.	%	3.5	4.4
T	Limits of flammability: % gas in gas–air mixture			
	lower	%	1.9	2.0
	higher	%	8.5	11.0
U	Gross calorific value (E × M/1000)	GJ/tonne (MJ/kg)	49.5	~~512.7~~ 51.7

[a] LPG composition varies according to source, and there are consequently small variations in properties.
[b] Commercial butane consists of n-butane and isobutane (2-methylpropane), both C_4H_{10}, together with small amounts of propane, C_3H_8.
[c] Commercial propane consists of propane and propylene (propene), C_3H_6.

Table A6 based on information provided by Calor Gas Ltd.

A7 Typical properties of liquid fuels[a].

Values stated are at 15 °C and 101.3 kPa

	Oils				
	Kerosene Class C2	Gas oil Class D	Light oil Class E	Medium oil class F	Heavy oil Class G
Gross calorific value (MJ/kg)	46.5	45.2	43.7	43.4	42.5
Gross calorific value (MJ/l)	36.7	38.0	40.6	40.8	41.2
Viscosity (centistoke)[b]	2 (40 °C)	4 (40 °C)	120 (25 °C) 14 (80 °C)	230 (40 °C) 35 (80 °C)	240 (60 °C) 85 (80 °C)
C:H ratio	6.3	6.4	6.6	7.4	7.7
Water content (max.) (%)	–	0.05	0.5	0.75	1.0
Sulphur content (max.) (%)	0.06	<0.75	3.2	3.5	3.5
Ash content (max.) (%)	–	0.01	0.05	0.12	0.20
Density at 15 °C (kg/l)	0.79	0.84	0.93	0.95	0.97
Storage temperature (°C)$_u$	A	A	10	25	40
Handling temperature (°C)	A	A	10	30	50

	Coal tar fuels					
	CTF50	CTF100	CTF200	CTF250	CTF300	CTF400
Gross calorific value (ave) (MJ/kg)	39.5	39.5	38.6	38.3	37.9	37.2
Density at 15 °C (ave) (kg/l)	1.0	1.0	1.15	1.15	1.20	1.25
Handling temperature (°C)†	A	30	30	60	85	130

[a] British Standard 2869 (1983) covers detailed properties of burner fuels from C2 to H; it also gives viscosity – temperature relationships and the relationships of Gross calorific value/nett calorific value to density, sulphur content, and moisture and ash content.

[b] Viscosity being temperature-dependent, values shown are approximate for the range of temperatures indicated.

[c] A = normal ambient temperature.

A8 Properties of solid fuels

Calorific values (dry, ash-free for coals)

Fuel	GCV (MJ/kg)	Fuel	GCV (MJ/kg)
Dry hardwood	18–19	Peat	22
Coke breeze	26	Lignite	26
Large coke	29	Bituminous coals	33–35
Low volatile coals	33–37	Anthracites	35–37

Analyses of coals (information by courtesy of National Coal Board)

Rank code no.	Gray–King coke type	Basis[a]	Volatile matter (%)	GCV (MJ/kg)	Ultimate analysis (%)						C:H ratio
					Moisture	Ash	C	H	N	S	Oxygen, others, errors	
102	A	a	8.5	31.2	2.2	8.7	81.1	3.3	1.6	0.9	2.2	24.6
		b	9.5	35.0								
		c	8.5	35.4			92.1	3.7	1.9	0.8	1.6	24.9
202	C	a	14.1	33.5	0.8	6.6	84.2	4.6	1.2	0.9	2.3	18.3
		b	14.8	36.2								
		c	14.3	36.6			92.0	4.2	1.3	0.8	1.7	21.9
202	D	a	13.4	35.4	1.1	2.1	89.3	4.3	1.4	0.7	1.1	20.8
		b	13.8	36.6								
		c	13.6	36.8			92.6	4.4	1.4	0.7	0.9	21.0
202	A	a	14.4	33.6	1.7	4.7	84.6	4.1	1.7	0.6	2.6	20.6
		b	15.4	35.9								
		c	15.0	36.1			90.9	4.3	1.9	0.7	2.2	21.6
204	G3	a	18.8	34.9	0.5	3.3	87.3	4.5	1.2	0.7	2.5	19.4
		b	19.5	36.3								
		c	19.3	36.5			91.2	4.6	1.3	0.7	2.2	19.8
204	G3	a	17.0	34.8	0.8	3.6	86.8	4.4	1.3	0.7	2.4	19.7
		b	17.6	36.5								
		c	17.4	36.7			91.3	4.5	1.4	0.7	2.1	20.3
206	B	a	12.5	33.0	1.4	6.4	84.0	3.6	1.9	1.2	1.5	23.3
		b	13.6	35.8								
		c	12.7	36.1			92.1	3.9	2.1	0.8	1.1	23.6
206	A	a	15.7	34.5	1.8	1.8	86.2	4.3	2.0	0.6	3.3	20.0
		b	16.2	35.8								
		c	16.2	35.8			89.7	4.4	2.1	0.7	3.1	20.4
601	G2	a	32.0	31.2	4.8	5.4	76.0	4.8	1.8	1.4	5.8	15.8
		b	35.6	34.7								
		c	34.0	35.4			86.0	5.4	2.0	0.8	5.4	15.9
602	G3	a	35.5	31.6	4.6	4.4	75.4	4.9	1.8	2.5	6.4	15.4
		b	39.0	34.7								
		c	38.2	35.2			84.2	5.4	2.1	1.1	6.9	15.6
802	D	a	33.8	30.9	7.9	1.5	75.4	4.8	1.6	0.6	8.2	15.7
		b	37.1	34.1								
		c	37.0	34.1			83.4	5.3	1.8	0.7	8.6	15.7
802	C-D	a	37.1	29.4	9.9	2.4	72.0	4.6	1.2	0.5	9.4	15.6
		b	42.2	33.4								
		c	42.2	33.5			82.2	5.2	1.4	0.8	10.4	15.8
802/902	B-C	a	35.1	28.1	10.1	3.2	69.6	4.4	1.2	1.3	10.2	15.8
		b	40.4	32.5								
		c	40.2	32.7			81.0	5.1	1.3	0.9	11.6	15.9
802	C-D	a	35.2	29.1	10.1	3.2	71.0	4.5	1.4	0.9	8.9	15.8
		b	40.5	33.6								
		c	40.1	33.8			82.4	5.2	1.6	0.8	9.8	15.8
802	C	a	32.8	28.4	11.1	3.4	70.3	4.4	1.2	0.7	8.9	16.0
		b	38.3	33.3								
		c	37.3	33.4			82.6	5.2	1.4	0.8	9.4	15.9
802	C	a	33.1	28.3	11.2	3.7	69.2	4.5	1.6	0.6	9.2	15.4
		b	38.8	33.2								
		c	38.3	33.4			81.8	5.2	1.9	0.8	10.1	15.7

[a] Air-dried basis; [b] dry, ash free; [c] dry, mineral matter free.

15

A9 Coal classification system used by National Coal Board

(1) Main class is by volatile matter content: anthracites, 100; low volatile steam coals, 200; medium volatile coals, 300; high volatile coals, 400–900.

(2) Class divisions (101, 102 etc) reflect Gray–King coke types: A, B, non-caking to G9, G10, very strongly caking.

(3) Full lines on chart indicate boundaries of classes; broken lines indicate limits found in practice.

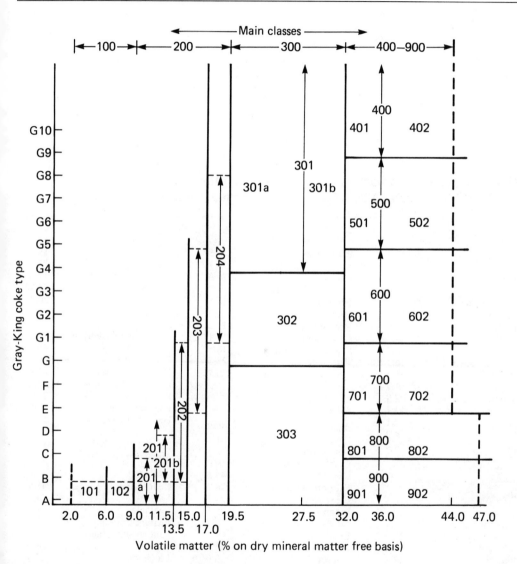

Figure A9 *Coal classification system used by National Coal Board*

A10 Effect of ash and moisture content on calorific values of coals

The calorific values quoted in **A8** cover (a) air dried, (b) dry ash free, (c) dry, mineral matter free coals. In practice, all coals contain both ash and moisture with percentages of either or both up to 15. The presence of these incombustibles reduces the calorific value of the coal, and the table below serves as a guide to calorific values reached in practice

Rank	Ash (% w/w)	Moisture (%w/w)	GCV (MJ/kg)	Size groups for low volatile coals			
100	<10	<2	35				
200–300	<6	<2	36				
200–300	10	10	28.5				
400	<5	<2	32.5	*Welsh anthracite*	*Size (mm)*	*Welsh dry steam coal*	*Size (mm)*
400	10	10	27				
600	<5	<5	35	Cobbles	80–125	Cobbles	80–125
600	10	10	26.5	French nuts	63–80	Large nuts	56–80
800	<4	<8	34	Stove nuts	36–63	Small nuts	18–56
800	10	10	26	Stovesse	16–36	Beans	16–18
800	15	15	22	Beans	10–20	Peas	10–18
900	<3	<10	33	Peas	10–16	Washed duff	<10
900	15	15	21	Grains	5–10		
				Washed duff	<5		

A11 Calorific values of wastes (MJ/kg), based on information supplied by Thorn–EMI Industrial Boilers, General Engineering, Radcliffe (1979) Ltd

Agriculture and foods

Animal fats	39.5
Barley dust	24.0
Citrus rinds	4.0
Cocoa waste	18.0
Coffee grounds	22.8
Corn cobbs	17.5
Corn shelled	19.9
Cotton hulls	24.7
Cotton husks	18.4
Dry food waste	18.1
Furfural	33.3
Grass (lawn)	4.7
Grape stalks	7.9
Rice husks	13.7
Starch	18.1
Straw (dry)	14.0
Sugar	16.0
Sunflower husks	17.7

Wood

Dry hardwood	18–19
Sawdust and chips	
dry	17.4
30% wet	15.1
40% wet	12.8
50% wet	10.5

Paper

Brown paper	16.5
Buttered paper[a]	20.3
Cardboard[a]	20.0
Corrugated boxes	17.7
Fibre board	16.5
Food cartons	11.2
Magazines	11.2
Newspapers	18.1
Plastic coated paper	16.5
Tar paper	23.5
Waxed milk cartons	25.1

Solvents

Acetone	30.8
Benzene	41.6
Chloroform	3.1
Dichlorobenzene	19.0
Diethyl ketone	40.5
Diisopropyl ketone	38.4
Ethyl acetate	24.7
Ethanol	29.5
Ethylene dichloride	10.9
Ethylene glycol	19.1
Heptane	46.5
Isopropyl alcohol	33.0
Methanol	22.6
Methyl butyl ketone	37.4
Methyl ethyl ketone	34.0
Methyl isopropyl ketone	35.6
Toluene	42.3
Xylene	42.8

Synthetics

Algea	15.6
Cellulose	17.0
Gelatin	18.4
Gluten	24.2
Melamine	20.1
Naphthene	40.2
Nylon	27.0
Phenol–formaldehyde	35.1
Polyethylene	46.5
Polypropylene	46.5
Polystyrene foam	41.9
Polyurethane	30.2
Polyvinyl acetate	22.8
Polyvinyl chloride	20.0
Urea–formaldehyde	17.7

Miscellaneous

Asphalt	39.8
Cellophane	41.9
Latex	23.3
Leather	18.9
Leather trimmings	17.8
Leather shreddings	19.8
Lignin	14.0
Linoleum	17.9
Lubricants (spent)	27.9
Paint (waste)	19–29
Pigbristles	21.4
Pitch	35.1
Rubber	23.3
Tyres	34.9
Average household refuse	12–14

[a] Granulated

A12 Properties of gases commonly involved in combustion

All values are at 101.3 kPa and at 0 °C except where otherwise indicated

Gas	(1) Relative molecular mass	(2) Ideal gas density[a] (kg/m³)	(3) Molar enthalpy (kJ/mol)	(4) Gross CV[b] (MJ/kg)	(5) Gross CV[c] (MJ/m³)	Specific heat capacity (kJ/kg K) At 0°C	At 15°C	At 100°C	At 200°C	At 300°C	Relative density[d]	Boiling point (°C)	Melting point (°C)	Ratio of specific heats (γ)	Thermal conductivity (W/m K)
Oxygen (O_2)	32.00	1.428				0.901	0.907	0.937	0.973	1.002		−183	−218	1.4	0.024
Nitrogen (N_2)	28.01	1.250				1.082	1.055	1.030	1.051	1.085		−196	−210	1.41	0.024
Atmospheric nitrogen[e]	28.17	1.257													0.024
Air	28.98	1.293				0.985	0.988	1.008	1.030	1.052				1.4	0.024
Carbon dioxide (CO_2)	44.01	1.964	283			0.830	0.850	0.914	1.000	1.066		−78		1.3	0.014
Carbon monoxide (CO)	28.01	1.250	286	10.10	12.63	1.083	1.068	1.043	1.064	1.094		−192		1.4	0.023
Hydrogen (H_2)	2.02	0.090		142.0	12.76	14.47	14.43	14.39	14.44	14.53		−253	−259	1.41	0.17
Water vapour (H_2O)	18.02	0.804				1.87	1.90	2.06	2.95	5.28		100	0		0.60
Sulphur dioxide (SO_2)	64.06	2.858				0.607	0.617	0.664	0.711	0.749		−10		1.26	
Hydrogen sulphide (H_2S)	34.08	1.521				0.99	1.00	1.03	1.08	1.13		−61	−86	1.3	0.012
Methane (CH_4)	16.04	0.716	892	55.6	39.8	2.08	2.15	2.48	2.86	3.21	0.42	−164	−182	1.32	0.030
Ethane (C_2H_6)	30.07	1.342	1562	51.9	69.7	1.63	1.70	2.06	2.47	2.87	0.55	−89	−183	1.22	0.017
Propane (C_3H_8)	44.10	1.968	2221	50.4	99.1	1.54	1.62	2.02	2.44	2.83	0.58	−42	−190	1.15	0.015
n-Butane (C_4H_{10})	58.12	2.596	2876	49.5	128.3	1.55	1.63	2.01	2.43	2.81	0.58	−0.5[g]	−138	1.11	0.013
n-Pentane (C_5H_{12})[f]	72.15	3.220	3536	49.1	157.8						0.63	36	−130		
Ethane (C_2H_4)	28.05	1.251	1412	50.3	63.0	1.44	1.49	1.81	2.16	2.46	0.57	−104	−169	1.40	
Propene (C_3H_6)	42.08	1.877	2059	48.9	91.9	1.40	1.49	1.77	2.15	2.47	0.84	−47	−185	1.22	
Benzene (C_6H_6)[f]	78.12	3.485	3268	41.8	145.8	0.94	1.00	1.34	1.67	1.95	0.88	80	5.5	1.14	
Ethyne (C_2H_2)[h]	26.04	1.162	1300	49.9	58.0	1.62	1.67	1.88	2.06	2.21	0.62	−84	−81		
n-Hexane (C_6H_{14})[f]	86.18	3.845	4198	48.7	187.3	1.63	1.69	2.04	2.41	2.76	0.66	69	−95	1.08	0.018

[a] Column (1)/22.41.
[b] Column (3)/column (1).
[c] Column (2)/22.41.
[d] Formerly called specific gravity.
[e] This is air without oxygen.
[f] These substances are liquid at ambient temperatures.
[g] Isobutane (2-methylpropane) boils at −12°C.
[h] Ethyne = acetylene.

A13 Mass relationships of common combustion reactions

Values are in kg per kg of fuel, stoichiometric combustion being assumed

Kilomole reaction	Combustion			Flue gases				
	O_2	Air	N_2	CO_2	CO	H_2O	Total	CO_2 (%)
$C(12) + O_2(32) \rightarrow CO_2(44)$	2.67	11.50	8.83	3.67			12.50	29.4
$C(12) + \frac{1}{2}O_2(16) \rightarrow CO(28)$	1.33	5.73	4.40		2.33		6.73	–
$CO(28) + \frac{1}{2}O_2(16) \rightarrow CO_2(44)$	0.57	2.46	1.89	1.57			3.46	45.4
$H_2(2) + \frac{1}{2}O_2(16) \rightarrow H_2O(18)$	8.00	34.5	26.5			9.0	35.5	–
$CH_4(16) + 2O_2(64) \rightarrow CO_2(44) + 2H_2O(36)$	4.00	17.2	13.2	2.75		2.25	18.2	15.1
$C_2H_2(26) + 2\frac{1}{2}O_2(80) \rightarrow 2CO_2(88) + H_2O(18)$	3.08	13.3	10.2	3.38		0.69	14.3	23.6
$C_2H_4(28) + 3O_2(96) \rightarrow 2CO_2(88) + 2H_2O(36)$	3.43	14.8	11.4	3.14		1.29	15.8	19.9
$C_2H_6(30) + 3\frac{1}{2}O_2(112) \rightarrow 2CO_2(88) + 3H_2O(54)$	3.73	16.1	12.4	2.93		1.80	17.1	17.1
$C_3H_6(42) + 4\frac{1}{2}O_2(144) \rightarrow 3CO_2(132) + 3H_2O(54)$	3.43	14.8	11.4	3.14		1.29	15.8	19.9
$C_3H_8(44) + 5O_2(160) \rightarrow 3CO_2(132) + 4H_2O(72)$	3.64	15.7	12.1	3.00		1.64	16.7	18.0
$C_4H_{10}(58) + 6\frac{1}{2}O_2(208) \rightarrow 4CO_2(176) + 5H_2O(90)$	3.59	15.5	11.9	3.03		1.55	16.5	18.4
$C_5H_{12}(72) + 8O_2(256) \rightarrow 5CO_2(220) + 6H_2O(108)$	3.56	15.3	11.8	3.06		1.5	16.3	18.8
$C_6H_6(78) + 7\frac{1}{2}O_2(240) \rightarrow 6CO_2(264) + 3H_2O(54)$	3.08	13.3	10.2	3.38		0.69	15.0	22.5
$S(32) + O_2(32) \rightarrow SO_2(64)$	1.00	4.31	3.31		SO_2 2.00		5.31	–
$H_2S(34) + 1\frac{1}{2}O_2(48) \rightarrow H_2O(18) + SO_2(64)$	1.41	6.08	4.67		1.88	0.53	7.08	–

A14 Volumetric relationships of common combustion reactions

The values shown assume stoichiometric combustion, and are in m^3 per m^3 of fuel except where the fuel is carbon or sulphur; in these cases the values are m^3 per m^3 of carbon dioxide or sulphur dioxide respectively. Volumes are related to 101.3 kPa and 0 °C; pentane and benzene are liquids at 0 °C but the volumes are those of the gaseous state corrected to 0 °C

Fuel	Products	Combustion			Flue gases					
		O_2	Air	N_2	CO_2	CO	SO_2	H_2O	Total	CO_2 (%)
Carbon	CO_2	1.00	4.76	3.76	1.0				4.76	21.0
Carbon	CO	0.50	2.38	1.88		1.00			2.88	–
Carbon monoxide	CO^2	0.50	2.38	1.88	1.00				2.88	3.5
Hydrogen	Water vapour	0.50	2.38	1.88				1.00	2.88	–
Methane (CH_4)		2.00	9.52	7.52	1.00			2.00	10.52	9.5
Ethane (C^2H_6)		3.50	16.7	13.2	2.00			3.00	18.2	11.0
Propane (C_3H_8)		5.00	23.8	18.8	3.00			4.00	25.8	11.6
Butane (C_4H_{10})	Water vapour and CO_2	6.50	31.0	24.5	4.00			5.00	33.5	11.9
Pentane (C_5H_{12})		8.00	38.1	30.1	5.00			6.00	41.1	12.2
Ethene (C_2H_4)		3.00	14.3	11.3	2.00			2.00	15.3	13.1
Propene (C_3H_6)		4.50	21.4	16.9	3.00			3.00	22.9	13.1
Benzene (C_6H_6)		7.50	35.7	28.2	6.00			3.00	37.2	16.1
Ethyne (C_2H_2)		2.50	11.9	9.4	2.00			1.00	12.4	16.1
Sulphur	SO_2	1.00	4.76	3.76			1.00		4.76	
Hydrogen Sulphide	Water vapour and SO_2	1.50	7.14	5.64			1.00	1.00	7.64	

A15 Specific heat capacity of gases

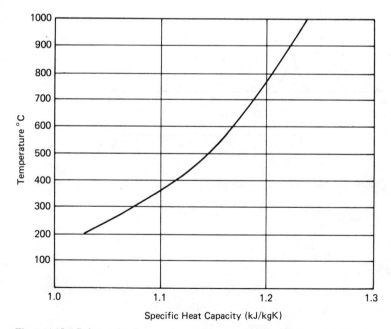

Figure A15.1 *Relationship between specific heat capacity and temperature for average flue gases*

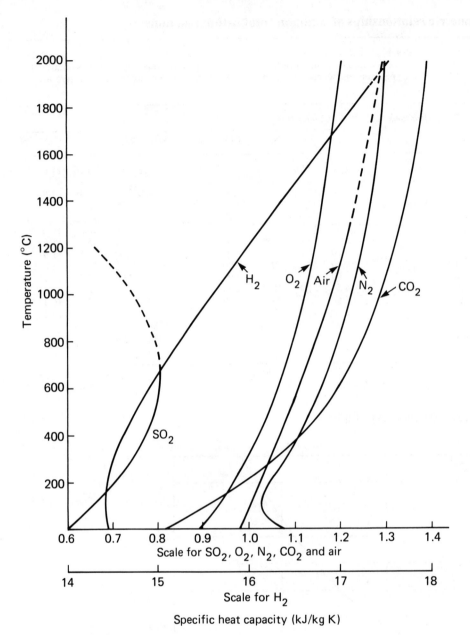

Figure A15.2 *Relationship of specific heat capacity of gases to temperature. The broken line indicates limit of normal temperature range*

A16 Variation with temperature of the physical properties of dry air

Values are based on a pressure of 101 325 N/m^2

Temp. (°C)	Specific heat capacity (kJ/kg K)		Absolute viscosity, μ (N s/m^2) (kg/m s)	Thermal conductivity, κ (W/m K)
	Constant pressure, C_p	Constant volume, C_v		
0	0.985	0.704	17.1×10^{-6}	0.0241
5	0.986	0.704	17.3×10^{-6}	0.0245
10	0.987	0.705	17.5×10^{-6}	0.0249
15	0.988	0.706	17.7×10^{-6}	0.0253
20	0.989	0.706	18.0×10^{-6}	0.0257
25	0.990	0.707	18.3×10^{-6}	0.0261
30	0.992	0.709	18.6×10^{-6}	0.0265
40	0.994	0.710	19.0×10^{-6}	0.0272
50	0.996	0.711	19.4×10^{-6}	0.0280
60	0.998	0.713	19.9×10^{-6}	0.0289
70	1.001	0.715	20.4×10^{-6}	0.0296
80	1.003	0.717	20.8×10^{-6}	0.0304
90	1.005	0.719	21.2×10^{-6}	0.0311
100	1.008	0.721	21.6×10^{-6}	0.0318
110	1.010	0.724	22.0×10^{-6}	0.0325
120	1.012	0.727	22.5×10^{-6}	0.0331
130	1.014	0.730	23.0×10^{-6}	0.0336
140	1.017	0.733	23.4×10^{-6}	0.0345
160	1.021	0.736	24.2×10^{-6}	0.0359
180	1.026	0.739	25.0×10^{-6}	0.0373
200	1.030	0.743	25.7×10^{-6}	0.0387
220	1.034	0.746	26.5×10^{-6}	0.0400
240	1.039	0.750	27.3×10^{-6}	0.0412
260	1.043	0.755	28.0×10^{-6}	0.0425
280	1.048	0.758	28.7×10^{-6}	0.0438
300	1.052	0.762	29.3×10^{-6}	0.0450

(1) The ratio of specific heats C_p/C_v is substantially constant: 1:40 at 0 °C to 1.38 at 300 °C.
(2) The Prandtl number

$$\frac{\text{Specific heat capacity at constant pressure} \times \text{viscosity}}{\text{Thermal conductivity}}$$

is also substantially constant at 0.70–0.69.
(3) The density of air at constant pressure is inversely proportional to absolute temperature, with a value at 0 °C of 1.293 kg/m^3.

22

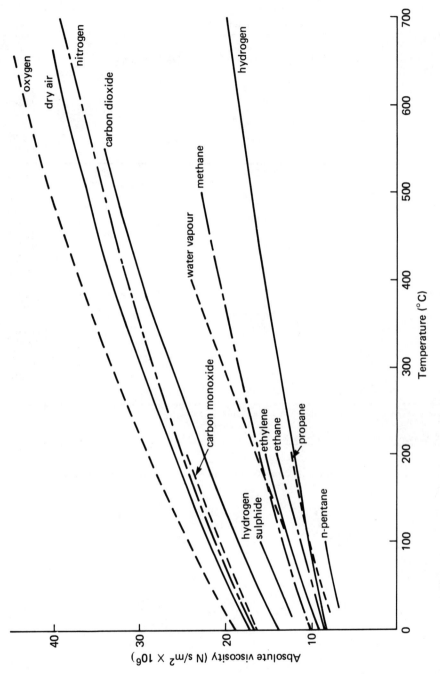

Figure A17 *Variation with temperature of the viscosity of some gases*

A18 Variation with temperature of some physical properties of water (data shown graphically in A19)

Temp. (°C)	Specific heat capacity at constant pressure (kJ/kg K)	Density (kg/m³)	Absolute viscosity (N s/m²) (kg/m s)	Thermal conductivity (W/m K)	Saturation vapour pressure (N/m²)	Prandtl number[a]
0.01	4.224	999.8	0.001 79	0.566	610	13.36
2	4.216	999.9	0.001 66	0.570	710	12.28
4	4.210	1000	0.001 55	0.574	810	11.37
6	4.204	999.9	0.001 46	0.578	930	10.62
8	4.199	999.8	0.001 39	0.582	1 070	10.02
10	4.195	999.7	0.001 31	0.586	1 230	9.37
12	4.192	999.6	0.001 24	0.589	1 410	8.83
14	4.188	999	0.001 18	0.593	1 600	8.33
16	4.185	999	0.001 12	0.597	1 820	7.85
18	4.183	998	0.001 06	0.600	2 060	7.39
20	4.180	998	0.001 01	0.603	2 330	7.00
25	4.177	997	0.000 90	0.612	3 160	6.14
30	4.176	996	0.000 80	0.620	4 240	5.39
35	4.175	994	0.000 73	0.627	5 620	4.86
40	4.174	992	0.000 66	0.633	7 380	4.35
45	4.175	990	0.000 60	0.639	9 580	3.92
50	4.175	988	0.000 55	0.644	12 300	3.57
55	4.176	986	0.000 51	0.649	15 700	3.28
60	4.178	983	0.000 47	0.654	19 900	3.00
65	4.181	981	0.000 44	0.658	25 000	2.80
70	4.184	978	0.000 41	0.663	31 100	2.59
75	4.188	975	0.000 38	0.666	38 500	2.39
80	4.192	972	0.000 36	0.670	47 400	2.25
85	4.197	969	0.000 34	0.673	57 800	2.12
90	4.201	965	0.000 32	0.676	70 100	1.99
95	4.207	962	0.000 30	0.679	84 500	1.86
100	4.211	958	0.000 28	0.682	101 000	1.72
110	4.222	951	0.000 26	0.685	143 000	1.60
120	4.237	943	0.000 232	0.686	199 000	1.43
130	4.254	935	0.000 215	0.686	270 000	1.33
140	4.277	926	0.000 195	0.685	361 000	1.22
150	4.301	917	0.000 181	0.683	476 000	1.18
160	4.326	907	0.000 168	0.682	618 000	1.07
170	4.352	897	0.000 157	0.679	792 000	1.01
180	4.382	887	0.000 148	0.676	1 000 000	0.96
190	4.416	876	0.000 140	0.671	1 260 000	0.93
200	4.452	864	0.000 134	0.666	1 560 000	0.90
250	4.870	800	0.000 109	0.620	3 980 000	0.86
300	5.790	713	0.000 091	0.540	8 590 000	0.98
350		574			16 540 000	
374.15		316			22 120 000	

[a] Obtained by multiplying the specific heat capacity of column 2 by the viscosity of column 4, dividing by the thermal conductivity of column 5, and multiplying by 1000.

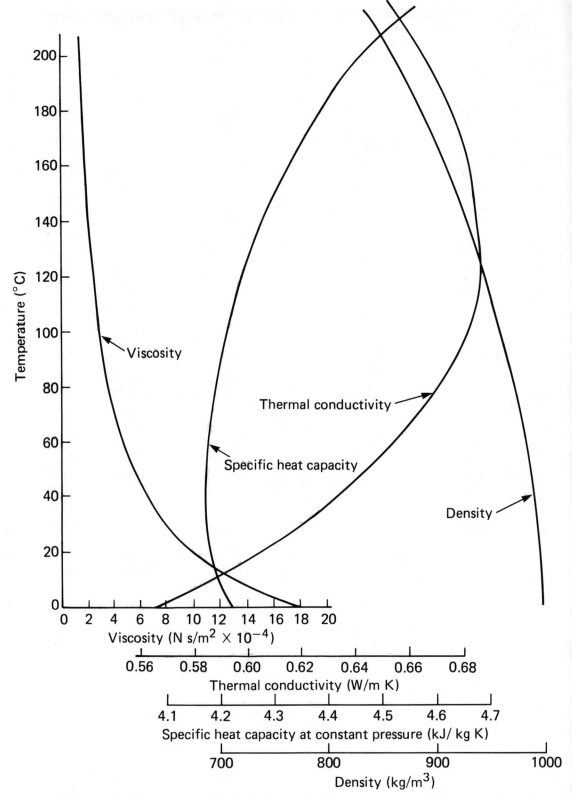

Figure A19 *Variation with temperature of the physical properties of water*

A20 Thermal conductivity of building materials, concrete and brickwork

A20.1 Condensed table of thermal conductivity values for building materials

Values of thermal conductivity shown are for general guidance and have been extracted from many sources, notably Section A3 of the CIBS guide

Material	Density ρ (kg/m³)	Thermal conductivity κ (W/m K)	Material	Density ρ (kg/m³)	Thermal conductivity κ (W/m K)
General building materials			Plastic linoleum	1750	0.35
			Inlaid linoleum	1150	0.22
Granite	2600	2.5	Cork floortiles	540	0.085
Marble	2500	2.0	Cork linoleum	510	0.070
Limestone	2180	1.5	Cellular rubber		
Sandstone	2000	1.3	underlay	270	0.065
Artificial stone	1750	1.3	Wilton carpet	190	0.058
Asbestos cement sheet	1600	0.36	Felt underlay	160	0.045
Perlite plasterboard	800	0.18	Cork board	145	0.042
Gypsum plasterboard	950	0.16			
Hardboard			*Rigid translucents*		
Standard	600	0.08			
Medium	900	0.13	Window glass	2500	1.05
Asbestos board	–	0.11	Solid sheet		
Resin bonded jute			Polypropylene	915	0.24
board	430	0.065	Epoxy glass fibre	1500	0.23
Fibre insulating board	300	0.057	Polycarbonate	1150	0.23
Glass fibre	12–150	0.040	Polystyrene	1050	0.17
Polystyrene expanded			Polyvinyl chloride	1350	0.17
board	15	0.037			
Phenolic foam board	50	0.036	*Roofing*		
Polyurethane board	30	0.025			
			Steel sheeting	8000	48.0
Building finishing materials			Slate	2700	2.00
			Asphalt heavy mastic,		
Wall tiles	–	0.13	20% grit	2325	1.15
Thermalite block	500	0.13	Concrete tiles	2100	1.10
Wood wool building			Clay tiles	1900	0.85
slab	500	0.10	PVC asbestos tiles	2000	0.85
Rigid mineral wool slab	155	0.050	Plastic tiles	1050	0.50
Mineral wool felted			Asphalt roofing	1900	0.50
mat	180	0.042	Screed roof	120	0.41
Render			Sarking felt	1100	0.20
Internal	1600	0.73	Roofing felt	960	0.19
External	1300	0.50	Timber sarking board	630	0.13
Dense plaster	1300	0.50	Straw thatch	240	0.07
Gypsum plaster	1280	0.46			
Vermiculite plaster	800	0.26	*Insulation*		
Perlite plaster	610	0.19			
Timber, chipboard,			Polyurethane foam		
plywood	530–800	0.13–0.15	(aged)	30	0.026
(Denser timbers have			Mineral fibre slab	30	0.035
slightly higher values of			Kapok quilt	20	0.035
κ)			PVC rigid foam	25	0.035
Woodwool building			Urea–formaldehyde	10–15	0.032–0.040
slabs	500	0.10	Felted mineral wool	50	0.039
			Glass fibre quilt	12	0.040
Floors			Perlite expanded		
			granules	65	0.042
Composition flooring	1600–2200	0.44–0.80	Vermiculite granules	100	0.065
Bitumen floors	1700	0.75			
Polyvinyl chloride	1000	0.40			

26

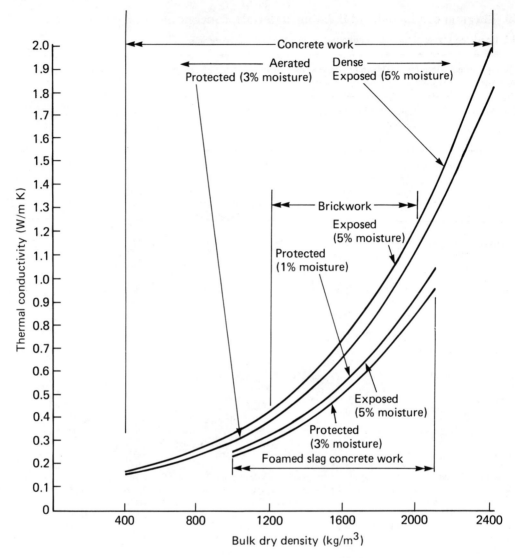

Figure A20.2 *Concrete and brickwork: variation of thermal conductivity with density*
(Based on data in Section A3 of CIBS guide with permission)

A21 Tables of values for thermal resistance and thermal transmittance

Tables **A21.1**, **A21.2**, **A21.3** and **A21.4** are reproduced from Section A3 of the CIBS guide, by permission of the Chartered Institution of Building Services

A21.1 Inside surface resistance

Building element	Heat flow	Surface resistance (m² K/W)	
		High emissivity factor[a]	Low emissivity factor[a]
Walls	Horizontal	0.12	0.30
Ceilings or roofs, flat or pitched, floors	Upward	0.10	0.22
Ceilings and floors	Downward	0.14	0.55

[a] For detailed definition of high and low emissivity factor, refer to CIBS guide A3.

A21.2 Outside surface resistance for stated exposure

Building element	Emissivity of surface	Surface resistance (m² K/W)		
		Sheltered[a]	Normal[a]	Severe[a]
Wall	High	0.08	0.06	0.03
	Low	0.11	0.07	0.03
Roof	High	0.07	0.04	0.02
	Low	0.09	0.05	0.02

[a] Refer to **B11.3** for definitions of sheltered, normal and severe.

A21.3 Standard thermal resistances for unventilated airspaces

Type of airspace		Thermal resistance (m² K/W) for heat flow in stated direction		
Thickness	Surface emissivity	Horizontal	Upward	Downward
5 mm	High	0.10	0.10	0.10
	Low	0.18	0.18	0.18
25 mm or more	High	0.18	0.17	0.22
	Low	0.35	0.35	1.06
High emissivity plane and corrugated sheets in contact		0.09	0.09	1.11
Low emissivity multiple foil insulation with airspace one side		0.62	0.62	1.76

A21.4 Standard thermal resistances for ventilated airspaces

Type of airspace (thickness 25 mm minimum)		Thermal resistance (m² K/W)
Airspace between asbestos cement or black metal cladding with unsealed joints	High emissivity lining	0.16
	Low emissivity surface facing airspace	0.30
Loft space above flat ceiling	Pitched roof with unsealed asbestos cement sheets or black metal cladding	0.14
	Pitched roof with aluminium cladding or low emissivity upper surface on ceiling	0.25
	Pitched roof lined with felt or building paper	0.18
Airspace between tiles and roofing felt or building paper		0.12
Airspace behind tiles on tile-hung wall (including tile resistance)		0.12
Airspace in cavity wall		0.18

A21.5 'U' values for solid and suspended floors[a] (from Section A3 of CIBS Guide, with permission)

Length (m)	Breadth (m)	'U' value (W/m² K)		
		4 edges exposed	2 perpendicular edges exposed	Suspended floor
V. long	100	0.06	0.03	0.07
V. long	40	0.12	0.07	0.15
V. long	20	0.22	0.12	0.26
100	40	0.15	0.09	0.18
100	20	0.24	0.14	0.28
60	40	0.17	0.10	0.20
60	20	0.26	0.15	0.30
40	20	0.28	0.16	0.31
40	10	0.43	0.25	0.47
20	10	0.48	0.28	0.51
10	10	0.62	0.36	0.59

[a] Note that for areas less than 10 m × 10 m 'U' values increase steeply.

A21.6 'U' values for glazing (without frames) (from Section A3 of CIBS Guide, with permission)

Construction	'U' value for stated exposure (W/m² K)		
	Sheltered	Normal (standard)	Severe
Single window glazing	5.0	5.6	6.7
Double window glazing (12 mm airspace)	2.8	3.0	3.3
Triple window glazing (12 mm airspace)	2.0	2.1	2.2
Roof glazing skylight	5.7	6.6	7.9

A21.7 U-values for typical windows (from Section A3 of CIBS Guide, with permission)

Window type	Fraction of area occupied by frame	U-value for stated exposure (W/m² K)		
		Sheltered	Normal	Severe
Single glazing				
Wood frame	10%	4.7	5.3	6.3
	20%	4.5	5.0	5.9
	30%	4.2	4.7	5.5
Aluminium frame				
(no thermal break)	10%	5.3	6.0	7.1
	20%	5.6	6.4	7.5
	30%	5.9	6.7	7.9
Aluminium frame				
(with thermal break)	10%	5.1	5.7	6.7
	20%	5.2	5.8	6.8
	30%	5.2	5.8	6.8
Double glazing				
Wood frame	10%	2.8	3.0	3.2
	20%	2.7	2.9	3.2
	30%	2.7	2.9	3.1
Aluminium frame				
(no thermal break)	10%	3.3	3.6	4.1
	20%	3.9	4.3	4.8
	30%	4.4	4.9	5.6
Aluminium frame				
(with thermal break)	10%	3.1	3.3	3.7
	20%	3.4	3.7	4.0
	30%	3.7	4.0	4.4

Note: Where the proportion of the frame differs appreciably from the above values, particularly with wood or plastic, the *U*-value should be calculated (metal members have a *U*-value similar to that of glass).

A21.8 Some values of emissivity

Material	Temperature basis (°C)	Emissivity[a]
Polished brass	250–350	0.028–0.031
Polished aluminium	40	0.04
Polished steel	400	0.14
Aluminium roofing sheets	40	0.22
Galvanized sheet	25	0.25
Lead	25	0.28
Metal based paint	25	0.27–0.52
Oxidized brass	200	0.61
Paint on rough metal	25	0.91
Plaster	10–80	0.91
Wood	20	0.90
Asbestos	90	0.93
Brickwork	20	0.93
Oil paint (colours)	25	0.92–0.96
Black sheet metal	40	0.94
Water	20	0.95
Dull black lacquer	25	0.96–0.98

[a] See **Section A2.6** for definition of emissivity.

A22 Properties of furnace insulating materials

A22.1 Analyses and properties of insulating refractories (reproduced by courtesy of MPK Insulation)

Product ref.	Classification temp.[a] (°C)	Chemical analysis (%)						Bulk density[b] (kg/m³)	Cold crushing strength[b] (MN/m²)	Thermal conductivity (W/m K at mean temp. °C)					
		Al₂O₃	SiO₂	Fe₂O₃	TiO₂	CaO	MgO			200	400	600	800	1000	1200
SUPRA	850	11	75	6.0	1.0	1.0	2.0	750	6.0	0.17	0.18	0.20			
ROTOL	900	11	79	4.0	1.0	0.5	2.0	800	8.0	0.20	0.21	0.23			
100	1000	20	53	4.0	1.2	1.4	16.4	400	1.3	0.14	0.16	0.19	0.22		
115	1150	19	45	2.9	1.0	17.3	11.4	580	2.9	0.20	0.23	0.27	0.30		
125	1250	36	46	1.1	0.2	14.6	0.4	670	1.6	0.18	0.20	0.22	0.24	0.26	
130LW	1300	39	44	0.6	1.1	14.6	0.1	500	1.0	0.14	0.16	0.18	0.21	0.24	
130HSR	1300	36	54	1.7	0.5	0.3	5.8	1200	18.0	0.56	0.59	0.62	0.65	0.68	
130HD	1300	37	54	1.2	0.2	0.2	4.8	1450	27.0	0.74	0.76	0.77	0.79	0.81	
135	1350	40	43	0.6	1.0	13.8	0.2	700	2.1	0.19	0.21	0.24	0.27	0.30	
135HSR	1350	37	54	1.7	0.5	0.3	5.1	960	10.0	0.40	0.44	0.48	0.52	0.56	
140	1400	41	54	1.3	0.6	0.3	0.3	860	4.8	0.28	0.31	0.34	0.36	0.39	
150	1500	46	52	1.0	1.4	0.3	0.1	820	1.5	0.31	0.33	0.37	0.41	0.46	0.52
155HA	1550	61	36	0.4	1.0	0.2	0.1	880	4.0	0.38	0.38	0.38	0.39	0.41	0.43
165HA	1650	72	26	0.3	0.7	0.2	0.1	1070	4.8	0.44	0.44	0.45	0.47	0.49	0.51
180	1800	93	6.6	0.08	0.01	0.08	0.01	1410	16.0			1.02	1.03	1.10	1.18
185	1850	98.9	0.7	0.05	0.01	0.07	0.01	1440	18.0			1.34	1.24	1.19	1.17
185HP	1850	99.5	0.2	0.05	0.01	0.05	0.03	1760	17.0			1.71	1.57	1.48	1.45

[a] Classification temperature is the temperature giving 3% shrinkage during 24 hours heat soak.
[b] Bulk density and cold crushing strength expressed at classification temperature.

A22.2 Analyses and properties of castables and mouldables (reproduced by courtesy of Morgan Refractories Ltd)

Grade	Product name	Max. service temp. (°C)	Chemical analysis				Bulk density (Fired) (kg/m³)	Cold crushing strength^a (MN/m²)	Thermal conductivity at mean 600 °C (W/m K)
			Al_2O_3	SiO_2	CaO	Fe_2O_3			
Insulating castables	Coolcast	1100	22.7	38.9	26.7	3.3	705	0.52	0.22
	Insucast	1200	31.3	42.1	12.2	9.4	1320	13.1	0.46
	Insulite	1370	35.8	45.0	9.5	5.8	1345	4.8	0.35
Dense castables	Standard	1400	38.0	48.0	5.8	4.7	1850	9.3	0.72
	H.S.	1250	38.7	40.2	11.5	6.1	1907	15.9	0.72
	Emcast	1550	47.0	42.0	5.5	3.0	2080	41	0.79
	Midcast	1550	76.0	6.7	8.8	4.6	2500	59	0.87
	H.T. cast	1700	61.0	33.2	2.3	0.9	2130	59	0.79
	1800 cast	1800	96.2	0.1	2.7	0.1	2788	41	1.73
Castables designed for gunning	Guncrete B.F.	1500	46.4	42.0	6.8	1.4	2115	38	0.79
	Extra H.S.	1250	39.9	40.3	11.4	5.8	2150	22.8	0.79
	Guncrete 130	1300	38.8	44.4	8.6	5.4	1850	17.2	0.72
	Guncrete 160	1600	50.3	40.8	5.3	1.2	2020	35	0.79
	Guncrete 170	1700	65.5	26.9	4.2	1.0	2115	35	0.79
Mouldables	Plastic standard	1450	39.0	55.0	0.2	2.4	2160	20.7	0.79
	Plastic super	1600	42.9	51.1	0.2	1.2	2290	19.3	0.87
	Plastic H.T.	1700	60.6	35.1	0.2	1.5	2516	15.9	0.87
	Plastic extra C.B.	1720	83.5	11.4	0.1	1.3	2800	52	1.45
Mouldables designed for casting	Plascast super	1600	47.0	46.6	–	1.7	2180	38	0.79
	Plascast H.T.	1700	64.6	31.1	–	1.5	2320	38	0.87
	Plascast extra	1750	81.2	13.9	–	1.6	2500	38	0.87

^a Cold crushing strength expressed at maximum service temperature.

A22.3 Heat flow data for range of ceramic fibre modules[a]

Type		Standard duty fibre						High temperature fibre					
		Hot face temperature (°C)											
		600°C	750°C	900°C	1000°C	1100°C	1200°C	1300°C	1350°C	1400°C	1450°C	1500°C	1550°C
P3	Cold face (°C)	68	91	113	132	152	171						
	Heat loss (MJ/m² hr)	2.03	3.26	4.62	6.03	7.56	9.69						
	Heat storage (MJ/m²)	2.17	2.71	3.28	3.66	4.04	4.41						
P4	Cold face (°C)	57	77	96	113	129	146						
	Heat loss (MJ/m² hr)	1.52	2.51	3.58	4.62	5.80	7.11						
	Heat storage (MJ/m²)	2.84	3.58	4.30	4.79	5.29	5.78						
P5	Cold face (°C)	52	68	85	99	113	127	129	136	143	150	157	164
	Heat loss (MJ/m² hr)	1.28	2.03	3.00	3.76	4.62	5.79	5.80	6.39	6.96	7.50	8.08	8.69
	Heat storage (MJ/m²)	3.53	4.41	5.32	5.91	6.53	7.15	9.56	9.93	10.32	10.70	11.03	11.45
P6	Cold face (°C)	46	60	77	88	102	116	103	109	116	120	126	131
	Heat loss (MJ/m² hr)	1.07	1.70	2.51	3.18	3.98	4.87	4.17	4.52	4.88	5.27	5.67	6.10
	Heat storage (MJ/m²)	4.22	5.26	6.34	7.06	7.78	8.50	13.2	13.7	14.2	14.8	15.3	15.8
P8	Cold face (°C)	41	52	63	71	82	93	96	101	106	111	117	122
	Heat loss (MJ/m² hr)	0.82	1.28	1.77	2.23	2.90	3.51	3.58	3.97	4.30	4.63	4.96	5.37
	Heat storage (MJ/m²)	5.60	6.97	8.36	9.26	10.26	11.19	15.0	15.6	16.2	16.8	17.3	17.9
P10	Cold face (°C)		46	56	63	74	82	84	89	93	97	102	107
	Heat loss (MJ/m² hr)		1.07	1.45	1.77	2.43	2.90	2.93	3.28	3.51	3.75	3.98	4.35
	Heat storage (MJ/m²)		8.63	10.39	11.53	12.7	13.9	18.6	19.3	20.0	20.8	21.5	22.2
P12	Cold face (°C)			50	57	63	74	76	79	84	89	92	95
	Heat loss (MJ/m² hr)			1.21	1.52	1.77	2.43	2.49	2.69	2.93	3.15	3.36	3.61
	Heat storage (MJ/m²)			12.4	13.7	15.1	16.5	22.1	23.0	23.9	24.8	25.6	26.5

[a] The information which is theoretical and for general guidance only, is reproduced by courtesy of MPK Insulation Ltd.

A22.4 Specific heat capacities

Materials	Temperature range (°C)	kJ/kg K
Castables	200–400	1.045
	400–1000	1.090
	above 1000	1.130
Refractories	500	1.060
	1000	1.100
	1500	1.130

A22.5 Specific heat capacity of ceramic fibre

Fibre	kJ/kg K
Triton[a] Kaowool[a]	1.07
Saffil[b]	1.0

[a] Triton is a trade name of Morgan Refractories and Kaowool is a trade name of Babcock and Wilcox.
[b] Saffil is a trademark of Imperial Chemical Industries PLC for alumina fibre.

A22.6 Thermal conductivity of Unifelt[a] sheet and veneering modules (reproduced by courtesy of Morgan Ceramic Fibres Ltd)

Grade	Nominal density (kg/m³)	Service temp. (°C)		Thermal conductivity at mean temp.[b] (W/m K)						
		Continuous	Intermittent	200 °C	400 °C	600 °C	800 °C	1000 °C	1200 °C	1400 °C
13U	140	1200	1260	0.070	0.10	0.15	0.21	0.29	0.39	
14U	125	1300	1400	0.070	0.10	0.14	0.20	0.28	0.38	
15U	120	1400	1500	0.070	0.10	0.14	0.195	0.275	0.37	0.50
16U	110	1500	1600	0.070	0.095	0.135	0.185	0.25	0.35	0.48
17U	100	1600	1700	0.070	0.095	0.13	0.175	0.245	0.325	0.45

[a] Unifelt is a Morganite Ceramic Fibres product and is a blend of Triton Kaowool and Saffil.
[b] Values of thermal conductivity are as stacked blocks. Values in the thickness direction would be 25–30% lower.

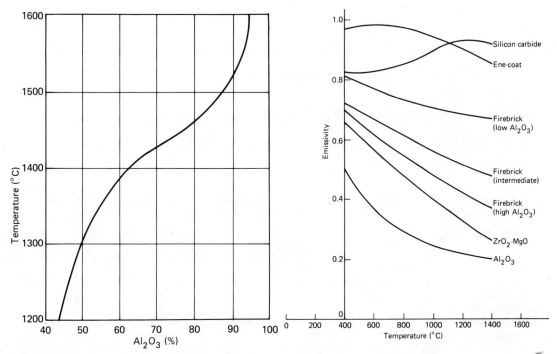

Figure A22.7 *Classification temperature of SAFFIL alumina fibre, aluminosilicate fibre and various mixed fibre blends*

(Reproduced by courtesy of Imperial Chemical Industries PLC)

Figure A22.8 *Refractory materials: variation of emissivity with temperature*

(Reproduced by courtesy of MPK Insulation Ltd.)

A23 Tables of ventilation requirements

(Refer to **Section B19**).

A23.1 Minimum ventilation rates where density of occupation is known (reproduced from Section B2 of CIBS guide, with permission)

Air space per person[a] (m^3)	Outdoor air supply per person[b] (l/s)		
	Minimum	Recommended minima	
		Smoking not permitted	Smoking permitted
3	11.3	17.0	22.6
6	7.1	10.7	14.2
9	5.2	7.8	10.4
12	4.0	6.0	8.0

[a] The statutory minimum volume per person in factories and offices is 11.5 m^3.
[b] The corresponding minimum outdoor air supply is 4.27 l/s per person.

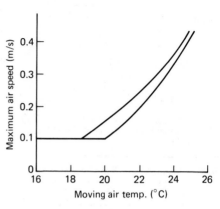

Figure A23.2 *Criteria for air movement*

A23.3 Empirical values of air changes per hour for natural infiltration[a] (based on data in Section A4 of CIBS guide)

Location[b]	Changes per hour
Department stores, storage space and warehouses	¼
Lecture halls, churches, dining/banqueting rooms, exhibition halls, bedrooms (residences), libraries, large shops	½
Museums, bars, canteens, living rooms, hotel bedrooms and public rooms, offices, restaurants, lecture rooms	1
Lavatories, staircases, corridors	1½
Bathrooms, classrooms, hospital waiting rooms	2

[a] Ventilation loss (W/m^3 K) = one third air change value.
[b] Air change rates may be increased by up to 50% for exposed sites or large area of openable windows, and reduced by up to 35% for sheltered positions.

A23.4 Empirical values of air change rates for infiltration to factories (based on data in Section A4 of CIBS guide)

Construction	Floor area (m^3)			
	up to 300	300–3000	3000–10 000	over 10 000
Single storey, brick or concrete, unpartitioned	1½	¾	½	¼
Curtain wall or sheet construction, lined	1¾	1	¾	½
Sheet construction, unlined	2¼	1½	1	¾
Multistorey brick or concrete	1–1½ depending on roof structure			

A23.5 Mechanical ventilation rates for various types of building (from Section B2 of CIBS guide, with permission)

Room	Recommended air changes/hr[a]	Notes
Boiler houses, engine rooms	15–30	Rate includes combustion air and should be checked against particular equipment installed
Canteens	8–12[b]	
Hotel and industrial kitchens	20–60[c]	Rates should be checked against requirements of equipment installed
Laboratories	4–6	Inlet air change rate must be checked against extraction rate via fume cupboard
Lavatories, toilets	6–8[c]	Rates apply to public toilet areas. For congested areas 10 litre per m² of floor area is better. Check extract and inlet positions
Offices	4–6[b]	London Building Act calls for 6 l/s per occupant or per 4.7 m² of floor area, whichever provides the greater ventilation rate

[a] The recommended rates do not apply to warm air heating; check against heating requirement.
[b] Not necessarily all outdoor air; check against rate per person and number of occupants.
[c] Refers to extract ventilation.

A23.6 Recommended outdoor air supply rates for air conditioned spaces (from Section 2 of CIBS guide, with permission)

Type of space	Smoking	Outdoor air supply (l/s)		
		Recommended per person	Minimum per person[a]	Minimum per m² of floor area[a]
Factories[b, c]	None	8	5	0.8
Offices, open plan	Some	8	5	1.3
Laboratories[c]	Some	12	8	–
Offices (private)	Heavy	12	8	1.3
Conference rooms (average)	Some	18	12	–
Boardrooms	V. heavy	25	18	6.0
Executive offices and conference rooms	V. heavy	25	18	6.0

[a] Take greater of the two.
[b] See statutory requirements and local bye-laws.
[c] Rate of extract may be over-riding factor. The outdoor air supply rates take account of the likely density of occupation and of the type and amount of smoking.

A24 Tables of contaminants, poisons, toxic vapours and dusts

Refer to Section B19.

A24.1 Tables of some threshold limit values (TLV)

Gases and vapours	Effectiveness of carbon filters (%)[a]	TLV (ppm)	TLV (mg/m³)	Toxic fumes and dusts	TLV (mg/m³)
Ozone	20–50	0.1	0.2	Beryllium	0.002
Phosgene	10–25	0.1	0.4	Silver (soluble compounds)	0.01
Bromine	20–50	0.1	0.7	Mercury (except Alkyl compounds)	0.05
Hydrogen peroxide		1.0	1.5	Chromic acid	0.1
Chlorine	2	1.0	3.0	Nickel (soluble compounds)	0.1
Formaldehyde	2	2.0[b]	3.0	Phosphorus (yellow)	0.1
Nitric acid	10–25	2.0	5.0	Tetraethyllead	0.1
Sulphur dioxide	2	2.0	5.0	Tin (organic compounds)	0.1
Hydrogen chloride	2	5.0[b]	7.0	Paraquat	0.1
Carbon tetrachloride	20–50	5	30	Strychnine	0.15
Hydrogen sulphide	2	10	14	Lead (inorganic compounds and dusts)	0.15
Benzene	20–50	10	30	Coal tar pitch volatiles	0.2
Ammonia	2	25	18	Arsenic and compounds	0.2
Carbon monoxide	1	50	55	Cotton dust (raw)	0.2
Trichlorethylene	20–50	50	270	Lindane	0.5
Turpentine	20–50	100	560	Nicotine	0.5
Methyl alcohol	10–25	200	260	Sulphuric acid	1.0
Octane	20–50	300	1450	Copper dusts	1.0
Ethyl ether	10–25	400	1200	Carbon black	3.5
Carbon dioxide	1	5000	9000	Oil mist (mineral)	5.0

[a] Mass of adsorbed contaminant expressed as a percentage of the mass of carbon.

[b] These are TLV ceiling values which should not be exceeded; others are TLV-time weighted average values based on 8 hr day or 40 hr week. For TLV's of quartz, silica, silicates and all forms of asbestos, refer to data published by the American Conference of Governmental Industrial Hygienists, from whose 1983 publication information on TLV is derived and published with their permission.

A24.2 Exhaust gas analysis for road vehicles (ppm)[a]

Gas	Petrol engines	Diesel engines
Carbon monoxide	30 000	200–1000
Carbon dioxide	132 000	90 000
Aldehydes	40	20
Formaldehyde	7	11
Oxides of nitrogen	600	400
Sulphur dioxide	60	200

[a] Values are in parts per million and because of wide varietions in engine and fuel conditions are for guidance only. They are based on data in Section B2 of the CIBS guide.

A24.3 Some limits for mineral dusts

Dust	Millions of particles per m^3	mg/m^3
Coal dust (bituminous)	–	2
Talc (non-asbestiform)	–	2
Natural graphite	530	–
Mica, soapstone	700	–
Inert or nuisance particles (**A24.4**)	1060	10

A24.4 Some nuisance dusts

Alundum (Al_2O_3)	Synthetic graphite	Pentaerythritol	Sucrose
Calcium carbonate	Gypsum	Plaster of Paris	Titanium dioxide
Cellulose (paper fibre)	Kaolin	Portland cement	Vegetable oil mists (except
Corundum	Limestone	Rouge	castor, cashew or similar
Emery	Magnesite	Silicon carbide	irritant oils)
Glycerine mist	Marble	Starch	

A24.5 Some simple asphyxiants[a]

Acetylene (ethyne)	Ethane	Hydrogen	Nitrogen
Argon	Ethylene	Methane	Propane
Butane	Helium	Neon	Propylene

[a] Limiting factor is minimum oxygen content of 18% by volume.

A25 Thermal properties of pure metals

Metal	Density at 20 °C (kg/m³)	Melting point (°C)	Specific heat capacity (kJ/kg K)		Thermal conductivity at 25 °C (W/m K)	Linear expansivity[a] (K⁻¹ × 10⁶)
			At 25 °C	Liquid at 2000 K		
Aluminium	2 700	660	0.90	1.09	237	25.0
Antimony	6 684	631	0.21	0.26	–	9.0
Beryllium	1 850	1283	1.82	3.26	218	12.0
Bismuth	9 750	271	0.13	0.15	8.4	13.0
Cadmium	8 650	321	0.23	0.26	93	30.0
Chromium	7 180–7200[b]	1857 ±20	0.23	0.94	91	6.0
Cobalt	8 900	1495	0.42	0.49	69	12.0
Copper	8 960	1083	0.38	0.49	398	16.6
Gold	19 320	1064	0.13	0.15	315	14.2
Iron	7 870	1535	0.45	0.82	80	12.0
Lead	11 350	328	0.13	0.14	35	29.0
Magnesium	1 738	649	1.02	1.34	159	25.0
Manganese	7 210–7440[b]	1244	0.48	0.84	–	22.0
Mercury	14 170[c]	−39	0.14	–	8.4	–
Molybdenum	10 220	2617	0.25	0.37	140	5.0
Nickel	8 900	1453	0.44	0.73	90	13.0
Platinum	21 450	1772	0.13	0.18	73	9.0
Potassium	862	64	0.75	–	9.9	83.0
Selenium	4 790[d]	217	0.32	–	0.05	37.0
Silicon	2 330	1420	0.70	0.91	8.4	3.0
Silver	10 500	962	0.24	0.28	427	19.0
Sodium	970	98	1.22	–	13.4	70.0
Tin	5 750–7310[e]	232	0.23	0.24	64	20.0
Titanium	4 540	1660	0.52	0.79	20	8.5
Tungsten	19 300	3410	0.13	0.17	178	4.5
Uranium	18 950	1132	0.12	0.20	2.5	13.4
Vanadium	6 110	1890	0.49	0.87	60	8.0
Zinc	7 130	420	0.39	–	115	35.0

[a] Formerly coefficient of linear expansion.
[b] Allotropic variation.
[c] At −39 °C.
[d] 4280 in vitreous state.
[e] Crystalline state changes at 13.2 °C (white to grey).

A26 Properties of commercial metals and alloys

A26.1 Properties of commercial metals and alloys

Substance	Density at 20 °C (kg/m³)	Melting point (°C)	Thermal conductivity (W/m K)	Linear expansivity (K⁻¹ × 10⁶)
Ingot iron	7 860	1538	133	12.2
Cast grey iron (ASTM A48)	7 200	1177	83	12.1
Malleable iron (ASTM A47)	7 320	1232		11.9
Ductile cast iron (ASTM A339–A395	7 200	1149	59	13.5
NI-resist cast iron (Type 2)	7 300	1232	71	17.3
Plain carbon steel (SAE 120)	7 860	1515	97	12.1
Stainless steel (Type 304)	8 020	1427	33	17.3
Hastelloy (C)	3 940	1288	17	11.3
Inconel X (annealed)	8 250	1399	1.7	12.1
Haynes Stellite alloy	9 150	1371	17	13.7
Aluminium 3003 (rolled)	2 730	649	283	23.2
Aluminium 2017 (annealed) (ASTM B221)	2 800	641	301	22.9
Aluminium alloy 389 (ASTM SC 84B)	11 600	566	176	48.6
Copper (ASTM B1/2/3/124/133/152)	8 910	1082	711	16.7
Cupronickel 30%	8 950	1227	54	15.3
Cupronickel (55–45 constantan)	8 900	1260	42	16.0
Nickel (ASTM B160/161/162)	8 890	1441	109	11.9
Beryllium copper (25 ASTM B194)	8 250	927	21	16.7
Yellow brass (ASTM B36/134/135)	8 470	932	221	18.9
Red brass cast (ASTM B30 No. 4A)	8 700	996	133	18.0
Aluminium bronze (ASTM B124/150/169)	7 800	1038	130	16.6
Nickel silver 18% alloy A (ASTM B122 No. 2)	8 800	1110	59	16.2
Chemical lead	11 350	327	62	29.5
Antimonial lead (hard)	10 900	290	54	27.2
Solder (50–50)	8 890	216	83	23.6
Magnesium alloy (ASTM A231B)	1 770	627	142	26.1
Monel K	8 470	1332	33	13.3
Commercial titanium	5 000	1816	33	9.0
Zinc (ASTM B69)	7 140	418	197	12.9
Commercial zirconium	6 500	1843	33	11.7

A26.2 Curie temperatures (°C) for ferromagnetic substances

Iron	1043	Manganese–bismuth	630
Cobalt	1394	Manganese–arsenic	318
Nickel	631	Fe_2O_3	893
Gadolinium	317		

A27 Commercial steel pipes (BS 1600)

A27.1 Summary of commercial steel pipe sizes (refer B16.4) (data from Crane Technical Paper B410 and based on British Standard BS 1600 Part 2 1970 (pipe sizes))

Wall thickness schedule (all dimensions in millimetre)

Ext. diam. (mm)	5 Stainless		10[a]		20		30		40[b,c]		60		80[d,e]		100		120		140		160		Double extra strong	
	Wall	i.d.	Wall	i.d.	Wall	i.d.	Wall	i.d.	Wall	i.d.	Wall	i.d.	Wall	i.d.	Wall	i.d.	Wall	i.d.	Wall	i.d.	Wall	i.d.	Wall	i.d.
10.3			1.24	7.8					1.73	6.8			2.41	5.5										
13.7			1.65	10.4					2.24	9.2			3.02	7.7										
17.1			1.65	13.8					2.31	12.5			3.20	10.7										
21.3	1.65	18.0	2.11	17.1					2.77	15.8			3.73	13.8							4.78	11.7	7.47	6.4
26.7	1.65	23.4	2.11	22.5					2.87	21.0			3.91	18.9							5.56	15.6	7.82	11.1
33.4	1.65	30.1	2.77	27.9					3.38	26.6			4.55	24.3							6.35	20.7	9.09	15.2
42.2	1.65	38.9	2.77	36.7					3.56	35.1			4.85	32.5							6.35	29.5	9.70	22.8
48.3	1.65	45.0	2.77	42.8					3.68	40.9			5.08	38.1							7.14	34.0	10.16	28.0
60.3	1.65	57.0	2.77	54.8					3.91	52.5			5.54	49.2							8.74	42.8	11.07	38.2
73.0	2.11	68.8	3.05	66.9					5.16	62.7			7.01	59.0							9.52	54.0	14.02	45.0
88.9	2.11	84.7	3.05	82.8					5.49	77.9			7.62	73.7							11.13	66.6	15.24	58.4
102	2.11	97.4	3.05	95.5					5.74	90.1			8.08	85.4										
114	2.11	110	3.05	108					6.02	102			8.56	97.2			11.13	92.0			13.49	87.3	17.12	80.1
141	2.77	136	3.40	135					6.55	128			9.52	122			12.70	116			15.88	110	19.05	103
168	2.77	163	3.40	162					7.11	154			10.97	146			14.27	140			18.26	132	21.95	124
219	2.77	214	3.76	212	6.35	206	7.04	205	8.18	203	10.31	199	12.70	194	15.09	189	18.26	183	20.62	178	23.01	173	22.22	175
273	3.40	266	4.19	265	6.35	260	7.80	257	9.27	255	12.70	248	15.09	243	18.26	237	21.44	230	25.40	222	28.58	216	25.40	222
324	3.96	316	4.57	315	6.35	311	8.38	307	10.31	303	14.27	295	17.47	289	21.44	281	25.40	273	28.58	267	33.34	257	25.40	273
356			6.35	343	7.92	340	9.52	337	11.13	333	15.09	325	19.05	318	23.82	308	27.79	300	31.75	292	35.71	284		
406			6.35	394	7.92	391	9.52	387	12.70	381	16.64	373	21.44	364	26.19	354	30.96	345	36.52	333	40.49	325		
457			6.35	445	7.92	441	11.13	435	14.27	429	19.05	419	23.82	410	29.36	399	34.92	387	39.69	378	45.24	367		
508			6.35	495	9.52	489	12.70	483	15.09	478	20.62	467	26.19	456	32.54	443	38.10	432	44.45	419	50.01	408		
610			6.35	597	9.52	591	14.27	581	17.48	575	24.61	560	30.96	548	38.89	532	46.02	518	52.39	505	59.54	491		
762			7.92	746	12.70	737	15.88	730																

a 356–762 mm Schedule 10 available only as stainless.
b Standard wall is as Schedule 40 except 324 mm which is 9.52 mm wall and 305 mm i.d. (not available 356 mm or above).
c Schedule 40 stainless dimensions as standard wall.
d Extra strong wall as Schedule 80 except 273 mm and 324 mm which both have 12.70 mm wall and 248 mm i.d. (not available 356 mm and above).
e Schedule 80 stainless dimensions as extra strong wall.

A27.2 Commercial steel pipes: values of factor C_2 for simplified flow formula (refer B16.4)

Ext. diam. (mm)	Schedule number										Double extra strong
	10	20	30	40	60	80	100	120	140	160	
10.3				13 940 000		46 100 000					
13.7				2 800 000		7 550 000					
17.1				561 000		1 260 000					
21.3				164 600		327 500				756 800	19 680 000
26.7				37 300		65 000				176 200	1 104 000
33.4				10 470		17 000				39 600	200 800
42.2				2 480		3 720				6 140	24 000
48.3				1 100		1 590				2 920	8 150
60.3				297		415				859	1 582
73.0				117		162				257	669
88.9				37.7		50.5				85	170
102				17.6		23.2					
114				9.10		11.88		15.73		20.77	32.72
141				2.798		3.590		4.734		6.318	8.677
168				1.074		1.404		1.786		2.422	3.275
219		0.234	0.243	0.257	0.287	0.326	0.371	0.444	0.509	0.586	0.558
273		0.069 9	0.074 1	0.078 7	0.090 5	0.100 1	0.114 8	0.132 5	0.159 3	0.185 2	
324		0.027 6	0.029 6	0.031 7	0.036 3	0.040 7	0.047 0	0.054 6	0.061 6	0.074 4	
356	0.016 70	0.017 53	0.018 41	0.019 34	0.021 89	0.024 92	0.029 16	0.033 40	0.038 37	0.044 35	
406	0.008 15	0.008 50	0.008 87	0.009 66	0.010 77	0.012 32	0.014 15	0.016 30	0.019 34	0.021 89	
457	0.004 35	0.004 51	0.004 86	0.005 24	0.005 90	0.006 64	0.007 66	0.008 87	0.010 08	0.011 77	
508	0.002 48	0.002 65	0.002 83	0.002 98	0.003 36	0.003 82	0.004 42	0.005 05	0.005 89	0.006 78	
610	0.000 940	0.000 994	0.001 081	0.001 146	0.001 304	0.001 470	0.001 711	0.001 970	0.002 242	0.002 600	

A28 Commercial steel pipes (BS3600)

A28.1 Summary of internal diameters of commercial steel pipe (refer B16.4)

Ext. diam. (mm)	Thickness (mm)																
	1.6	1.8	2.0	2.3	2.6	2.9	3.2	3.6	4.0	4.5	5.0	5.4	5.6	5.9	6.3	7.1	8.0
10.2	7.0	6.6	6.2	5.6													
13.5		9.9	9.5	8.9	8.3	7.7											
17.2			13.2	12.6	12.0	11.4	10.8										
21.3					16.1	15.5	14.9	14.1	13.3	12.3	11.3	10.5					
26.9					21.7	21.1	20.5	19.7	18.9	17.9	16.9	16.1	15.7	15.1	14.3	12.7	
33.7							27.3	26.5	25.7	24.7	23.7	22.9	22.5	21.9	21.1	19.5	17
42.4							36.0	35.2	34.4	33.4	32.4	31.6	31.2	30.6	29.8	28.2	26
48.3							41.9	41.1	40.3	39.3	38.3	37.5	37.1	36.5	35.7	34.1	32
60.3								53.1	52.3	51.0	50.3	49.5	49.1	48.5	47.7	46.1	44
76.1											66.1	65.3	64.9	64.3	63.5	61.9	60
88.9												78.1	77.7	77.1	76.3	74.7	72
102													90.4	89.8	89.0	87.4	85
114												103.1	102.5		101.7	100.1	98
140														128	127	126	124
168															156	154	152
219															207	205	203
273															260	259	257
324															311	310	308
356															343	341	340
406															394	392	390
457															444	443	441
508															495	494	492
610															597	596	594

Data derived from Crane Technical Paper B410 and based on BS 3600 (Pipe Sizes). The sizes in the box are excluded from BS 3600 (1973).

A28.2 Steel pipe to BS 3600: values of factor C_2 for simplified flow formula (refer B16.4)

Ext. diam. (mm)	Wall (mm)							
	1.6	**1.8**	**2.0**	**2.3**	**2.6**	**2.9**	**3.2**	**3.6**
10.2	12 700 000	17 500 000	24 600 000	42 800 000				
13.5		2 010 000	2 530 000	3 620 000	5 290 000	7 940 000		
17.2			436 000	562 000	732 000	967 000	1 300 000	
21.3					151 000	186 000	229 000	309 000
26.9					31 700	36 800	42 900	53 100

Ext. diam. (mm)	**3.2**	**3.6**	**4.0**	**4.5**	**5.0**	**5.4**	**5.6**	**5.9**
33.7	9390	11 000	13 000	15 400	19 400	23 000	25 000	29 300
42.4	2200	2 480	2 800	3 170	3 750	4 250	4 500	5 040
48.3	990	1 100	1 220	1 350	1 560	1 730	1 820	2 000
60.3		283	307	333	371	402	418	449
76.1					88.6	94.1	96.8	102

Ext. diam. (mm)	**5.4**	**5.6**	**5.9**	**6.3**	**7.1**	**8.0**	**8.8**	**10.0**
88.9	37.1	38.0	39.8	42.3	47.1	52.7	59.2	71.5
102		17.2	17.9	18.9	20.7	22.8	25.2	29.6
114		8.71	9.0	9.42	10.22	11.10	12.11	13.91
140			2.83	2.94	3.14	3.35	3.59	4.00
168				1.02	1.08	1.13	1.20	1.31
219				0.234	0.244	0.254	0.265	0.283

Ext. diam. (mm)	**6.3**	**7.1**	**8.0**	**8.8**	**10.0**	**11.0**	**12.5**	**14.2**
273	0.069 9	0.072 1	0.074 4	0.076 9	0.810	0.084 8	0.090 5	0.096 7
324	0.027 6	0.028 4	0.029 2	0.030 0	0.031 3	0.032 5	0.034 3	0.036 3
356	0.016 7	0.017 1	0.017 5	0.018 0	0.018 7	0.019 3	0.020 3	0.021 4
406	0.008 14	0.008 31	0.008 49	0.008 68	0.008 98	0.009 26	0.009 66	0.010 09
457	0.004 34	0.004 42	0.004 51	0.004 59	0.004 74	0.004 86	0.005 05	0.005 24
508	0.002 48	0.002 52	0.002 56	0.002 61	0.002 68	0.002 74	0.002 83	0.002 93
610	0.000 939	0.000 952	0.000 966	0.000 980	0.001 002	0.001 021	0.001 051	0.001 08

Ext. diam. (mm)	**40**	**45**	**50**	**55**	**60**
406	0.021 48	0.025 30			
457	0.010 14	0.011 67	0.013 50		
508	0.005 27	0.005 97	0.006 69	0.007 74	
610	0.001 746	0.001 930	0.002 137	0.002 372	0.002 624

Thickness (mm)

.8	10.0	11.0	12.5	14.2	16.0	17.5	20.0	22.2	25	28	30	32	36	40	45	50	55	60
16.1																		
24.8	22.4																	
30.7	28.3																	
42.7	40.3	38.3																
58.5	56.1	54.1	51.1	47.7														
71.3	68.9	66.9	63.9	60.5	56.9													
84.0	81.6	79.6	76.6	73.2	69.6	66.6												
96.7	94.3	92.3	89.3	85.9	82.3	79.3	74.3											
.22	120	118	115	111	108	105	100											
51	148	146	143	140	136	133	128	124										
02	199	197	194	191	187	184	179	175	169									
55	253	251	248	245	241	238	233	229	223	217	213							
06	304	302	299	296	292	289	280	274	268	264	260	252						
38	336	334	331	327	324	321	316	311	306	300	296	292	284					
89	386	384	381	378	374	371	366	362	356	350	346	342	334	326	316			
39	437	435	432	429	425	422	417	413	407	401	397	393	385	377	367	357		
90	488	486	483	480	476	473	468	464	458	452	448	444	436	428	418	408	398	
92	590	588	585	582	578	575	570	566	560	554	550	546	538	530	520	510	500	490

Wall (mm)

.0	4.5	5.0	5.4	5.6	5.9	6.3	7.1
23 000	591 000	955 000	1 380 000				
66 400	83 800	116 000	148 000	166 000	208 000	289 000	539 000

.3	7.1	8.0	8.8	10.0	11.0	12.5	14.2
6 700	55 400	86 400	143 000				
5 910	7 850	10 600	14 800	26 300			
2 290	2 900	3 730	4 880	7 720			
496	592	711	864	1 190	1600		
110	125	144	166	209	258	354	495

1.0	12.5	14.2	16.0	17.5	20.0	22.2	25.0
4.9	109.4	143.1	191.2				
4.2	42.3	52.8	66.9	85.5			
5.77	18.88	22.80	27.86	34.30	48.61		
4.41	5.08	5.87	6.84	8.01	10.37		
1.42	1.59	1.79	2.02	2.28	2.79	3.35	
0.300	0.326	0.355	0.388	0.425	0.490	0.559	0.677

6.0	17.5	20.0	22.2	25.0	28.0	30.0	32.0	36.0
.103 6	0.111 0	0.124 1	0.137 3	0.159 1	0.179 6	0.198 3		
.038 4	0.040 7	0.044 6	0.048 4	0.054 5	0.060 1	0.065 1	0.070 5	0.083 2
.022 5	0.023 7	0.025 7	0.027 7	0.030 8	0.033 6	0.036 1	0.038 8	0.044 9
.010 55	0.011 03	0.011 83	0.012 61	0.013 83	0.014 90	0.015 83	0.016 83	0.019 06
.005 45	0.005 67	0.006 03	0.006 38	0.006 91	0.007 38	0.007 78	0.008 21	0.009 14
.003 03	0.003 14	0.003 32	0.003 49	0.003 75	0.003 97	0.004 16	0.004 37	0.004 80
.001 112	0.001 144	0.001 198	0.001 248	0.001 324	0.001 388	0.001 442	0.001 500	0.001 621

A29 Volume/velocity relationship for compressed air in steel pipes

A29.1 Volumes (m³/min) of compressed air at 700 kPa flowing at various velocities in steel pipes of different bores

Velocity (m/s)	Pipe bore (mm)																						
	6.8	9.2	12.5	15.8	21.0	26.6	35.1	40.9	52.5	62.7	77.9	90.1	102	128	154	203	255	303	381	429	478	575
3.0	0.0065	0.0120	0.022	0.035	0.062	0.100	0.174	0.24	0.39	0.56	0.86	1.15	1.47	2.3	3.4	5.8	9.2	13.0	21	26	32	47
3.5	0.0076	0.0140	0.026	0.041	0.073	0.117	0.20	0.28	0.45	0.65	1.00	1.34	1.72	2.7	3.9	6.8	10.7	15.1	24	30	38	55
4.0	0.0087	0.0160	0.029	0.047	0.083	0.133	0.23	0.32	0.52	0.74	1.14	1.53	1.96	3.1	4.5	7.8	12.3	17.3	27	35	43	62
4.5	0.0098	0.0179	0.033	0.053	0.094	0.150	0.26	0.35	0.58	0.83	1.29	1.71	2.2	3.5	5.0	8.7	13.8	19.5	31	39	48	70
5.0	0.0109	0.0200	0.037	0.059	0.104	0.167	0.29	0.39	0.65	0.83	1.43	1.91	2.5	3.9	5.6	9.7	15.3	22	34	43	54	78
5.5	0.0120	0.022	0.040	0.065	0.114	0.183	0.32	0.43	0.71	1.02	1.57	2.1	2.7	4.2	6.1	10.7	16.9	24	38	48	59	86
6.0	0.0131	0.024	0.044	0.071	0.125	0.200	0.35	0.47	0.78	1.11	1.72	2.3	2.9	4.6	6.7	11.7	18.4	26	41	52	65	93
6.5	0.0142	0.026	0.048	0.076	0.135	0.22	0.38	0.51	0.84	1.20	1.86	2.5	3.2	5.0	7.3	12.6	19.9	28	44	56	70	101
7.0	0.0153	0.028	0.052	0.082	0.145	0.23	0.41	0.55	0.91	1.30	2.00	2.7	3.4	5.4	7.8	13.6	21	30	48	61	75	109
7.5	0.0163	0.030	0.055	0.088	0.156	0.25	0.44	0.59	0.97	1.39	2.1	2.9	3.7	5.8	8.4	14.6	23	32	51	65	81	117
8.0	0.0174	0.032	0.059	0.094	0.166	0.27	0.46	0.63	1.04	1.48	2.3	3.1	3.9	6.2	8.9	15.5	25	35	55	69	86	125
8.5	0.0185	0.034	0.063	0.100	0.177	0.28	0.49	0.67	1.10	1.57	2.4	3.3	4.2	6.6	9.5	16.5	26	37	58	74	92	132
9.0	0.0196	0.036	0.066	0.106	0.187	0.30	0.52	0.71	1.17	1.67	2.6	3.4	4.4	6.9	10.1	17.5	28	39	62	78	97	140

A30 Pressure drop for compressed air in steel pipe

A30.1 Table of pressure drop for compressed air at 700 kPa in steel pipe

Free air (m³/min)	Compressed air at 700 kPa (m³/min)	Pressure drop (kPa/m) of Schedule 40 pipe for pipe bore as indicated[a]														
		6.8 mm	9.2 mm	12.5 mm	15.8 mm	21.0 mm	26.6 mm	35.1 mm	40.9 mm	52.5 mm	62.7 mm	77.9 mm	90.1 mm	102 mm	128 mm	154 mm
0.1	0.0127	0.86	0.184	0.039	0.013											
0.2	0.0253	3.36	0.698	0.146	0.047	0.011	0.0035									
0.3	0.0379	7.55	1.57	0.319	0.099	0.024	0.0073									
0.4	0.0506		2.71	0.548	0.170	0.041	0.012									
0.5	0.0632		4.10	0.842	0.257	0.062	0.018									
0.6	0.0759		5.90	1.19	0.370	0.088	0.026	0.0066								
0.7	0.0885		8.03	1.62	0.494	0.117	0.035	0.0086	0.0041							
0.8	0.101			2.12	0.634	0.150	0.044	0.011	0.0053							
0.9	0.114			2.64	0.803	0.187	0.055	0.014	0.0065							
1.0	0.126			3.26	0.991	0.231	0.067	0.017	0.0079							
1.25	0.158			4.99	1.55	0.353	0.102	0.026	0.012							
1.50	0.190			7.20	2.19	0.499	0.147	0.036	0.017	0.0048						
1.75	0.221			9.79	2.98	0.679	0.196	0.047	0.022	0.0064						
2.0	0.253				3.82	0.871	0.257	0.062	0.029	0.0082						
2.5	0.316				5.97	1.36	0.393	0.094	0.045	0.012	0.0051					
3.0	0.379				8.6	1.92	0.565	0.135	0.063	0.018	0.0073					
3.5	0.442					2.61	0.754	0.184	0.086	0.024	0.0097					
4.0	0.506					3.41	0.984	0.236	0.110	0.030	0.012					
4.5	0.569					4.32	1.25	0.298	0.136	0.038	0.016	0.0051				
5.0	0.632					5.34	1.54	0.368	0.164	0.046	0.019	0.0063				
6	0.759					7.68	2.17	0.518	0.236	0.066	0.027	0.0090				
7	0.885						2.95	0.689	0.321	0.090	0.036	0.012	0.0059			
8	1.011						3.85	0.900	0.419	0.115	0.047	0.015	0.0075			
9	1.138						4.88	1.14	0.530	0.145	0.058	0.019	0.0094			
10	1.264						6.02	1.41	0.640	0.179	0.072	0.023	0.011			
11	1.391						7.29	1.71	0.774	0.217	0.085	0.028	0.014	0.0073		
12	1.517						8.67	2.02	0.921	0.252	0.101	0.033	0.016	0.0085		
13	1.643							2.38	1.08	0.295	0.119	0.039	0.019	0.0098		
14	1.770							2.76	1.25	0.343	0.138	0.045	0.022	0.011		
15	1.896							3.13	1.44	0.393	0.158	0.051	0.025	0.013		
16	2.02							3.57	1.64	0.443	0.178	0.058	0.028	0.015		
17	2.15							4.01	1.85	0.500	0.200	0.065	0.031	0.016		
18	2.28							4.49	2.07	0.558	0.223	0.072	0.035	0.018		
19	2.40							5.01	2.31	0.618	0.247	0.081	0.039	0.020		
20	2.53							5.49	2.53	0.685	0.266	0.089	0.043	0.022	0.0072	
22	2.78							6.65	3.07	0.825	0.328	0.107	0.052	0.027	0.0086	
24	3.03							7.91	3.61	0.982	0.388	0.126	0.061	0.032	0.010	
26	3.29							9.28	4.22	1.15	0.455	0.148	0.071	0.037	0.012	
28	3.54								4.86	1.33	0.525	0.171	0.082	0.043	0.014	0.0054
30	3.79								5.62	1.52	0.603	0.197	0.094	0.049	0.016	0.0061

A30.1 Table of pressure drop for compressed air at 700 kPa in steel pipe (Continued)

Free air (m³/min)	Compressed air at 700 kPa (m³/min)	Pressure drop (kPa/m) of Schedule 40 pipe for pipe bore as indicated[a]								
		62.7 mm	77.9 mm	90.1 mm	102 mm	128 mm	154 mm	203 mm	40.9 mm / 255 mm	52.5 mm / 303 mm
32	4.05	0.682	0.222	0.106	0.055	0.018	0.0069		6.39	1.73
34	4.30	0.770	0.251	0.119	0.062	0.020	0.0078		7.22	1.94
36	4.55	0.863	0.280	0.134	0.070	0.022	0.0087		8.09	2.17
38	4.80	0.957	0.312	0.148	0.077	0.024	0.0096			2.41
40	5.06	1.05	0.346	0.164	0.086	0.027	0.011			2.67
50	6.32	1.65	0.534	0.254	0.132	0.042	0.016			4.15
60	7.59	2.37	0.765	0.363	0.188	0.059	0.023	0.0058		5.98
70	8.85	3.23	1.03	0.495	0.254	0.080	0.031	0.0077	255 mm	8.14
80	10.11	4.22	1.35	0.639	0.332	0.104	0.040	0.010		
90	11.38	5.34	1.70	0.808	0.418	0.130	0.051	0.013	0.0041	
100	12.64	6.59	2.10	0.992	0.513	0.160	0.062	0.015	0.0050	
110	13.91	7.97	2.54	1.19	0.621	0.192	0.075	0.019	0.0060	
120	15.17	9.49	3.02	1.42	0.739	0.228	0.089	0.022	0.0071	
130	16.43		3.55	1.67	0.862	0.267	0.103	0.026	0.0082	
140	17.70		4.12	1.93	1.00	0.308	0.120	0.029	0.0095	303 mm
150	18.96		4.73	2.22	1.15	0.353	0.138	0.034	0.011	0.0045
200	25.3		8.4	3.94	2.03	0.628	0.243	0.059	0.019	0.0078
250	31.6			6.16	3.17	0.975	0.378	0.090	0.029	0.012
300	37.9			8.88	4.56	1.40	0.540	0.129	0.041	0.017
350	44.3				6.21	1.90	0.735	0.174	0.056	0.023
400	50.6				8.11	2.48	0.960	0.227	0.072	0.030
450	56.9					3.14	1.22	0.286	0.091	0.037
500	63.2					3.88	1.50	0.352	0.112	0.046
550	69.5					4.69	1.82	0.424	0.134	0.056
600	75.9					5.58	2.16	0.504	0.160	0.066
650	82.2					6.55	2.54	0.592	0.188	0.076
700	88.5					7.60	2.94	0.686	0.218	0.089
750	94.8					8.72	3.38	0.788	0.248	0.101
800	101.1						3.84	0.896	0.282	0.115
850	107.5						4.34	1.01	0.319	0.130

[a] For internal diameters greater or less than those shown, decrease or increase the pressure drop in ratio $(d_1/d_2)^5$

A31 Flow of water in steel pipes

A31.1 Flow of water at 15 °C in Schedule 40 steel pipe

The numerals marked ∗ are external/internal diameter (mm).

Volume (m³/min)	Pressure drop (kPa per 100 m)							
	∗10.3/6.8	∗13.7/9.2	∗17.1/12.5	∗21.3/15.8	∗26.7/21.0	∗33.4/26.6	∗42.2/35.1	∗48.3/40.9
0.001	72.6	17						
0.002	259	60	13.6	4.4				
0.003	559	122	29	9.1	2.3			
0.004	957	209	48	15.1	3.8	1.2		
0.005	1445	318	70	22.3	5.7	1.7		
0.006	2029	446	98	30.9	7.7	2.4		
0.008	3516	736	169	52.4	12.9	4.1	1.1	
0.010		1181	252	79.8	19.3	6.1	1.5	0.8
0.015		2567	537	169	40.3	12.4	3.2	1.5
0.020			924	284	68.3	21.0	5.4	2.6
	∗60.3/52.5							
0.030	1.6	∗73.0/62.7		617	145	44.2	11.4	5.3
0.040	2.7	1.0		1072	250	75.8	19.3	9.1
0.050	3.9	1.7			383	114	29.0	13.5
0.060	5.5	2.3			541	161	40.0	18.7
0.070	7.8	3.1			727	215	54.1	24.8
			∗88.9/77.9					
0.080	9.5	3.9	1.4	∗102/90.1	927	276	69.0	31.5
0.090	11.5	4.8	1.7	0.8		347	86.2	39.7
0.100	14.1	5.9	2.0	1.0	∗114/102	425	105	48.8
0.150	29.5	12.5	4.2	2.1	1.1	930	226	103
0.200	51.2	21.2	7.2	3.6	1.9		391	181
0.25	77.3	32.2	10.8	5.3	2.8	∗141/128		274
0.30	110	44.9	15.2	7.4	4.0	1.4		382
0.35	147	60.6	20.3	9.9	5.3	1.8	∗168/154	518
0.40	192	78.0	26.4	12.8	6.8	2.3	0.9	669
0.45	239	97.9	32.9	16.1	8.4	2.8	1.2	845
0.50	295	120	40.3	19.6	10.1	3.4	1.4	
0.55	355	144	47.9	23.2	12.2	4.1	1.6	
0.60	420	169	56.6	27.3	14.6	4.7	1.9	
0.65	488	197	65.8	31.9	16.9	5.5	2.2	
0.70	563	228	75.9	36.8	19.4	6.3	2.5	
0.75	644	260	86.3	42.0	21.8	7.2	2.9	
0.80		295	97.7	47.3	24.6	8.1	3.2	∗219/203
0.85		331	109	52.8	27.7	9.1	3.6	0.9
0.90			122	58.5	30.8	10.0	4.1	1.0
0.95			135	64.9	34.2	11.1	4.5	1.2
1.0			150	71.4	37.7	12.2	4.9	1.3
1.1			175	86.0	45.2	14.7	5.9	1.5
1.2			214	102	53.4	17.2	6.9	1.8
1.3				119	62.7	20.0	8.0	2.1
1.4				137	72.2	23.2	9.1	2.4
1.5				156	81.8	26.4	10.5	2.7
1.6				178	92.4	29.7	11.8	3.1
1.7	∗273/255			199	104	33.1	13.2	3.5
1.8	1.2				116	36.9	14.7	3.9
1.9	1.4				128	41.0	16.3	4.2
2.0	1.5				141	45.2	18.1	4.6
2.2	1.8				170	54.5	21.7	5.6
2.4	2.1	∗324/303				64.5	25.3	6.5
2.6	2.5	1.0				74.9	29.6	7.6
2.8	2.8	1.2				85.9	33.9	8.7

continued

A31.1 Flow of water at 15 °C in Schedule 40 steel pipe (Continued)

Volume (m³/min)	Pressure drop (kPa per 100 m)							
			*356/333					
3.0	3.2	1.3	0.8			98.2	38.7	9.9
3.5	4.3	1.8	1.1	*406/381		133	52.6	13.4
4.0	5.5	2.3	1.4	0.9		172	67.3	17.2
5.0	8.4	3.4	2.2	1.1			104	26.2
6.0	11.8	4.9	3.1	1.6			147	37.3
					*457/429			
7.0	15.8	6.5	4.2	2.1	1.2		200	49.9
8.0	20.4	8.5	5.4	2.7	1.5		259	65.0
9.0	25.6	10.7	6.7	3.3	1.9			81.6
10.0	31.3	13.0	8.1	4.1	2.3	*508/478		99.2
12.0	44.7	18.4	11.4	5.7	3.2	1.9		141
14	60.0	24.6	15.3	7.7	4.4	2.5	*610/575	191
16	77.6	31.7	19.8	9.9	5.6	3.2	1.3	248
18	97.5	39.8	24.6	12.4	6.9	4.0	1.6	
20	119	48.7	30.2	15.2	8.4	4.9	2.0	
25	183	75.8	46.9	23.4	13.0	7.6	3.0	
30		108	66.9	33.2	18.3	10.8	4.3	
35		146	90.3	44.6	24.8	14.4	5.7	
40		190	117	57.8	31.9	18.6	7.4	
45		239	147	72.6	40.0	23.3	9.2	
50			181	88.8	49.1	28.4	11.3	
55				107	59.4	34.3	13.6	
60				127	70.8	41.1	16.1	
65				149	82.2	47.5	18.9	
70				170	95.5	55.2	21.6	
75				198	110	62.8	24.6	

Velocity is calculated from $\dfrac{\text{Volume}}{(\pi/4 \times (\text{int. diam})^2)}$ m/s

A31.2 Equivalent length (m) of pipe fittings for resistance calculations

Pipe bore (mm) Fitting	6.8	9.2	12.5	15.8	21.0	26.6	35.1	40.9	52.5	62.7	77.9	90.1	102	128
Elbow	0.14	0.18	0.25	0.35	0.47	0.62	0.86	1.0	1.4	1.7	2.2	2.6	3.2	4.4
Long 90° bend	0.08	0.11	0.14	0.17	0.24	0.33	0.46	0.54	0.74	0.88	1.1	1.3	1.6	2.1
Return bend	0.25	0.32	0.48	0.60	0.75	1.0	1.3	1.6	2.1	2.5	3.3	4.0	4.7	6.3
Globe valve	0.48	0.62	0.80	1.1	1.4	1.9	2.6	3.2	4.0	4.8	6.2	7.8	9.3	13.0
Gate valve	0.05	0.07	0.10	0.13	0.18	0.25	0.34	0.40	0.55	0.68	0.83	1.0	1.2	1.6
Tee-run	0.06	0.09	0.13	0.17	0.24	0.33	0.45	0.53	0.73	0.9	1.1	1.3	1.6	2.1
Tee-outlet	0.25	0.40	0.52	0.7	1.0	1.3	1.7	2.1	2.9	3.7	5.0	6.3	7.4	10.0

A31.3 Velocity head loss factors for ducts (Refer B16.1)

Restriction	Factor	Restriction	Factor
90° rounded elbow	0.65	Abrupt enlargement	0.30
90° square elbow	1.25	Gradual enlargement	0.15
90° bend (r = 2d)	0.10	Abrupt reduction	0.30
Branch	0.6–1.3	Gradual reduction	0.04
Open damper	0.30	Diffuser	0.60
Wire mesh	0.40	Outlet	1.00

A32 Loss of heat from pipes

A32.1 Heat emission (W/m) from bare steel pipes in still air at 10–20° C ambient

Temp diff. steam to air (K)	Nominal bore of pipe (mm)									
	15	20	25	32	40	50	65	80	100	150
50	46	54	68	92	97	116	136	162	206	292
60	58	70	87	113	121	147	174	207	261	364
70	72	86	108	134	149	181	212	255	316	443
80	87	103	128	157	178	214	252	305	375	527
90	101	122	150	179	205	248	297	353	438	615
100	115	140	172	210	235	285	345	404	503	706
110	133	160	196	238	268	330	390	463	580	810
120	150	182	222	269	302	373	440	522	657	910
130	168	203	248	303	337	418	492	585	737	1015
140	186	224	276	337	377	465	548	650	820	1135
150	205	250	303	374	416	512	607	715	905	1265
160	226	276	342	413	460	565	670	785	995	1395
170	248	302	368	454	507	620	732	860	1085	1530
180	273	329	405	498	559	675	807	955	1180	1660
190	297	363	437	546	610	735	880	1050	1275	1800
200	323	400	475	594	661	795	945	1140	1385	1940

Factors for reduction of pipe emission

A32.2 Vertical pipes

Bore (mm)	Factor	Bore (mm)	Factor
15	0.76	50	0.88
20	0.80	65	0.91
25	0.82	80	0.93
32	0.84	100	0.95
40	0.86	150	1.00

A32.3 Banked pipes

No. high	Factor	No. high	Factor
10	0.63	5	0.82
9	0.67	4	0.86
8	0.70	3	0.91
7	0.74	2	0.96
6	0.78	1	1.00

A32.4 Factors for increase in emission for air movement

Air velocity (m/s)	Factor	Air velocity (m/s)	Factor
1.5	1.4	7.0	3.5
2.0	1.7	8.0	3.7
3.0	2.2	9.0	4.0
4.0	2.6	10.0	4.2
5.0	2.9	12.0	4.6
6.0	3.2	15.0	5.2

A32.5 Evaporation rates from one m² of water surface[a]

Water temp (°C)	Evaporation (kg/hr)	Water temp (°C)	Evaporation (kg/hr)	Water temp (°C)	Evaporation (kg/hr)
20	0.02	70	0.54	94	1.33
30	0.05	75	0.66	96	1.42
40	0.11	80	0.80	98	1.54
50	0.20	85	0.94	99	1.59
60	0.35	90	1.16	100	1.65
65	0.44	92	1.23		

[a] These figures are based on 15°C, 101.3 kN/m² and R.H. 55% for still air. Increase rates by 30% for slight air movement and by 60% for rapid air movement.

A33 Variation with pressure-temperature of the physical properties of saturated steam

Absolute pressure (kPa)	Temp (°C)	Specific enthalpy[a] (kJ/kg)			Specific heat at constant pressure (kJ/kg K)	Density (kg/m³)	Absolute viscosity ×10⁶ (Ns/m²)	Thermal conductivity (W/m K)	Prandtl no. for saturated vapour	Pressure factor[b]
		Saturated liquid	Evaporation increment	Saturated vapour						
0.6	0	0	2502	2502	1.86	0.0048	8.00	0.0176	0.85	
0.7	1.8	8	2497	2505	1.86	0.0052	8.09	0.0178	0.85	
0.8	4.0	17	2492	2509	1.86	0.0060	8.19	0.0180	0.85	
0.9	5.5	25	2487	2512	1.86	0.0068	8.28	0.0181	0.85	
1.0	7.0	34	2482	2516	1.86	0.0076	8.37	0.0182	0.86	
1.5	13	54	2473	2527	1.87	0.011	8.54	0.0184	0.87	
2.0	17	71	2461	2532	1.87	0.015	8.76	0.0186	0.88	
2.5	21	88	2452	2540	1.87	0.019	8.96	0.0189	0.89	
3	24	101	2444	2545	1.88	0.023	9.07	0.0191	0.89	
4	29	122	2432	2554	1.89	0.030	9.28	0.0194	0.90	
5	33	138	2423	2561	1.89	0.037	9.44	0.0196	0.91	0.0031
6	36	150	2416	2566	1.89	0.044	9.56	0.0199	0.91	0.0042
7	39	164	2408	2572	1.90	0.051	9.68	0.0201	0.92	0.0062
8	42	175	2402	2577	1.90	0.057	9.81	0.0203	0.92	0.0078
9	44	184	2397	2581	1.90	0.063	9.88	0.0204	0.92	0.0097
10	46	192	2393	2585	1.90	0.069	9.96	0.0206	0.92	0.0115
12	50	209	2383	2592	1.91	0.082	10.12	0.0208	0.93	0.0162
14	53	222	2375	2597	1.91	0.094	10.20	0.0211	0.93	0.0218
16	56	234	2368	2602	1.92	0.106	10.30	0.0213	0.93	0.0288
18	58	242	2364	2606	1.92	0.118	10.40	0.0215	0.93	0.0356
20	60	252	2358	2610	1.92	0.131	10.50	0.0217	0.93	0.0442
22	62	259	2354	2613	1.93	0.143	10.57	0.0218	0.94	0.0540
24	64	268	2349	2617	1.93	0.155	10.64	0.0220	0.94	0.0640
26	66	276	2345	2621	1.94	0.167	10.71	0.0221	0.94	0.075
28	67	283	2340	2623	1.94	0.179	10.79	0.0222	0.94	0.086
30	69	289	2336	2625	1.95	0.191	10.86	0.0223	0.95	0.097
32	71	297	2332	2629	1.95	0.203	10.90	0.0224	0.95	0.110
34	72	303	2328	2631	1.96	0.215	10.95	0.0226	0.95	0.124
36	74	309	2324	2633	1.96	0.227	11.00	0.0227	0.95	0.138
38	75	314	2321	2635	1.96	0.239	11.05	0.0228	0.95	0.153
40	76	318	2319	2637	1.97	0.250	11.10	0.0229	0.95	0.169
45	79	330	2312	2642	1.97	0.280	11.21	0.0231	0.96	0.213
50	81	341	2304	2646	1.97	0.309	11.32	0.0233	0.96	0.261
55	84	351	2299	2650	1.98	0.337	11.41	0.0235	0.96	0.314
60	86	360	2294	2654	1.98	0.366	11.50	0.0237	0.96	0.372

65	88	368	2288	2656	1.99	0.394	11.58	0.0238	0.97	0.434
70	90	377	2283	2660	2.00	0.422	11.66	0.0240	0.97	0.501
80	94	392	2274	2666	2.01	0.479	11.78	0.0244	0.97	0.649
90	97	405	2266	2671	2.02	0.535	11.90	0.0247	0.97	0.815
100	100	418	2258	2675	2.03	0.59	12.02	0.0249	0.98	1.025
110	102	429	2251	2680	2.03	0.65	12.12	0.0251	0.98	1.230
120	105	439	2244	2683	2.04	0.70	12.21	0.0254	0.98	1.453
130	107	449	2238	2687	2.05	0.75	12.30	0.0256	0.98	1.694
140	109	459	2232	2690	2.06	0.81	12.40	0.0258	0.99	1.953
150	111	467	2226	2693	2.07	0.86	12.47	0.0260	0.99	2.23
160	113	475	2221	2696	2.08	0.92	12.56	0.0261	1.00	2.53
170	115	483	2216	2699	2.09	0.97	12.64	0.0262	1.01	2.84
180	117	491	2211	2702	2.10	1.02	12.69	0.0264	1.01	3.34
190	119	498	2206	2704	2.11	1.08	12.74	0.0265	1.01	3.51
200	120	504	2202	2706	2.12	1.13	12.80	0.0267	1.02	3.88
220	123	518	2193	2711	2.14	1.23	12.91	0.0270	1.02	4.66
240	126	530	2185	2715	2.15	1.34	13.01	0.0272	1.03	5.51
260	129	541	2177	2718	2.17	1.44	13.10	0.0276	1.03	6.43
280	131	551	2170	2721	2.19	1.55	13.19	0.0279	1.04	7.42
300	134	562	2163	2725	2.21	1.65	13.29	0.0282	1.04	8.47
320	136	571	2157	2728	2.22	1.75	13.38	0.0284	1.05	9.60
340	138	580	2150	2730	2.23	1.86	13.46	0.0286	1.05	10.79
360	140	589	2144	2733	2.25	1.96	13.55	0.0289	1.05	12.05
380	142	597	2138	2735	2.26	2.06	13.65	0.0291	1.06	13.37
400	144	605	2133	2738	2.28	2.16	13.75	0.0293	1.07	14.76
420	145	612	2128	2740	2.29	2.27	13.80	0.0295	1.07	16.22
440	147	620	2122	2742	2.30	2.37	13.85	0.0296	1.08	17.75
460	149	627	2117	2744	2.31	2.47	13.89	0.0298	1.08	19.34
480	150	634	2112	2746	2.32	2.57	13.94	0.0300	1.08	21.00
500	152	640	2107	2747	2.34	2.67	13.98	0.0302	1.08	22.72
520	153	647	2103	2749	2.35	2.77	14.02	0.0304	1.09	24.51
540	155	653	2098	2751	2.36	2.87	14.06	0.0305	1.09	26.36
560	156	659	2094	2753	2.37	2.97	14.10	0.0307	1.09	28.28
580	158	665	2089	2754	2.39	3.07	14.14	0.0309	1.09	30.27
600	159	670	2085	2755	2.40	3.17	14.18	0.0311	1.10	32.32
620	160	676	2081	2757	2.41	3.27	14.22	0.0312	1.10	34.44
640	161	682	2077	2758	2.42	3.37	14.26	0.0314	1.10	36.62
660	163	687	2073	2760	2.43	3.47	14.30	0.0316	1.10	38.86
680	164	692	2069	2761	2.44	3.57	14.34	0.0317	1.10	41.17
700	165	697	2065	2762	2.45	3.67	14.38	0.0319	1.10	43.54
720	166	702	2061	2763	2.46	3.77	14.43	0.0321	1.11	45.98
740	167	707	2057	2764	2.47	3.87	14.47	0.0322	1.11	48.48
760	168	712	2054	2765	2.48	3.96	14.51	0.0324	1.11	51.05

continued

A33 (Continued)

Absolute pressure (kPa)	*Temp* (°C)	*Specific enthalpy*[a] (kJ/kg)			*Specific heat at constant pressure* (kJ/kg K)	*Density* (kg/m^3)	*Absolute viscosity* $\times 10^6$ (Ns/m^2)	*Thermal conductivity* (W/m K)	*Prandtl no. for saturated vapour*	*Pressure factor*[b]
		Saturated liquid	*Evaporation increment*	*Saturated vapour*						
780	169	716	2050	2766	2.50	4.06	14.55	0.0325	1.12	53.68
800	170	721	2047	2768	2.51	4.16	14.59	0.0326	1.12	56.38
820	171	725	2043	2769	2.52	4.26	14.63	0.0327	1.12	59.13
840	172	730	2040	2769	2.53	4.36	14.67	0.0329	1.13	61.96
860	173	734	2036	2770	2.54	4.46	14.71	0.0330	1.13	64.84
880	174	739	2033	2771	2.56	4.56	14.75	0.0332	1.13	67.79
900	175	743	2030	2772	2.57	4.66	14.79	0.0334	1.14	70.80
920	176	747	2026	2773	2.58	4.75	14.84	0.0336	1.14	73.88
940	177	751	2023	2774	2.59	4.85	14.88	0.0337	1.14	77.02
960	178	755	2020	2775	2.60	4.95	14.92	0.0338	1.15	80.22
980	179	759	2017	2775	2.61	5.05	14.96	0.0340	1.15	83.49
1 000	180	763	2014	2776	2.62	5.15	15.00	0.0341	1.15	86.81
1 020	181	767	2011	2777	2.63	5.25	15.03	0.0343	1.15	90.20
1 040	182	770	2008	2778	2.64	5.34	15.07	0.0344	1.16	93.66
1 060	183	774	2005	2778	2.65	5.44	15.10	0.0346	1.16	97.18
1 080	183	777	2002	2779	2.66	5.54	15.13	0.0347	1.16	100.75
1 100	184	781	1999	2780	2.67	5.64	15.17	0.0349	1.16	104.4
1 150	186	790	1991	2781	2.70	5.88	15.23	0.0352	1.17	113.8
1 200	188	798	1984	2783	2.72	6.13	15.29	0.0356	1.17	123.5
1 250	190	807	1977	2784	2.75	6.37	15.36	0.0359	1.18	133.7
1 300	192	815	1971	2785	2.78	6.62	15.42	0.0361	1.19	144.3
1 350	193	823	1964	2787	2.81	6.86	15.48	0.0363	1.20	155.2
1 400	195	830	1958	2788	2.83	7.11	15.54	0.0365	1.20	166.5
1 450	197	838	1951	2789	2.85	7.35	15.59	0.0368	1.21	178.2
1 500	198	845	1945	2790	2.88	7.60	15.64	0.0371	1.21	190.3
1 600	201	859	1933	2792	2.92	8.08	15.76	0.0376	1.22	215.6
1 700	204	872	1921	2793	2.96	8.57	15.87	0.0381	1.23	242.5
1 800	207	885	1910	2795	3.01	9.06	15.97	0.0386	1.24	270.8
1 900	210	897	1899	2796	3.06	9.56	16.07	0.0392	1.25	300.7
2 000	212	908	1889	2797	3.10	10.05	16.16	0.0398	1.26	
2 100	215	920	1878	2798	3.15	10.54	16.24	0.0404	1.27	
2 200	217	931	1868	2799	3.19	11.03	16.32	0.0410	1.27	
2 300	220	942	1858	2800	3.23	11.52	16.40	0.0414	1.28	
2 400	222	952	1848	2800	3.27	12.02	16.47	0.0419	1.29	
2 500	224	962	1839	2801	[illegible]	[illegible]	[illegible]	[illegible]	[illegible]	

2 700	228	981	1820	2801	3.40	13.5	16.68	0.0434	1.31
2 800	230	991	1811	2802	3.45	14.0	16.75	0.0439	1.32
2 900	232	1000	1802	2802	3.49	14.5	16.82	0.0444	1.32
3 000	234	1008	1794	2802	3.53	15.0	16.89	0.0449	1.33
3 200	237	1025	1777	2802	3.61	16.0	17.02	0.0458	1.34
3 400	241	1042	1760	2802	3.69	17.0	17.15	0.0468	1.35
3 600	244	1058	1744	2802	3.77	18.0	17.27	0.0478	1.36
3 800	247	1073	1728	2801	3.85	19.1	17.39	0.0487	1.37
4 000	250	1087	1713	2800	3.93	20.1	17.51	0.0496	1.39
4 200	253	1101	1698	2799	4.00	21.1	17.63	0.0505	1.40
4 400	256	1115	1683	2798	4.08	22.2	17.75	0.0514	1.41
4 600	259	1129	1668	2797	4.17	23.2	17.86	0.0523	1.42
4 800	261	1142	1654	2796	4.26	24.3	17.97	0.0533	1.44
5 000	264	1154	1640	2794	4.36	25.4	18.07	0.0543	1.45
5 200	266	1167	1626	2793	4.45	26.4	18.17	0.0552	1.46
5 400	269	1179	1612	2791	4.54	27.5	18.27	0.0561	1.48
5 600	271	1191	1598	2789	4.63	28.6	18.37	0.0571	1.49
5 800	273	1202	1585	2787	4.73	29.7	18.47	0.0580	1.51
6 000	276	1214	1571	2785	4.83	30.8	18.57	0.0590	1.52
6 200	278	1225	1558	2783	4.94	31.9	18.66	0.0600	1.54
6 400	280	1236	1545	2781	5.04	33.1	18.75	0.0609	1.55
6 600	282	1246	1532	2778	5.14	34.2	18.84	0.0618	1.57
6 800	284	1257	1519	2776	5.23	35.4	18.93	0.0628	1.58
7 000	286	1267	1506	2773	5.32	36.5	19.02	0.0638	1.59
7 200	288	1278	1493	2771	5.41	37.7	19.11	0.0647	1.60
7 400	290	1288	1480	2768	5.51	38.9	19.21	0.0657	1.61
7 600	291	1298	1468	2766	5.61	40.1	19.31	0.0667	1.62
7 800	293	1308	1455	2763	5.71	41.3	19.41	0.0678	1.63
8 000	295	1317	1443	2760	5.82	42.5	19.50	0.0689	1.65
8 200	297	1327	1430	2757	5.93	43.7	19.59	0.0699	1.66
8 400	298	1336	1418	2754	6.03	45.0	19.67	0.0710	1.67
8 600	300	1345	1406	2751	6.13	46.3	19.75	0.0720	1.68
8 800	302	1355	1393	2748	6.23	47.5	19.83	0.0731	1.69
9 000	303	1364	1381	2745	6.32	48.8	19.89	0.0741	1.70
9 200	305	1373	1368	2741	6.42	50.1	19.96	0.0752	1.70
9 400	306	1382	1356	2738	6.52	51.4	20.02	0.0763	1.71
9 600	308	1391	1344	2735	6.62	52.7	20.07	0.0774	1.72
9 800	309	1399	1332	2731	6.73	54.0	20.1	0.0786	1.72
10 000	311	1408	1320	2728	6.83	55.4	20.2	0.0798	1.73
10 400	314	1425	1295	2720	7.04	58.2	20.4	0.0823	1.75
10 800	317	1442	1271	2713	7.4	61.0	20.6	0.0848	1.81
11 200	319	1459	1246	2705	7.9	63.9	20.8	0.0875	1.87
11 600	322	1475	1222	2697	8.3	66.9	21.0	0.0903	1.93

continued

54

A3.3 (Continued)

Absolute pressure (kPa)	Temp (°C)	Specific enthalpy[a] (kJ/kg)			Specific heat at constant pressure (kJ/kg K)	Density (kg/m³)	Absolute viscosity ×10⁶ (Ns/m²)	Thermal conductivity (W/m K)	Prandtl no. for saturated vapour	Pressure factor[b]
		Saturated liquid	Evaporation increment	Saturated vapour						
12 000	325	1492	1197	2689	8.7	70.0	21.2	0.0932	1.98	
12 400	327	1508	1173	2681	9.1	73.2	21.4	0.0962	2.03	
12 800	330	1524	1148	2672	9.6	76.5	21.6	0.0997	2.08	
13 200	332	1540	1122	2662	10.0	80.0	21.8	0.103	2.13	
13 600	334	1556	1097	2653	10.5	83.4	22.0	0.106	2.18	
14 000	337	1571	1071	2642	11.0	87.0	22.2	0.109	2.24	
14 400	339	1588	1044	2632	11.6	90.8	22.4	0.112	2.32	
14 800	341	1603	1018	2621	12.3	94.7	22.7	0.116	2.41	
15 200	343	1619	990	2609	13.0	98.8	23.0	0.120	2.49	
15 600	345	1635	962	2597	13.9	103	23.4	0.124	2.62	
16 000	347	1651	934	2585	15.0	107	23.7	0.128	2.78	
16 400	349	1667	905	2572	16.1	112	24.1	0.132	2.94	
16 800	351	1683	873	2556	17.2	117	24.6	0.137	3.09	
17 200	353	1700	843	2543	18.3	122	25.0	0.142	3.22	
17 600	355	1718	811	2529	20.5	128	25.4	0.147	3.54	
18 000	357	1735	778	2513	23.0	133	25.8	0.152	3.90	
18 500	359	1756	736	2492	27.3	141	26.5	0.159	4.55	
19 000	361	1779	692	2471	32.5	150	27.2	0.166	5.3	
19 500	363	1801	644	2445	40	159	27.9	0.173	6.5	
20 000	366	1826	592	2418	53	170	28.8	0.181	8.4	
20 500	368	1853	533	2386	74	183	29.7	0.190	11.6	
21 000	370	1887	459	2349	107	199	30.6	0.201	16.3	
21 500	371	1935	385	2310	190	222	32.0	0.216	28.1	
22 000	374	2011	185	2196		268	38.0	0.237		
22 120	374.15	2107	0	2107		315	41.4	0.240		

[a] Refer to **A2.5** and **B17.4.1**. Specific enthalpy saturated liquid is the 'water' component, while evaporational increment is the 'latent heat' component.
[b] Refer to **B17.4.3**.

A34 Superheated steam: variation with temperature of total specific enthalpy (kJ/kg) (top lines) and specific volume (m³/kg) (bottom lines)

Absolute pressure (bar)	Saturation temp. (°C)	Total temperature (°C)															
		50	100	120	140	160	180	200	250	300	350	400	450	500	550	600	650
0.01	7	2595	2689	2723	2762	2802	2841	2880	2978	3077	3177	3280	3384	3492	3599	3708	3818
		149	172	182	191	200	209	218	241	264	287	311	334	356	381	404	427
0.05	33	2594	2688	2722	2761	2801	2841	2880	2978	3077	3177	3280	3384	3491	3598	3708	3818
		29.8	34.4	36.2	38.1	39.9	41.8	43.7	48.3	52.9	57.5	62.1	66.7	71.2	76.0	80.6	85.2
0.5	81		2683	2718	2758	2798	2839	2878	2976	3076	3177	3279	3383	3490	3597	3707	3817
			3.42	3.60	3.80	3.98	4.17	4.36	4.82	5.28	5.75	6.21	6.67	7.12	7.60	8.06	8.52
1.0	100					2796	2836	2875	2975	3075	3176	3278	3382	3489	3597	3707	3816
						1.98	2.08	2.17	2.41	2.64	2.87	3.10	3.33	3.56	3.80	4.02	4.26
1.5	111					2793	2833	2873	2973	3073	3175	3277	3381	3489	3595	3706	3815
						1.33	1.41	1.46	1.62	1.78	1.94	2.09	2.25	2.41	2.56	2.73	2.91
2	120					2789	2830	2871	2971	3072	3174	3277	3381	3488	3595	3706	3815
						0.99	1.04	1.08	1.20	1.32	1.43	1.55	1.67	1.80	1.92	2.03	2.15
3	134					2782	2824	2866	2968	3070	3172	3275	3380	3487	3594	3705	3814
						0.65	0.68	0.72	0.80	0.88	0.95	1.03	1.11	1.19	1.26	1.34	1.42
4	144					2774	2818	2860	2965	3067	3170	3274	3379	3486	3593	3704	3814
						0.48	0.51	0.53	0.60	0.65	0.71	0.77	0.83	0.89	0.95	1.00	1.06
5	152					2766	2811	2855	2961	3065	3168	3272	3377	3485	3592	3703	3813
						0.38	0.40	0.42	0.47	0.52	0.57	0.62	0.66	0.71	0.76	0.81	0.85
6	159					2758	2805	2849	2958	3062	3166	3271	3376	3484	3591	3702	3812
						0.32	0.33	0.35	0.39	0.43	0.47	0.51	0.55	0.59	0.63	0.67	0.71
7	165						2798	2844	2954	3060	3164	3269	3375	3483	3590	3701	3811
							0.28	0.30	0.34	0.37	0.41	0.44	0.47	0.50	0.54	0.58	0.61
8	170						2791	2838	2950	3057	3162	3268	3373	3482	3589	3700	3811
							0.247	0.26	0.29	0.32	0.35	0.38	0.41	0.44	0.47	0.50	0.53
9	175						2783	2832	2947	3055	3161	3266	3372	3480	3588	3699	3810
							0.218	0.230	0.26	0.29	0.31	0.34	0.37	0.39	0.42	0.44	0.47
10	180						2777	2827	2943	3052	3159	3264	3371	3479	3587	3698	3809
							0.194	0.206	0.233	0.26	0.28	0.31	0.33	0.35	0.38	0.39	0.42

A34 Superheated steam: variation with temperature of total specific enthalpy (kJ/kg) (top lines) and specific volume (m³/kg) (bottom lines) (Continued)

Absolute pressure (bar)	Saturation temp. (°C)	Total temperature (°C) 200	220	240	260	280	300	320	340	360	380	400	450	500	550	600	650
11	184	2820	2870	2916	2961	3006	3050	3093	3135	3178	3220	3263	3370	3478	3586	3697	3809
		0.186	0.196	0.206	0.215	0.224	0.234	0.243	0.25	0.26	0.27	0.28	0.30	0.32	0.34	0.36	0.39
12	188	2814	2865	2912	2958	3003	3047	3090	3133	3176	3218	3261	3368	3477	3585	3696	3808
		0.169	0.179	0.188	0.196	0.205	0.214	0.223	0.231	0.239	0.247	0.25	0.27	0.29	0.31	0.33	0.35
13	192	2808	2859	2907	2954	2999	3044	3087	3131	3175	3217	3260	3367	3475	3584	3695	3807
		0.155	0.164	0.172	0.181	0.189	0.197	0.205	0.212	0.220	0.227	0.235	0.25	0.27	0.29	0.31	0.33
14	195	2801	2854	2905	2952	2997	3042	3085	3128	3173	3215	3258	3366	3474	3583	3695	3806
		0.143	0.152	0.160	0.167	0.175	0.182	0.189	0.196	0.203	0.210	0.218	0.235	0.25	0.27	0.29	0.30
16	201		2843	2891	2939	2988	3036	3080	3125	3169	3212	3255	3363	3472	3581	3692	3805
			0.131	0.138	0.145	0.152	0.159	0.165	0.172	0.178	0.184	0.190	0.205	0.220	0.235	0.25	0.27
18	207		2831	2882	2932	2982	3031	3075	3121	3165	3208	3252	3360	3470	3580	3691	3804
			0.115	0.122	0.128	0.134	0.140	0.146	0.151	0.157	0.162	0.168	0.182	0.196	0.209	0.222	0.235
20	212		2820	2873	2925	2976	3025	3070	3116	3160	3205	3249	3358	3468	3578	3690	3802
			0.102	0.108	0.114	0.120	0.126	0.131	0.136	0.141	0.146	0.151	0.163	0.176	0.188	0.200	0.211
22	217		2808	2864	2918	2969	3019	3065	3111	3156	3201	3246	3355	3466	3576	3689	3801
			0.092	0.097	0.103	0.108	0.113	0.118	0.122	0.127	0.132	0.137	0.148	0.159	0.170	0.181	0.192
24	222			2856	2910	2963	3013	3061	3107	3153	3197	3242	3353	3464	3574	3687	3799
				0.089	0.094	0.099	0.104	0.108	0.113	0.117	0.121	0.125	0.136	0.146	0.156	0.166	0.176
26	226			2848	2903	2957	3007	3056	3103	3149	3194	3239	3350	3461	3572	3685	3798
				0.081	0.086	0.090	0.095	0.099	0.103	0.107	0.111	0.119	0.126	0.134	0.144	0.153	0.162
28	230			2837	2894	2950	3001	3051	3099	3145	3190	3236	3347	3458	3570	3683	3796
				0.075	0.079	0.083	0.088	0.092	0.095	0.099	0.103	0.110	0.118	0.125	0.133	0.142	0.150
30	234			2827	2885	2942	2995	3045	3094	3141	3187	3232	3344	3456	3568	3682	3795
				0.069	0.073	0.077	0.081	0.085	0.089	0.093	0.096	0.103	0.110	0.116	0.124	0.132	0.140
32	237			2817	2876	2934	2989	3040	3089	3137	3183	3229	3341	3454	3566	3680	3794
				0.064	0.068	0.072	0.076	0.079	0.083	0.087	0.090	0.096	0.103	0.109	0.116	0.124	0.131
34	241				2866	2927	2982	3035	3084	3132	3180	3226	3339	3452	3564	3678	3792
					0.063	0.067	0.071	0.074	0.076	0.080	0.084	0.090	0.096	0.102	0.109	0.117	0.124
36	244				2856	2919	2976	3029	3080	3128	3176	3222	3336	3450	3562	3676	3791
					0.059	0.063	0.066	0.070	0.073	0.076	0.079	0.084	0.090	0.096	0.103	0.110	0.117
38	247				2846	2910	2969	3023	3075	3124	3172	3219	3334	3447	3561	3675	3789
					0.055	0.059	0.062	0.066	0.069	0.072	0.075	0.080	0.086	0.091	0.098	0.104	0.110
40	250				2836	2902	2962	3018	3070	3119	3168	3215	3331	3445	3559	3674	3788
					0.052	0.055	0.059	0.062	0.065	0.068	0.071	0.076	0.081	0.086	0.093	0.099	0.105
42	253				2825	2894	2955	3012	3065	3115	3165	3212	3328	3443	3557	3672	3786
					0.049	0.052	0.056	0.059	0.062	0.065	0.067	0.072	0.077	0.082	0.088	0.094	0.100
44	256				2814	2885	2948	3006	3060	3111	3161	3209	3325	3441	3555	3670	3785
					0.046	0.050	0.053	0.056	0.059	0.061	0.064	0.068	0.073	0.078	0.084	0.090	0.095
46	259				2802	2876	2941	3000	3055	3106	3157	3205	3323	3438	3553	3669	3784
						0.047	0.050										

A34 Superheated steam: variation with temperature of total specific enthalpy (kJ/kg) (top lines) and specific volume (m³/kg) (bottom lines) (Continued)

Absolute pressure (bar)	Saturation temp. (°C)	Total temperature (°C)														
		280	300	320	340	360	380	400	420	440	460	480	500	550	600	650
48	261	2866	2933	2993	3049	3101	3153	3201	3250	3297	3344	3390	3436	3551	3666	3782
		0.044	0.048	0.050	0.053	0.056	0.058	0.060	0.063	0.065	0.067	0.069	0.071	0.077	0.082	0.087
50	264	2857	2926	2987	3044	3096	3149	3198	3247	3294	3341	3388	3434	3549	3665	3781
		0.042	0.045	0.048	0.051	0.053	0.056	0.058	0.060	0.062	0.064	0.066	0.068	0.074	0.079	0.084
52	266	2847	2918	2981	3039	3092	3145	3194	3243	3291	3338	3385	3431	3547	3663	3779
		0.040	0.043	0.046	0.048	0.051	0.053	0.055	0.058	0.060	0.062	0.064	0.066	0.071	0.075	0.080
56	271	2827	2902	2968	3028	3082	3137	3187	3237	3285	3333	3380	3427	3543	3659	3776
		0.036	0.039	0.042	0.044	0.046	0.049	0.051	0.053	0.055	0.057	0.059	0.061	0.065	0.070	0.074
60	276	2805	2885	2954	3017	3073	3128	3179	3230	3279	3327	3375	3422	3539	3656	3774
		0.033	0.036	0.039	0.041	0.043	0.045	0.047	0.049	0.051	0.053	0.055	0.057	0.061	0.065	0.069
64	280	2782	2868	2940	3005	3063	3120	3172	3224	3273	3322	3370	3418	3535	3653	3771
		0.030	0.033	0.036	0.038	0.040	0.042	0.044	0.046	0.048	0.049	0.051	0.053	0.057	0.061	0.065
68	284		2849	2926	2993	3052	3111	3164	3217	3267	3317	3365	3413	3532	3650	3768
			0.031	0.033	0.035	0.037	0.039	0.041	0.043	0.045	0.046	0.048	0.050	0.055	0.058	0.061
72	288		2830	2911	2981	3041	3102	3156	3210	3260	3311	3359	3408	3528	3647	3765
			0.028	0.031	0.033	0.035	0.037	0.039	0.041	0.042	0.044	0.045	0.047	0.050	0.054	0.058
76	291			2897	2968	3033	3093	3150	3203	3255	3305	3354	3404	3524	3643	3762
				0.029	0.031	0.033	0.035	0.037	0.038	0.040	0.041	0.043	0.044	0.048	0.051	0.054
80	295			2882	2955	3023	3084	3142	3196	3249	3300	3348	3399	3520	3640	3759
				0.027	0.029	0.031	0.033	0.034	0.036	0.037	0.039	0.041	0.042	0.045	0.048	0.052
84	298			2867	2942	3012	3075	3134	3189	3242	3294	3344	3394	3516	3636	3756
				0.025	0.027	0.029	0.031	0.032	0.034	0.035	0.037	0.038	0.040	0.043	0.046	0.049
88	302				2928	3000	3065	3125	3182	3236	3288	3339	3389	3512	3633	3753
					0.026	0.027	0.029	0.031	0.032	0.034	0.035	0.036	0.038	0.041	0.044	0.047
92	305				2914	2989	3056	3117	3175	3229	3282	3333	3384	3508	3629	3751
					0.0241	0.026	0.028	0.029	0.031	0.032	0.033	0.034	0.036	0.039	0.042	0.045
96	308				2899	2977	3046	3109	3167	3223	3277	3328	3380	3504	3626	3748
					0.0227	0.0246	0.026	0.028	0.029	0.030	0.032	0.033	0.034	0.037	0.040	0.043

continued

A34 (Continued)

Absolute pressure (bar)	Saturation temp. (°C)	Total temperature (°C)											
		340	360	380	400	420	440	460	480	500	550	600	650
100	311	2883	2965	3036	3100	3160	3216	3271	3323	3375	3500	3623	3745
		0.0215	0.0233	0.0249	0.026	0.028	0.029	0.030	0.031	0.033	0.036	0.038	0.041
105	315	2863	2949	3023	3089	3150	3208	3263	3316	3368	3495	3619	3741
		0.0200	0.0218	0.0234	0.0249	0.026	0.028	0.029	0.030	0.031	0.034	0.036	0.039
110	318	2842	2933	3010	3078	3141	3199	3256	3309	3362	3490	3614	3738
		0.0186	0.0205	0.0221	0.0235	0.0248	0.026	0.027	0.028	0.030	0.032	0.035	0.037
115	321	2819	2916	2996	3066	3131	3191	3248	3302	3356	3485	3610	3734
		0.0174	0.0193	0.0208	0.0222	0.0235	0.0248	0.026	0.027	0.028	0.031	0.033	0.035
120	325	2795	2898	2982	3055	3121	3182	3240	3295	3350	3480	3606	3730
		0.0162	0.0181	0.0197	0.0211	0.0224	0.0235	0.0247	0.026	0.027	0.029	0.032	0.034
125	328	2769	2880	2968	3043	3111	3173	3232	3288	3343	3474	3601	3727
		0.0151	0.0170	0.0186	0.0200	0.0213	0.0224	0.0235	0.0246	0.026	0.028	0.030	0.032
130	331	2741	2860	2953	3031	3100	3164	3224	3280	3337	3469	3597	3723
		0.0140	0.0160	0.0176	0.0190	0.0203	0.0214	0.0225	0.0235	0.0245	0.027	0.029	0.031
135	334	2710	2840	2937	3018	3090	3155	3216	3273	3330	3464	3593	3719
		0.0130	0.0151	0.0167	0.0181	0.0193	0.0204	0.0215	0.0225	0.0235	0.026	0.028	0.030
140	337	2676	2818	2921	3006	3079	3146	3208	3266	3324	3459	3589	3716
		0.0120	0.0142	0.0159	0.0172	0.0184	0.0195	0.0206	0.0215	0.0225	0.0247	0.027	0.029
150	342		2771	2888	2979	3057	3127	3192	3251	3311	3448	3580	3708
			0.0126	0.0143	0.0157	0.0169	0.0179	0.0189	0.0199	0.0208	0.0229	0.0249	0.027
160	347		2717	2851	2951	3034	3108	3175	3236	3297	3438	3571	3701
			0.0110	0.0129	0.0143	0.0155	0.0165	0.0175	0.0184	0.0193	0.0213	0.0232	0.025
170	352		2652	2811	2922	3011	3088	3157	3221	3284	3427	3562	3694
			0.0096	0.0116	0.0130	0.0142	0.0153	0.0162	0.0171	0.0180	0.0199	0.0217	0.0234
180	357		2569	2767	2890	2986	3067	3139	3205	3270	3416	3553	3686
			0.0081	0.0104	0.0119	0.0131	0.0142	0.0151	0.0160	0.0168	0.0187	0.0204	0.0220
190	361			2717	2857	2960	3046	3121	3188	3255	3405	3545	3679
				0.0093	0.0109	0.0121	0.0131	0.0141	0.0149	0.0157	0.0176	0.0192	0.0208
200	366			2660	2821	2933	3024	3103	3172	3241	3394	3536	3671
				0.0082	0.0099	0.0112	0.0122	0.0132	0.0140	0.0148	0.0165	0.0182	0.0197
210	370			2593	2781	2905	3001	3084	3155	3227	3383	3527	3664
				0.0072	0.0091	0.0104	0.0114	0.0123	0.0131	0.0139	0.0156	0.0172	0.0187
220	374			2505	2739	2875	2978	3064	3138	3212	3372	3517	3656
				0.0061	0.0083	0.0096	0.0106	0.0116	0.0123	0.0131	0.0148	0.0163	0.0177

A35 Steam pipe capacities (kg/s)

A35.1 Gauge pressure (kPa) and velocity (m/s)

Gauge pressure (kPa)	Velocity (m/s)	External diameter and nominal bore of pipe (mm)										
		21.3 15.0	26.9 20.0	33.7 25.0	46.4 32.0	48.3 40.0	60.3 50.0	76.1 65.0	88.9 80.0	114.3 100	139.7 125	168 150
40	15	0.0017	0.0037	0.0064	0.0100	0.014	0.027	0.040	0.058	0.10	0.18	0.25
	25	0.0028	0.0066	0.011	0.017	0.025	0.044	0.072	0.10	0.18	0.26	0.40
	40	0.0046	0.010	0.017	0.028	0.039	0.073	0.11	0.16	0.28	0.46	0.63
70	15	0.0019	0.0043	0.0068	0.011	0.016	0.030	0.045	0.068	0.12	0.19	0.28
	25	0.0033	0.0072	0.012	0.019	0.027	0.049	0.077	0.12	0.19	0.31	0.43
	40	0.0048	0.010	0.019	0.029	0.046	0.081	0.12	0.17	0.30	0.47	0.66
100	15	0.0023	0.0052	0.0085	0.013	0.020	0.035	0.056	0.082	0.14	0.22	0.33
	25	0.0037	0.0095	0.014	0.023	0.032	0.059	0.093	0.13	0.23	0.36	0.51
	40	0.0059	0.012	0.023	0.035	0.054	0.10	0.15	0.21	0.37	0.57	0.81
200	15	0.0031	0.0066	0.012	0.019	0.027	0.050	0.077	0.11	0.20	0.31	0.43
	25	0.0052	0.012	0.019	0.031	0.044	0.081	0.12	0.18	0.33	0.48	0.69
	40	0.0081	0.017	0.031	0.048	0.075	0.13	0.20	0.28	0.52	0.80	1.14
300	15	0.0043	0.010	0.016	0.025	0.035	0.067	0.11	0.15	0.25	0.41	0.56
	25	0.0070	0.015	0.027	0.041	0.061	0.12	0.17	0.25	0.43	0.68	0.94
	40	0.011	0.024	0.043	0.068	0.098	0.16	0.28	0.40	0.70	1.11	1.63
400	15	0.0050	0.011	0.019	0.030	0.042	0.077	0.12	0.17	0.32	0.46	0.67
	25	0.0080	0.018	0.031	0.049	0.074	0.12	0.20	0.30	0.54	0.80	1.16
	40	0.013	0.032	0.054	0.081	0.13	0.22	0.34	0.50	0.86	1.36	1.94
500	15	0.0060	0.014	0.024	0.036	0.051	0.10	0.14	0.21	0.37	0.57	0.80
	25	0.010	0.022	0.038	0.059	0.082	0.15	0.25	0.35	0.61	0.98	1.42
	40	0.016	0.037	0.064	0.096	0.14	0.25	0.39	0.56	1.00	1.57	2.24
600	15	0.0070	0.016	0.029	0.042	0.061	0.12	0.17	0.25	0.43	0.69	0.93
	25	0.012	0.026	0.044	0.069	0.10	0.18	0.29	0.41	0.69	1.16	1.68
	40	0.019	0.043	0.074	0.11	0.16	0.28	0.44	0.62	1.15	1.78	2.55
700	15	0.0077	0.016	0.030	0.045	0.071	0.12	0.19	0.26	0.50	0.76	1.09
	25	0.013	0.031	0.052	0.079	0.12	0.21	0.33	0.48	0.83	1.32	1.88
	40	0.021	0.048	0.083	0.12	0.19	0.33	0.51	0.69	1.26	2.07	2.98
800	15	0.0087	0.019	0.034	0.051	0.077	0.13	0.22	0.30	0.54	0.82	1.23
	25	0.015	0.033	0.055	0.087	0.13	0.22	0.34	0.51	0.89	1.43	2.03
	40	0.023	0.052	0.089	0.14	0.20	0.37	0.56	0.85	1.40	2.23	3.41
1000	15	0.011	0.024	0.040	0.064	0.096	0.16	0.26	0.38	0.64	1.03	1.51
	25	0.018	0.039	0.070	0.11	0.15	0.27	0.42	0.60	1.05	1.72	2.46
	40	0.028	0.059	0.11	0.17	0.25	0.45	0.69	1.00	1.76	2.70	3.93
1400	15	0.014	0.033	0.050	0.084	0.13	0.22	0.35	0.51	0.88	1.43	2.02
	25	0.023	0.053	0.090	0.14	0.21	0.38	0.57	0.85	1.42	2.32	3.43
	40	0.035	0.083	0.15	0.22	0.33	0.60	0.94	1.30	2.33	3.57	5.11

A35.2 Steam pipe capacities (kg/s)
Showing relationship to pressure drop factor F (see B17.4.3)

External diameter and nominal bore of pipe (mm)

F	21.3 / 15.0	26.9 / 20.0	33.7 / 25.0	46.4 / 32.0	48.3 / 40.0	60.3 / 50.0	76.1 / 65.0	88.9 / 80.0	114.3 / 100	139.7 / 125	168 / 150	191 / 175	219 / 200	241 / 225	273 / 250	324 / 300
0.000 16						0.009	0.015	0.025	0.055	0.100	0.166	0.247	0.35	0.49	0.65	1.07
0.000 20					0.0045	0.010	0.017	0.029	0.063	0.113	0.184	0.279	0.40	0.55	0.73	1.18
0.000 25				0.0030	0.0050	0.011	0.019	0.031	0.069	0.124	0.200	0.303	0.43	0.60	0.79	1.28
0.000 30				0.0033	0.0054	0.012	0.021	0.034	0.075	0.134	0.216	0.327	0.47	0.65	0.86	1.38
0.000 35			0.0019	0.0035	0.0057	0.012	0.024	0.036	0.080	0.144	0.232	0.35	0.50	0.69	0.92	1.48
0.000 45		0.0010	0.0022	0.0040	0.0065	0.014	0.026	0.042	0.093	0.164	0.264	0.37	0.54	0.74	1.04	1.68
0.000 55		0.0011	0.0025	0.0045	0.0073	0.016	0.029	0.047	0.104	0.184	0.296	0.43	0.61	0.84	1.17	1.88
0.000 65		0.0012	0.0028	0.0049	0.0081	0.017	0.031	0.051	0.114	0.204	0.33	0.48	0.69	0.95	1.29	2.08
0.000 75		0.0014	0.0031	0.0054	0.0088	0.019	0.034	0.056	0.124	0.224	0.36	0.53	0.76	1.06	1.41	2.28
0.000 85		0.0015	0.0033	0.0059	0.0096	0.021	0.038	0.061	0.135	0.244	0.39	0.58	0.83	1.17	1.54	2.48
0.001 00	0.0005	0.0016	0.0036	0.0065	0.0106	0.023	0.041	0.068	0.150	0.269	0.44	0.66	0.94	1.31	1.73	2.79
0.001 25	0.0006	0.0018	0.0039	0.0072	0.0118	0.025	0.046	0.076	0.167	0.30	0.49	0.76	1.04	1.45	1.92	3.1
0.001 50	0.0007	0.0020	0.0042	0.0079	0.0131	0.028	0.050	0.083	0.184	0.34	0.54	0.85	1.15	1.60	2.11	3.4
0.001 75	0.0007	0.0022	0.0046	0.0086	0.014	0.030	0.055	0.091	0.201	0.37	0.59	0.95	1.26	1.74	2.30	3.8
0.002 00	0.0008	0.0024	0.0051	0.0093	0.016	0.033	0.060	0.099	0.218	0.40	0.64	1.05	1.36	1.88	2.49	4.1
0.002 5	0.0009	0.0027	0.0056	0.010	0.017	0.037	0.067	0.110	0.241	0.45	0.71	1.11	1.52	2.10	2.80	4.6
0.003 0	0.0010	0.0030	0.0062	0.011	0.019	0.041	0.074	0.121	0.265	0.49	0.78	1.18	1.68	2.31	3.1	5.0
0.004 0	0.0011	0.0035	0.0073	0.013	0.022	0.048	0.087	0.143	0.31	0.57	0.92	1.40	2.00	2.75	3.7	6.0
0.005 0	0.0013	0.0039	0.0082	0.015	0.025	0.054	0.098	0.160	0.35	0.64	1.03	1.59	2.27	3.1	4.1	6.8
0.006 0	0.0014	0.0044	0.010	0.017	0.027	0.060	0.109	0.180	0.39	0.71	1.15	1.74	2.52	3.4	4.6	7.5

0.008 0	0.0017	0.0051	0.011	0.019	0.032	0.070	0.126	0.208	0.46	0.83	1.35	2.04	2.9	4.0	5.3	8.7
0.010 0	0.0019	0.0058	0.012	0.022	0.036	0.079	0.143	0.235	0.52	0.92	1.52	2.31	3.3	4.5	6.0	9.8
0.012 5	0.0021	0.0065	0.013	0.024	0.040	0.088	0.160	0.258	0.58	1.03	1.70	2.56	3.7	5.0	6.7	10.9
0.015 0	0.0023	0.0071	0.015	0.027	0.044	0.096	0.177	0.290	0.63	1.14	1.87	2.82	4.0	5.5	7.3	12.0
0.017 5	0.0025	0.0077	0.016	0.030	0.048	0.105	0.194	0.31	0.69	1.25	2.05	3.07	4.4	6.0	8.1	13.0
0.020	0.0027	0.0084	0.017	0.032	0.052	0.114	0.210	0.34	0.75	1.36	2.22	3.33	4.8	6.5	8.7	14.0
0.025	0.0030	0.0094	0.019	0.036	0.059	0.127	0.232	0.38	0.83	1.51	2.47	3.71	5.3	7.3	9.7	15.7
0.030	0.0033	0.0104	0.021	0.040	0.065	0.140	0.255	0.42	0.92	1.66	2.72	4.09	5.9	8.0	10.6	17.4
0.040	0.0039	0.0123	0.025	0.047	0.078	0.167	0.30	0.50	1.09	1.97	3.22	4.84	7.0	9.6	12.6	20.8
0.050	0.0045	0.0138	0.028	0.053	0.087	0.188	0.34	0.56	1.22	2.23	3.62	5.4	7.9	10.9	14.3	23.7
0.060	0.0049	0.0148	0.031	0.058	0.095	0.208	0.38	0.62	1.35	2.45	3.98	5.9	8.7	12.0	15.9	
0.080	0.0057	0.0174	0.037	0.068	0.112	0.242	0.44	0.72	1.58	2.84	4.62	6.8	10.1			
0.100	0.0065	0.0196	0.041	0.077	0.126	0.272	0.50	0.82	1.78	3.20	5.24	7.6				
0.12	0.0071	0.0214	0.045	0.085	0.139	0.300	0.55	0.90	1.95	3.52	5.78					
0.15	0.0079	0.0241	0.050	0.094	0.154	0.334	0.61	1.00	2.18							
0.20	0.0093	0.0285	0.060	0.111	0.183	0.396	0.72	1.17	2.58							
0.25	0.0104	0.0316	0.067	0.124	0.204	0.440	0.80	1.29								
0.30	0.0114	0.0348	0.074	0.137	0.225	0.484	0.88	1.41								
0.35	0.0125	0.0379	0.081	0.150	0.246	0.528	0.95									
0.40	0.0135	0.0411	0.088	0.163	0.267	0.571	1.03									
0.45	0.0144	0.0439	0.093	0.174	0.285	0.607										
0.50	0.0153	0.0466	0.098	0.185	0.302	0.643										
0.60	0.0169	0.0518	0.109	0.204	0.333											
0.70	0.0183	0.0562	0.118	0.222												
0.80	0.0196	0.0603	0.126	0.238												
0.90	0.0208	0.0641	0.135													

62

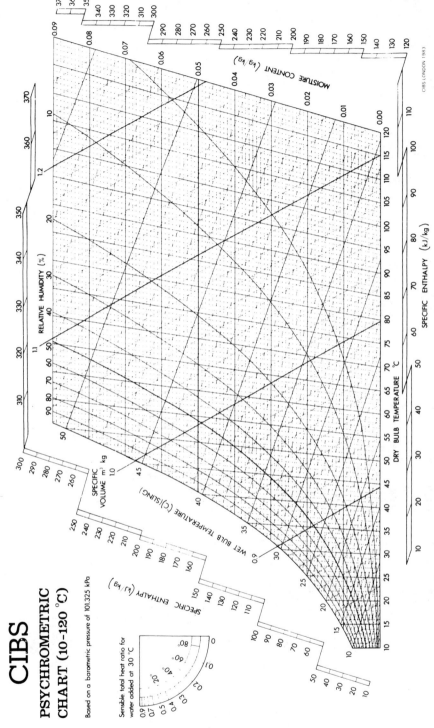

CIBS

PSYCHROMETRIC CHART (10–120 °C)

Based on a barometric pressure of 101.325 kPa

A36 CIBS Psychrometric chart (Reproduced by permission of the Chartered Institution of Building Services. Pads of charts for record purposes can be obtained from CIBS, Delta House, 222 Balham High Road, London SW12 9BS)

A37 Standard steel radiators

A37.1 Watts output[a] of standard steel radiators: S = single, D = double, F = finned (reproduced by courtesy of Stelrad Group Ltd)

Nominal length (mm)	300 mm high		440 mm high				590 mm high				740 mm high			
	S	D	S	D	SF	DF	S	D	SF	DF	S	D	SF	DF
480	210	332	289	460	359	661	372	591	471	860	453	722	583	1058
640	280	443	385	613	484	893	496	789	635	1161	604	962	786	1429
800	350	554	481	766	608	1125	620	986	798	1462	755	1203	989	1800
960	420	665	578	920	732	1357	744	1183	962	1764	906	1443	1192	2170
1120	490	776	674	1073	857	1589	868	1380	1126	2065	1058	1684	1395	2541
1280	560	886	770	1226	981	1821	992	1577	1289	2367	1209	1925	1598	2912
1440	630	997	867	1379	1105	2053	1116	1774	1453	2668	1360	2165	1802	3283
1600	700	1108	963	1533	1230	2285	1240	1971	1617	2970	1511	2406	2005	3654
1760	770	1219	1059	1686	1354	2517	1364	2169	1780	3271	1662	2646	2208	4025
1920	840	1330	1155	1839	1479	2749	1488	2366	1944	3572	1813	2887	2411	4395
2080	910	1440	1252	1992	1603	2981	1612	2563	2108	3874	1964	3127	2614	4766
2240	980	1551	1348	2146	1727	3213	1735	2760	2271	4175	2115	3368	2817	5137
2400	1050	1662	1444	2299	1852	3445	1859	2957	2435	4477	2266	3608	3020	5508
2560	1120	1773	1541	2452	1976	3677	1983	3154	2599	4778	2417	3849	3223	5879
2720	1190	1883	1637	2605	2101	3909	2107	3351	2762	5079	2568	4090	3426	6250
2880	1260	1994	1733	2759	2225	4141	2231	3549	2926	5381	2719	4330	3629	6620
3040	1330	2105	1829	2912	2349	4373	2355	3746	3090	5682	2871	4571	3832	6991
3200	1400	2216	1926	3065	2474	4605	2479	3943	3253	5984	3022	4811	4035	7362

[a] Outputs are based on mean water temperature of 76.7° C and room temperature of 21.1° C as defined in BS 3528 (1974). For other temperature conditions apply factors as **Table 37.2**.

A37.2 Factors for alternative temperatures

Room temp. (°C)	Mean water temperature (u C)					
	65	70	75	80	85	90
10	0.93	1.04	1.15	1.27	1.39	1.51
15	0.85	0.96	1.07	1.18	1.30	1.43
18	0.80	0.90	1.02	1.13	1.25	1.37
20	0.76	0.87	0.98	1.10	1.21	1.34
22	0.73	0.84	0.95	1.06	1.18	1.30
25	0.68	0.79	0.89	1.01	1.13	1.25

A37.3 Abridged table for estimating: S = single, D = double, F = finned

Height (mm)	Output[a] watts per nominal m²				Conversion factor: nominal area / heating area	
	S	D	SF	DF	S	D
300	1457	2306	–	–	2.32	4.64
440	1366	2175	1747	3260	2.27	4.54
590	1310	2083	1713	3154	2.24	4.48
740	1276	2033	1699	3100	2.23	4.46

[a] Outputs per m² fall below these values for the shorter radiator lengths

A38 Induction motors

A38.1 Typical performance values for standard three-phase 415 V 50 Hz T.E.F.C. squirrel cage induction motors[a]

Output (kw)	Speed rev/min	Efficiency			Power factor			Motor input (full load)		
		Full load	0.75 load	0.5 load	Full load	0.75 load	0.5 load	Amp	kW	kVA
0.75	1425	71	68	62	0.71	0.63	0.51	2.06	1.06	1.49
	910	68	67	62	0.66	0.60	0.50	2.32	1.10	1.67
	715	65	63	54	0.54	0.47	0.37	2.97	1.15	2.14
1.10	1410	75	74	69	0.75	0.67	0.53	2.72	1.47	1.96
	910	76	76	73	0.69	0.61	0.48	2.92	1.45	2.10
	705	69	66	60	0.59	0.50	0.39	3.76	1.59	2.70
1.50	2880	77	75	71	0.88	0.78	0.72	3.08	1.95	2.21
	1400	76	76	73	0.79	0.72	0.61	3.48	1.97	2.50
	950	75	73	70	0.68	0.58	0.46	4.09	2.00	2.94
	720	75	74	69	0.56	0.46	0.36	4.97	2.00	3.57
2.20	2875	79	78	75	0.89	0.84	0.77	4.35	2.78	3.13
	1435	79	76	72	0.78	0.69	0.56	4.97	2.78	3.57
	955	80	79	77	0.73	0.65	0.51	5.24	2.75	3.77
	720	79	77	73	0.62	0.54	0.42	6.25	2.78	4.49
3.00	2840	80	79	76	0.86	0.81	0.69	5.64	3.49	4.06
	1440	81	80	77	0.77	0.69	0.56	6.69	3.70	4.81
	960	82	81	77	0.72	0.66	0.54	7.07	3.66	5.08
	725	78	77	75	0.71	0.63	0.51	7.54	3.85	5.42
5.50	2900	83	83	80	0.86	0.83	0.75	11.1	6.63	7.98
	1440	85	85	83	0.75	0.67	0.55	12.0	6.47	8.63
	975	86	85	83	0.77	0.72	0.59	11.6	6.40	8.31
	730	84	83	81	0.79	0.75	0.61	11.5	6.55	8.29
7.50	2885	85	85	84	0.90	0.90	0.86	13.6	8.82	9.80
	1440	86	86	85	0.82	0.80	0.71	14.8	8.72	10.64
	975	84	83	79	0.76	0.71	0.57	16.3	8.93	11.75
	725	84	84	82	0.79	0.73	0.62	15.7	8.93	11.30
15.0	2940	85	84	80	0.88	0.85	0.80	27.9	17.6	20.1
	1460	87	87	85	0.86	0.82	0.76	27.9	17.2	20.0
	975	88	88	87	0.82	0.78	0.70	28.9	17.0	20.8
30.0	2955	87	85	82	0.91	0.89	0.84	52.7	34.5	37.9
	1475	90	90	88	0.86	0.84	0.78	53.9	33.3	38.8
	980	90	90	89	0.82	0.79	0.71	56.5	33.3	40.6
55.0	2960	91	90	86	0.90	0.90	0.88	93.4	60.4	67.2
	1475	91	91	90	0.88	0.86	0.84	97.8	60.4	70.3
	985	91	91	89	0.84	0.80	0.71	105	60.4	75.5
75.0	2960	91	90	86	0.92	0.91	0.88	125	82.4	89.5
	1475	92	91	89	0.88	0.86	0.81	129	81.5	92.6
	985	92	91	90	0.85	0.83	0.76	133	81.5	95.9
90.0	2970	91	90	86	0.91	0.90	0.84	151	98.9	109
	1480	92	91	90	0.87	0.85	0.80	156	97.8	112
	985	92	91	90	0.85	0.84	0.80	160	97.8	115
110.0	2970	92	91	88	0.91	0.90	0.87	183	120	131
	1480	92	91	89	0.87	0.83	0.77	191	120	137
	985	92	91	90	0.86	0.83	0.77	193	120	139
132.0	2955	92	92	89	0.93	0.93	0.91	215	143	154
	1480	92	91	89	0.86	0.83	0.77	232	143	167
	985	92	91	89	0.83	0.81	0.74	240	143	173

[a] These values should be used as a guide as they vary with manufacturer and can be improved, but usually at a price premium.

A39 Abridged tables of lamp data and standard service illuminance

A39.1 Lighting design lumens (LDL) for fluorescent tubes

	Tube details				Ordinary phosphors						Triphosphors		
Length (mm)	Diam. (mm)	Filling	Nominal watts	Circuit watts	Warm white 3000 K	White 3400 K	Plus white 3600 K	Natural 4000 K	Daylight 4300 K	North light 6500 K	Warm 3000 K	White 3400 K	Cool 4100 K
2400	38	Ar	125	138	8800	8900	8400	6500	8500	5600			
2400	38	Ar	85	103	6750	6850	6500	5000	6500	4100			
2400	38	Kr	100	112	7900	8000	–	–	7600	3200		8900	
1800	38	Ar	75	91	5650	5750	5500	4000	5450	–			
1800	26	Kr	70	80	5600	5700	–	–	5400	–	8900	8900	8900
1500	38	Ar	65	78	4600	4750	4500	3400	4450	–			
1500	26	Ar	50	63	3550	3600	–	2400	–	–	6300	6300	6300
1500	26	Kr	58	71	4550	4700	–	–	4400	–			
1200	38	Ar	40	51	2700	2800	2700	2100	2650	1700	5100	5100	5100
1200	26	Kr	36	47	2700	2800	–	–	2650	–			
600	38	Ar	40	51	1700	1700	–	–	1600	–	3200	3200	3200
600	38	Ar	20	30	1100	1100	–	800	1050	700			
600	26	Kr	18	28	1100	1100	–	–	1050	–	1325	1325	1325

A39.2 Lighting design lumens (LDL) for discharge lamps

Lamp watts	Mercury fluorescent MBF	Mercury tungsten MBTF	Mercury reflector MBFR	Mercury metal halide MBIF	High pressure sodium		Low pressure sodium SOX	
					SON	SON DL	Watts	Output
50	1 900				3 100			
70					5 300		18	1 750
80	3 650							
125	5 800						35	4 300
150					15 000	12 500		
160		2 560					55	7 500
250	12 500	4 840	10 500	16 000	25 500	22 000		
400		18 000		24 000	45 000		90	12 500
500		11 500						
700	38 000		32 500				135	21 500
1000	58 000		48 000	85 000	110 000			

A39.3 Light outputs (average through life) for tungsten filament lamps

Wattage	Lumens	Wattage	Lumens
40	350–400	300	4 300
60	625–675	500	7 700
100	1250–1300	750	12 400
150	2100	1000	17 300
200	2700		

Note: LDL shown is based on a life of 2000 hours. For 100 hours percentage increases are, approximately:

Fluorescents: All krypton filled and 1500–2400 mm argon filled 7–9%
               600–1200 mm argon filled up to 20%
Mercury: 7–10%
Sodium: 2–5% but Higher on 50 W and 70 W SON

A39.4 Values of standard service illuminance[a] (from CIBS code)

Location	lux	Location	lux	Location	lux
Storage areas	150	Casual work	200	Rough work	300
Routine work	500	Demanding work	750	Fine work	1000
Very fine work	1500	Minute work	3000		

[a] Refer to **B27** for details of lamp types. These are recommended levels and represent good current practice. CIBS code gives detailed definitions.

67

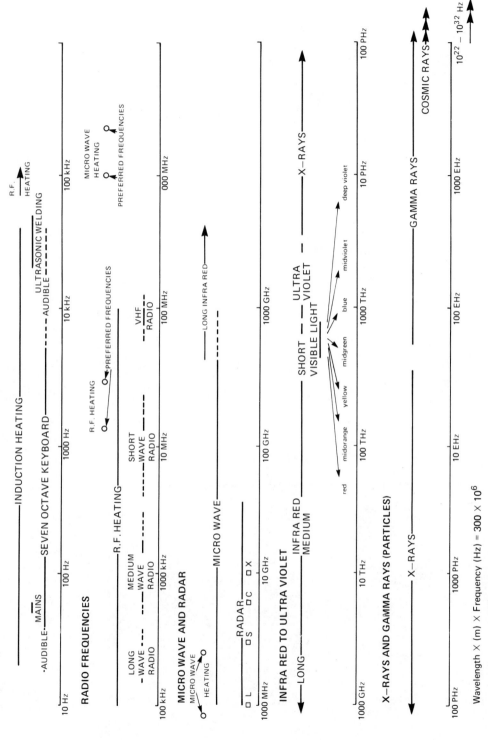

Figure A40 The electromagnetic radiation spectrum

Wavelength $\times$ (m) $\times$ Frequency (Hz) = 300 $\times$ 10^6

Section B

Calculation and analysis procedures

B1 The use of the gigajoule as the common unit of energy

The tables at **B1.1** illustrate the advantages of the gigajoule as the common unit (or depending on the scale of activity the terajoule or petajoule).

In making the change from supply units to common units the calorific value of the fuel must be known and when making comparisons of relative costs it must be remembered that as a heating fuel electricity has a conversion efficiency close to 100%, whereas all other fuels have combustion losses. The points brought out by these tables are

(1) An index of performance may be calculated relating total energy usage to factory size or to unit of factory output.
(2) A clear indication of relative costs enables informed decisions to be taken regarding fuel to be used.
(3) Care must be used in considering electricity costs bearing in mind that the 'night' charge per unit is a marginal charge and does not include a share of maximum demand and fuel surcharges. As a heating fuel electricity is inherently expensive unless supplied at an 'off peak' or 'night' rate; on the other hand it can be competitive in applications involving furnaces which are insulated to a high standard and closely controlled according to duty (this is because the flue loss at high temperatures can be very high with gas or oil fuels). The techniques of electric heating are covered in **B32**.
(4) Whenever any form of energy conservation is being considered it is absolutely essential to look at energy usage in common units.
(5) Both heating and process requirements can be expressed in gigajoule per annum and this enables an overall energy balance to be obtained.
(6) Allowance can be made for the variation in conversion efficiency of the different fuels and of the burners, boilers and distribution systems with which they are associated.

B1.1 Energy consumption expressed in gigajoule (GJ)

B1.1.1 10 000 m² medium engineering factory

Fuel	Consumption	Unit price	Cost (£)	Conversion factor to GJ	Total annual GJ	Cost per usable GJ
Electricity (day)	851 000 kWhr	3.95 p/kWhr	33 600	0.0036^a	3 064	£10.97^e
Electricity (night)	283 000 kWhr	1.27 p/kWhr	3 600	0.0036^a	1 019	£ 3.53^e
Class D oil	280 000 l	16.5 p/l	46 200	0.038^b	10 640	£ 5.43^f
Natural gas	81 000 therm	27.5 p/therm	22 300	0.1055^c	8 550	£ 3.48^g
Totals			105 700		23 273	

See **B1.1.2** for footnotes

B1.1.2 11 000 m² textile mill

Fuel	Consumption	Unit price	Cost (£)	Conversion factor to GJ	Total annual GJ	Cost per usable GJ
Electricity (day)	673 750 kWhr ⎫	3.52 p/kWhr	26 170	0.0036[a]	2 670	£9.80[e]
Electricity (night)	69 100 kWhr ⎭	average				
Coal	554 tonne	£43.6/tonne	24.210	29[d]	17 100	£2.56[h]
Class D oil	9 000 l	15.34 p/l	1 380	0.038[b]	340	£5.08[f]
Natural gas	51 600 therm	24.7 p/therm	12 790	0.1055[c]	5 440	£3.13[g]
Totals			64 550		25 560	

[a] 1 kWhr = 3.6 MJ (**A3**).
[b] Gross calorific value of Class D oil = 38.0 MJ/l (**A7**).
[c] 1 therm = 105.5 MJ (**A3**).
[d] Medium volatile coal with normal ash and moisture content: 29 MJ/kg (29 GJ/tonne) (**A10**).
[e] Assumes 100% conversion efficiency.
[f] Assumes 80% conversion efficiency.
[g] Assumes 75% conversion efficiency.
[h] Assumes 55% conversion efficiency.

B2 The assessment of process energy by analysis of summer and winter consumption

Charts B2.1, **B2.2**, **B2.3** and **B2.4** are all based on case studies and illustrate some graphical methods of analysis of consumption. Much valuable information can be obtained by plotting consumption figures from records available without having recourse to installing special meters or recording instruments.

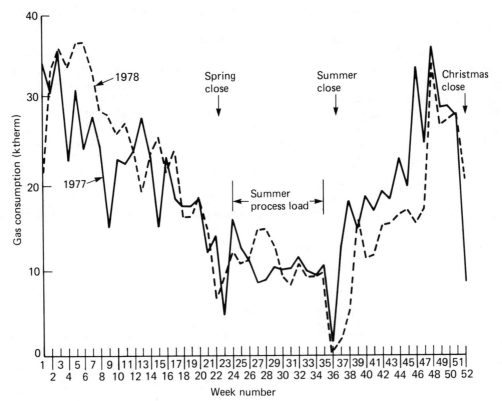

Figure B2.1 *Assessment of process energy from weekly fuel records*

B2.1 Summary of data for Chart B2.1

Total annual consumption (average 1977/78)

981 000 therm (103 500 GJ)

Weekly summer (process) load		10 700 therm	
Annual process consumption	10 700 × 47	= 503 000 therm	(53 070 GJ)
Annual space heating consumption	981 000–503 000	= 478 000 therm	(50 430 GJ)

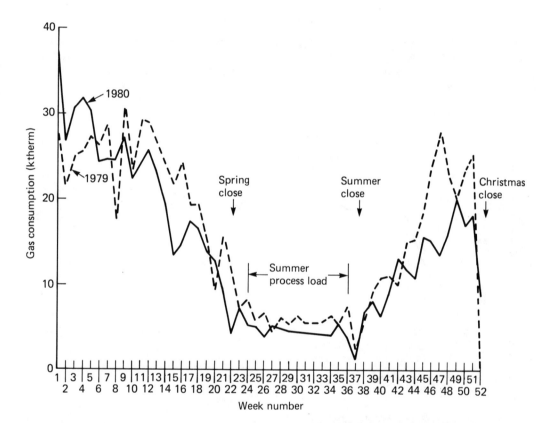

Figure B2.2 *Assessment of process energy from weekly fuel records*

B2.2 Summary of data for chart B2.2

Total annual consumption:	1979		816 000 therm (86 000 MJ)
	1980		668 000 therm (70 000 MJ)
Weekly summer (process load):	1979		6 100 therm
	1980		5 200 therm
Annual process consumption:	1979	47 × 6100	= 287 000 therm (30 250 MJ)
	1980	47 × 5200	= 244 000 therm (25 800 MJ)
Annual space heating consumption:	1979	816 000 − 287 000	= 529 000 therm (55 800 MJ)
	1980	668 000 − 244 000	= 424 000 therm (44 700 MJ)

B2.1/B2.2 *Annual load profiles for bulk propane usage*

The average summer consumption may be taken as the process load; as can be seen the 1977 and 1978 consumption for process work was 53 000 GJ and this declined to 30 200 GJ in 1979 and 25 800 GJ in 1980; these figures were shown to be consistent with reduced activity in the factory.

For space heating the 1977 and 1978 level was 50 500 GJ rising to 55 800 GJ in 1979 (a particularly cold November and December) and falling back to 44 700 GJ in 1980 (owing largely to a winter period of short-time working).

Spring, summer and winter factory closures show up clearly and account should be taken of these periods in assessing average figures.

B2.3 *Comulative annual consumption of steam boilers fired by class E oil*

When fuel deliveries are irregular and there is no record of stocks a clearer picture can be obtained by plotting on a cumulative basis. The 1979/80 graph shows the effect of reduced activity. Taking averages over the summer and winter periods enables the process load to be separated out.

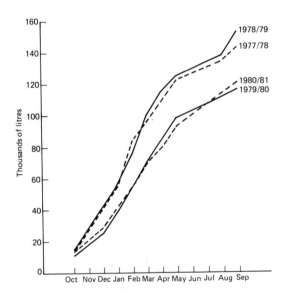

Figure B2.3 *Cumulative annual consumption of steam boilers fired by Class E oil*

B2.3 Summary of data for Chart B2.3

Average annual process consumption	75 000 l
Average annual heating consumption	60 000 l
Average annual total consumption	135 000 l

Month	Rounded average monthly consumption (kl)	Estimated monthly process load (kl)	Month	Rounded average monthly consumption (kl)	Estimated monthly process load (kl)
Oct	13	6	Apr	13	7
Nov	11	6	May	13	6
Dec	10	5	Jun	5	5
Jan	16	8	Jul	4	4
Feb	18	9	Aug	5	5
Mar	16	8	Sep	11	6

In this factory steam was essential for process work, but owing to old and inefficient boilers, and owing also to heavy losses in distribution, the heating system was shown to be very costly; this led to a change to gas fired radiant heating and the installation of a very much smaller steam boiler dedicated to the process work.

B2.4 Seven year variations in energy usage

In this case figures were available of total energy consumption (electricity, gas, class D oil and propane) going back over a seven year period. The graph shows the maximum and the minimum consumption in each of the twelve calendar months over the seven year period and is a striking illustration of the danger of drawing conclusions from short term observations. Variations over the seven years were caused by cold and mild winters, rise and fall in factory activity, major changes in factory lay-out, and vigilance in energy conservation.

A reminder of what has happened to energy costs is given by the table which gives data for the same factory; the year by year costs are shown using the 1979/80 consumption as a basis and calculating what the cost would have been for the same consumption in the seven previous years.

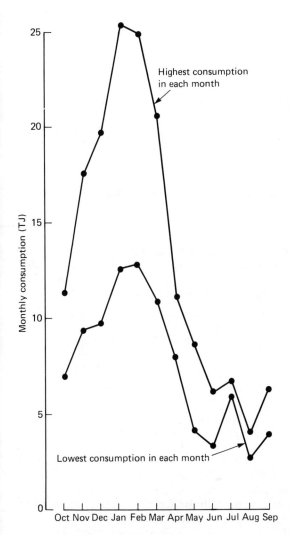

Figure B2.4 *Seven year variations in energy consumption*

B2.4 Summary of data for Chart B2.4

Consumption and cost in base year 1979/80

	Consumption (TJ)	Cost £000
Electricity	21.0	158
Propane	25.0	70
Gas	52.8	150
Class D oil	44.0	154
Total	*142.8*	*532*

Comparative costs of same consumption (£000), 1972/81

Year	Electricity	Propane	Gas	Oil	Total
72/73	53	20	34	15	122
73/74	66	22	38	47	173
74/75	78	29	52	52	211
75/76	106	30	67	64	267
76/77	119	46	74	73	312
77/78	131	51	83	74	339
78/79	137	52	85	74	348
79/80	158	70	150	154	532
80/81	185	85	170	180	640

B3 Analysis of electricity charges

Tables B3.1 and **B3.2** suggest a method of presentation which should enable monthly or quarterly accounts to be viewed against a background which is meaningful; simply to look at the amount payable has very limited value. Both these tables come from actual case studies and give a two year picture on a single sheet; they illustrate two very different tariffs but almost any tariff can be recorded in a similar manner. The information which can be gleaned from these tables is as follows:

Prices

The impact of price increases shows up clearly. The overall average price has more meaning than the price per kWhr (i.e. excluding maximum demand charges and fuel surcharge); also the impact of high winter maximum demand charges is clearly shown. The care needed in referring to the price per kWh is emphasized by the summary for 1980/81 in **Table B3.3**.

Trends

Forecasting can be made on a sound basis using the latest known tariffs and assessing the elements of the total charge separately.

B3.1 shows an increase in consumption and maximum demand year on year and the forecast indicates a continuation of this trend based on the expected future rate of activity.

B3.2 shows a rather more stable pattern year on year with almost identical maximum demand.

Seasonal changes

B3.1 represents a 365 day, 24 hour process with a normal increase in winter demand and consumption. **B3.2** shows clearly the effect of summer and year end factory closure; these 'short' months need to be allowed for in assessing 'normal' weekly consumption.

The higher winter consumption and maximum demand in **B3.2** is more clearly shown and this was, in fact, traced to the use of expensive 'top-up' electric heating.

B3.1 Electricity consumption and cost[a]

Month	Maximum demand			Units, step one			Units step two			Total	Fuel surcharge[e]	Total
	kVA	£/kVA	£	kWhr	p/kWhr	£	kWhr	p/kWhr	£	kWhr	(£)	(£)
Apr '79	1850	0.68/0.61/0.575[b]	1134	370 000	1.549	5731	419 600	1.419	5954	789 600	1993	14 812
May	1760	0.68/0.61/0.575[b]	1082	352 000	1.549	5452	500 650	1.419	7104	852 650	2578	16 217
Jun	1810	0.68/0.61/0.575[b]	1111	362 000	1.549	5607	438 570	1.419	6223	800 570	2522	15 463
Jul	1900	0.68/0.61/0.575[b]	1163	380 000	1.549	5886	498 270	1.419	7070	878 270	2803	16 923
Aug	1800	0.68/0.61/0.575[b]	1105	360 000	1.549	5576	498 600	1.419	7075	858 600	3462	17 220
Sep	1840	0.68/0.61/0.575[b]	1128	368 000	1.607	5914	440 830	1.472	6189	808 830	3533	17 064
Oct	1740	0.68/0.61/0.575[b]	1071	348 000	1.607	5592	505 530	1.472	7441	853 530	4087	18 191
Nov	1720	0.68/0.61/0.575[b]	3872	344 000	1.607	5528	473 530	1.472	6970	817 530	4052	20 422
Dec	1690	4.51/4.175/4.01[b]	7109	338 000	1.607	5432	144 480	1.472	2127	482 480	2553	17 221
Jan '80	1630	4.51/4.175/4.01[b]	6869	326 000	1.607	5239	676 780	1.472	9962	1 002 780	5307	27 377
Feb	1650	2.36/2.245/2.18[b]	3720	330 000	1.607	5303	496 500	1.472	7308	826 500	4373	20 705
Mar	1600	2.36/2.245/2.18[b]	3611	320 000	1.607	5142	433 580	1.472	6382	753 580	3988	19 123
Apr	1520	0.81/0.71[c]	1129	304 000	2.43	7387	526 340	2.17	11 422	830 340	1046	20 984
May	1490	0.81/0.71[c]	1108	298 000	2.43	7241	487 740	2.17	10 584	785 740	525	19 758
Jun	1580	0.81/0.71[c]	1172	316 000	2.43	7679	417 310	2.17	9056	733 310	739	18 645
Jul	1580	0.81/0.71[c]	1172	316 000	2.43	7679	389 340	2.17	8449	705 340	711	18 010
Aug	1860	0.81/0.71[c]	1361	372 000	2.52	9374	435 710	2.24	9760	807 710	678	21 183
Sep	1780	0.81/0.71[c]	1314	356 000	2.52	8971	551 180	2.24	12 346	907 180	762	23 393
Oct	1730	0.81/0.71[c]	1278	346 000	2.52	8719	520 790	2.24	11 666	866 790	874	22 537
Nov	1740	2.33/2.23[c]	3930	348 000	2.52	8770	480 460	2.24	10 762	828 460	870	24 332
Dec	1860	4.24/4.14[c]	7750	372 000	2.52	9374	477 420	2.24	10 694	849 420	910	28 729
Jan '81	2020	4.24/4.14[c]	8413	404 000	2.52	10 181	534 810	2.24	11 980	938 810	2366	32 939
Feb	1800	2.33/2.23[c]	4064	360 000	2.52	9072	510 000	2.24	11 424	807 000	2323	26 883
Mar	1970	2.33/2.23[c]	4443	399 000	2.52	9929	649 180	2.24	14 542	1 043 180	2804	31 718

1979/80 9 724 920 kWhr at average 2.27p = £220 738
1980/81 10 166 280 kWhr at average 2.84p = £289 111
1981/82 forecast 10 800 000 kWhr at average 3.29p = £355 000

[a] Southern Electricity Board Demand Tariff 3 High Voltage Supply. Units Step one applies to first 200 per kVA. Step two to remainder.
[b] First figure 0–500 kVA, second 500–1000 kVa, third above 1000 kVA.
[c] First figure 0–500 kVA, second above 500 kVA.

B3.2 Electricity consumption and cost[a]

Month	Day			Night			Maximum demand							Fuel sur-charge	Total
							kVA	kVA	Monthly			Winter			
	kWhr	p/kWhr	£	kWhr	p/kWhr	£	month	year	Charge (£)	Cost (£)		Charge (£)	Cost (£)	(£)	(£)
Jan '79	69 880	1.7	1188	8530	0.7	60	440	440	0.46/0.34[b]	180		2.55	1122	189	2738
Feb	57 770	1.7	982	4920	0.7	34	410	440	0.46/0.34[b]	180		2.55	1046	151	2393
Mar	63 750	1.7	1084	4810	0.7	34	420	440	0.46/0.34[b]	180		0.68	286	165	1748
Apr	54 470	1.84	1000	5070	0.7	35	410	440	0.48/0.36[b]	188		—		150	1376
May	49 330	1.84	908	4930	0.7	35	380	440	0.48/0.36[b]	188		—		164	1295
Jun	47 770	1.84	879	3150	0.7	22	370	440	0.48/0.36[b]	188		—		160	1250
Jul	52 090	1.84	958	4380	0.7	31	380	440	0.48/0.36[b]	188		—		180	1358
Aug	32 770	1.84	603	3720	0.7	26	370	440	0.48/0.36[b]	188		—		147	968
Sep	52 990	1.9	1007	6770	0.7	47	370	440	0.48/0.36[b]	188		—		261	1504
Oct	65 100	1.9	1237	8690	0.7	61	380	440	0.48/0.36[b]	188		—		353	1839
Nov	62 190	1.9	1182	8540	0.7	60	380	440	0.48/0.36[b]	188		0.73	277	351	2058
Dec	44 760	1.9	850	5590	0.7	39	410	440	0.48/0.36[b]	188		2.77	1136	266	2480
Jan '80	67 320	1.9	1279	8840	0.7	62	410	440	0.48/0.36[b]	188		2.77	1219	403	3151
Feb	60 470	1.9	1149	6350	0.7	44	430	440	0.48/0.36[b]	188		2.77	1191	354	2926
Mar	62 400	1.9	1186	6990	0.7	49	420	440	0.48/0.36[b]	188		0.73	307	367	2097
Apr	66 890	2.69	1798	6700	1.17	78	410	440	0.48/0.39[b]	194		—		93	2183
May	52 190	2.69	1380	5070	1.17	59	410	440	0.48/0.39[b]	194		—		59	1712
Jun	56 420	2.69	1518	4600	1.17	54	380	440	0.48/0.39[b]	194		—		62	1847
Jul	50 460	2.69	1357	4160	1.17	49	390	440	0.48/0.39[b]	194		—		55	1675
Aug	35 940	2.81	1010	4180	1.17	49	390	440	0.53/0.43[b]	214		—		34	1327
Sep	55 850	2.81	1569	5240	1.17	61	360	440	0.53/0.43[b]	214		—		51	1916
Oct	57 630	2.81	1619	5620	1.17	66	380	440	0.53/0.43[b]	214		—		64	1983
Nov	58 830	2.81	1653	6250	1.17	73	390	440	0.53/0.43[b]	214		0.79	308	68	2337
Dec	50 300	2.81	1913	5080	1.17	59	410	440	0.53/0.43[b]	214		3.04	1246	59	3013

1979 721 970 kWhr at average 2.91p = £21 004
1980 742 830 kWhr at average 3.52p = £26 167
1981 forecast 729 000 kWhr at average 4.05p = £29 550
[a] Midlands Electricity Board Maximum Demand Tariff Low Voltage.
[b] First 250 kVA charged at higher figure, balance at lower.

B3.3 Summary of 1980/81 prices for Tables B3.1 and B3.2

Table	Pence per kWhr (as tariff)	Inclusive average for year	Summer average	Winter average
B3.1	2.17–2.5	2.84	2.60	3.50
B3.2	2.69–2.81 (day) 1.17 (night)	3.50	3.30	5.44

Information from electricity meters

Much valuable information can be gleaned from observation of electricity meters and the estimation of Power Factor is fully described in **B30**. Most industrial installations have the following meters:

(1) kWhr meter recording units used; this meter normally enables the kW loading to be estimated by timing the rotating disc.
(2) kVAhr meter with maximum demand indicator. The maximum demand mechanism normally operates on a half-hour period during which the actual demand is integrated; if at the end of the period a new 'high' is reached the driving pointer moves the maximum demand indicating needle forward. By getting to know the time at which the half hour period ends (i.e. 10.14, 10.44, etc) it is possible to observe the actual M.D. at various times in the day.
(3) kVAR hr meter which is used solely as a check on Power Factor (*see* **B30**).

B4 Audit of electricity demand and consumption

Details of three case studies are given in **Tables B4.1** (office complex), **B4.2** (hotel) and **B4.2** (timber mill). In each of these studies the information was obtained without recourse to special long term instrumentation. Full use was made of information from electricity

B4.1 Office complex (maximum demand 350 kVA)

Load	Max demand (kW)	Hours per day	Days per year	Utilization (%)	kWhr/year
Lighting	158	15	240	100	570 000
Air handling	75	11	240	100	198 000
Air conditioning	82	15	240	50/80	193 000
Computer room	25	13	240	100	78 000
Compressor	9	10	240	100	22 000
Canteen	35	3	240	70	18 000
Heaters/humidifiers	22	10	240	30	16 000
Exterior lighting	6	5	365	100	11 000
Boiler auxiliaries	4	12	120	90	5000
Miscellaneous					39 000

Total 1 150 000

accounts, from limited metering available in addition to supply meters and by observation, enquiry and judgement. The useful data obtained are summarized below:

(1) Those functions making the major demands and with the highest consumption can be isolated and studied for their energy saving potential.
(2) The major effect of continuous operations is emphasized by the kiln load in the timber plant and by the 365 days per year service offered by the hotel complex.

(3) The office study illustrates the effect of a very high standard of lighting and in this case no means of control apart from block switching. Energy savings of a minimum of 30% were foreseen.

(4) The total installed power of electric motors on woodworking machinery (as also in the majority of machine shops) can be four or more times the average load and three or more times the maximum demand.

(5) A preliminary assessment may be made of the effects of switching off processes to reduce maximum demand charges.

(6) If detailed audit of electricity usage is subsequently called for the best locations for instruments can be established and the types of recording instruments required can be determined.

(7) The advantage of alternative tariffs can be studied noting that 'off peak' or 'night unit' tariffs normally require some sacrifice in the optimum normal rate per kWhr.

Table B4.2 Hotel complex (maximum demand 130 kVA)

Load	Max demand (kW)	Hours per day	Days per year	Utilization (%)	kWhr/year
Reception, corridors	18	24	365	65	102 000
Restaurant 1 (electric)	33	10	300	60	59 000
Restaurant 2	19	8	365	70	38 000
Restaurant 3	10	10	100	80	8000
Boiler etc.	21	24	200	50	50 000
Main kitchens	14	14	365	70	50 000
Cellar plant	10	24	365	50	44 000
Cold rooms	8	24	365	50	35 000
Bedrooms: kettles	22	½	300	100	35 000
light, ventilation	29	4	300	100	7000
Meeting rooms	12	8	200	80	15 000
Proprietor's house	8	8	300	60	11 000
Laundry	6	6	300	90	7000
No. 1 Bar	6	7	365	80	12 000
No. 2 Bar, Lounges	5	10	365	45	8000
Manager's flat	10				9000

Total 490 000

B4.3 Timber curing and processing plant (maximum demand 300 kVA in winter)

Department	kW installed	Max demand (kVA)	Average load (kW)	Hours per year	kWhr/year
Main mill					
power	360	130	80	2300	184 000
light	10				
New mill					
power	180	65	39	2300	90 000
light	4				
Compressor	25	30	22	2300	51 000
Structure shop					
power	25	30	19	2000	38 000
light	2				
Kilns	120	70	37	8650	320 000
Space heating	48	48	23	1500	35 000
Office lighting	5	5	4	1800	7000
Yard lighting	9	9	8	600	20 000
Impression	25	15	9	2300	20 000

Total 750 000

B5 Reconciliation of fuel used with heating requirements of building

In assessing the effectiveness with which energy is used to provide comfort conditions the following preliminary steps are necessary:

(1) Eliminate process energy.
(2) Translate fuel input to heat output from equipment expressed in gigajoules per year (or per heating seasoon).
(3) Establish occupation hours per week and per year.
(4) Assess length of heating season.
(5) As appliances are sized according to the maximum heating requirements a factor must be applied relating the actual average output of the heating plant to its maximum continuous rating.

The approach can be illustrated by the four examples in **B5.1–B5.4**.

B5.1 Small factory unit with single warm air heater, output 130 kW.

Actual fuel input 5000 therms gas per year.
Output per year from heater assuming 75% conversion efficiency:

$5000 \times 0.1055 \times 0.75 = 396\,GJ$
(0.1055 is conversion from therms to gigajoules, *see* **A5**).

If heater operates continuously 50 hours per week for a 28 week heating season its annual output will be

$$\frac{130\,kW \times 3.6}{1000} \times 50 \times 28 = 655\,GJ$$

($1\,kWhr = 3.6\,MJ$ or $0.036\,GJ$; *see* **A3**)
Utilization factor $= 396/655 = 60\%$

B5.2 Medium engineering factory

Two 3 750 000 BThU steam boilers
Fuel input 281 000 litres of Class D oil per year
Boiler efficiency 69%   Distribution loss 10%
GCV of fuel 38 MJ/litre (*see* **A7**)

Input in GJ $\dfrac{281\,000 \times 38}{1000} = 10\,680\,GJ/year$

Usuable output $10\,680 \times 0.69 \times 0.90 = 6600\,GJ/year$

This particular case study has been used to construct the Sankey diagram at **B6.1** and from this diagram the approximate split between four production shops was derived. As all the energy was used for space heating the study showed very clearly the high costs involved owing to low boiler efficiency, high distribution loss, poorly insulated buildings, and unplanned ventilation.

B5.3 Furniture factory

Three steam boilers, each 12 500 lb/hr

Fuel input 1 200 000 litres of Class G oil per year
Boiler efficiency 80.2%   Distribution loss 5%
GCV of fuel 41 MJ/l (*see* **A5**)

Input in GJ $\dfrac{1\,200\,000 \times 41}{1000} = 49\,200\,GJ/year$

Usable output 49 200 × 0.802 × 0.95 = 37 500 GJ/year.

By plotting fuel consumption on a cumulative basis as in **B2.3** the total annual process load was assessed at 9800 GJ or 820 GJ per month average, leaving 27 700 GJ or an average 3500 GJ per month for the heating season from October to May and a probable maximum total requirement in the mid/winter months of 7500 GJ. The output from the boilers per month assuming 200 hours running is

3 × 12.5 (1000 lb steam) × 1.0237 × 200 = 7500 GJ
(1.0237 converts 1000 lb steam to GJ)

B5.4 Hotel complex

Four gas fired boilers (see **B8.1** for efficiency calculation)
The input of fuel to the boiler is calculated from

$$\frac{\text{Output} \times 1055}{10^9 \times \text{efficiency}}$$

where
1055 converts BThU to joule (**A3**)
10^9 converts joule to GJ

Data are given in **Table B5.4**.

B5.4 Boilers in hotel complex

Boiler	Nominal output (BThU/hr)	Measured efficiency (%)	Input of fuel (calc.) (GJ/hr)	Gas consumption (therm)	Gas consumption (GJ)[a]
1	600 000	81.6	0.78	24 000	2500
2	440 000	79.3	0.59	24 000	2500
3	800 000	81.5	1.04	14 000	1200
4	800 000	81.3	1.04	14 000	1200

[a]BThU × 0.1055 = GJ (**A3**)

These figures can be reconciled by making reasonable assumptions with regard to heating requirements and boiler running hours, i.e.

Nos 1 and 2:   75 hr per week for 30 weeks heating season at average utilization 60% plus 30 hr per week for water heating only during 22 weeks of summer period at 75% utilization.
Total input required per year for Nos 1 and 2

{(75 × 30 × 0.6) + (30 × 22 × 0.75)} × (0.78 + 0.50) = 2530 GJ

Nos 3 and 4:   Assume that only one boiler is required to meet the demand as clearly there was excess capacity; also assume 45 hr per week at 60% in winter + 20 hr per week at 80% in summer. Total input per year for No 3

(45 × 30 × 0.6) + (20 × 22 × 0.8) × 1.04 = 1200 GJ

B6 Construction and use of Sankey diagrams

The Sankey diagram portrays the flow of energy in all its forms by the band width in the diagram, and in the examples which follow the scale is GJ per year. Its main value is that it enforces a discipline in taking count of all the energy supplied to a factory or other complex;

it also ensures that the requirements for process work and for power, lighting and space and water heating and are all accounted for. The most straightforward way of constructing the diagram is to consider the total energy supplied per year from all sources and this total, of course, embraces the seasonal variations.

B6.1 Energy distribution in a 10 000 m² factory

This case study has already been referred to in **B1.1** and **B5.2**. As can be seen all the class D oil is supplied to fire steam boilers and the output from these used entirely for factory heating; with large high bay shops separated from one another distribution loss was unavoidable. In the case of Natural Gas supplied, part of it was fed direct to heat treatment plant, part to gas fired warm air heating, part to gas fired radiant heating and the balance to

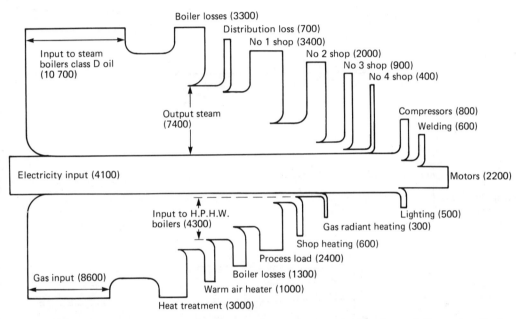

Figure B6.1 *Sankey diagram for 10 000 m² medium engineering factory with oil fired steam boilers and gas fired high pressure hot water boilers (all figures are in GJ/year)*

high pressure hot water boilers. The output from the boilers was used partly for space heating, but mainly for process work. The electrical energy was supplied to compressor drives (continuous operation during working week), welding plant, machine tool and other electric motor drives, and lighting.

The major lessons learned from this study concerned the need to utilize surplus heat from the heat treatment plant, the need to reduce heat loss from the fabric of the building and the need to reduce the heavy loss caused by the steam space heating system.

B6.2 and B6.3 'Before' and 'After' diagrams for a light engineering factory

This case history has already been referred to in **B2.3**; diagram **B6.2** shows the very heavy loss due to inefficient boilers and the high distribution loss in the steam based space heating system.

The main changes made were to introduce gas fired radiant as the main space heating system for the works and to replace the old boilers with a single much smaller unit dedicated to the process requirements. One of the old boilers was retained as a standby unit. Other

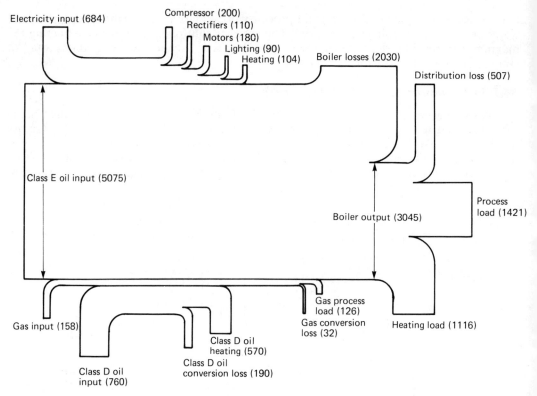

Figure B6.2 *Sankey diagram for light engineering factory: energy distribution before conservation measures (all figures are in GJ/year)*

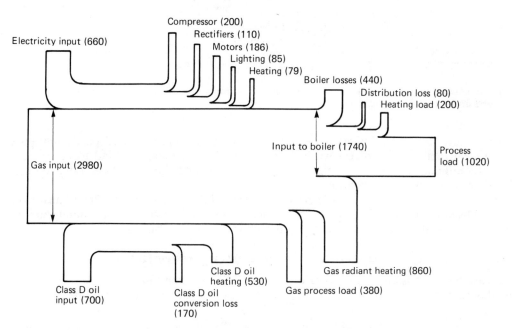

Figure B6.3 *Sankey diagram for light engineering factory: prospective energy distribution after conservation measures (all figures are in GJ/year)*

improvements were to reduce the wasteful use of process steam, eliminate electric space heating and improve the thermal properties of the buildings. There was also a change of fuel for some of the process work from steam to direct fired gas. As can be seen the gross annual energy input was reduced from 5917 GJ to 4340 GJ.

B7 The chemistry of combustion

B7.1 The basic principles

(1) Gas volume calculations are made on the 'Ideal Gas' basis using Molar Gas Volume as a measure (**A2.2**). all molecules of gas, regardless of chemical formula, are treated as having the same volume at the same temperature and pressure. All calculations rest on the supposition that 32 kg of oxygen at 0 °C and standard atmospheric pressure will occupy 22.4136 m^3. The mass of the *same volume* of other gases may then be taken in ratio of their relative molecular masses, i.e. 32 kg oxygen (O_2), 2 kg hydrogen (H_2), 16 kg methane (CH_4) etc. (Simple relative molecular masses suffice, i.e. $H_2 = 2$ exactly.)

(2) Mass calculations are related to the relative molecular mass (rmm) of the elements or compounds involved and the unit used is kilomole; in any chemical reaction the masses of the reactant compounds balance with the masses of the product compounds, i.e.

$$C \ (rmm \ 12) + O_2 \ (rmm \ 32) \rightarrow CO_2 \ (rmm \ 44)$$

(*see also* **A12** and **A13**)

(3) Volumes of gases must always be related to the temperature and pressure conditions prevailing; corrections may be made for changes using the gas constant as defined in **A2.3** such that

$$\frac{P \times V}{T} = R$$

where R, the gas constant, is 8314 J/K kmol
P is pressure in N/m^2
T is absolute temperature
V is kilomolar gas volume (22.4136 m^3 at 0 °C)

Thus, if the temperature of a gas is raised from 0 °C to 100 °C and its pressure raised from one atmosphere to two, the new kilomolar volume (V_1) becomes

$$V_1 = \frac{8314 \times (273.15 + 100)}{2 \times 101 \ 325} = 15.31 \ m^3$$

If only a change of temperature or a change of pressure occurs the new kilomolar volume may be calculated by ratio

$$V_1 \ (\text{at } 100 \ °C) = \frac{22.41 \times 373.15}{273.15} = 30.62 \ m^3$$

$$V_1 \ (\text{at } 100 \ °C \text{ and twice atmospheric pressure}) = \frac{30.62 \times 101.3}{202.6} = 15.31 \ m^3$$

In combustion calculations volume or mass may be used according to the nature of the fuel (solid, liquid or gas) and in the examples which follow both methods are illustrated.

B7.2 The combustion of methane

Combustion of fuels may be examined and analysed on a mass or volume basis and for complex fuels 'ultimate analysis' is used; the source of combustion energy is in the rearrangement of the forces holding the various molecules together (the'bond energies'). In the text which follows these concepts are examined in relation to methane which is the simplest of the hydrocarbon fuels and which constitutes 70–90% of natural gases.

Methane reacts with the oxygen in air according to the equation

$$CH_4 + 2O_2 \rightarrow CO_2 + 2H_2O + 892 \, MJ/kmol \qquad (7.2.1)$$

Since this book is written for engineers rather than chemists, we will continue to write thermochemical equations in this style rather than in the chemically conventional form

$$CH_4 + 2O_2 \rightarrow CO_2 + 2H_2O; \Delta H = -892 \, MJ/kmol$$

(The 892 MJ is the kilomolar enthalpy of combustion or the heat of reaction and arises from the difference in the stability of the molecular forces of the reactants and the products; *see also* **B7.3**).

All combustion processes are exothermic in nature and give out heat energy with reaction. Endothermic reactions take place when heat energy is supplied from an external source.

All fuels which have a hydrogen content will produce water vapour in their flue gases when burned. If the latent energy in the water vapour is included in the heat energy of combustion then the calorific value of the fuel is referred to as gross (GCV). If the value is quoted exclusive of the energy which will be obtained by condensing the water vapour in the flue gases then the value is referred to as nett (NCV). These two values are also referred to as 'wet' and 'dry' respectively.

Mass calculation

Substitution of relative molecular mass in equation (7.2.1) leads to:

[12 (carbon) + 4 (hydrogen)] + [2 × 16 × 2 (oxygen)] →
[12 + 32 (carbon dioxide)] + [2 × 18 (water vapour)] + 892 MJ/kmol   (7.2.2)

This can be restated as:

16 kg of methane will react with 64 kg of oxygen to produce 44 kg of carbon dioxide and 36 kg of water vapour with the release of 892 MJ of heat energy.

Reducing the reaction to the basis 'per kg of fuel':

$$1 \, kg \, CH_4 + 4 \, kg \, O_2 \rightarrow 2.75 \, kg \, CO_2 + 2.25 \, kg \, H_2O + 55.6 \, MJ \qquad (7.2.3)$$
(55.6 is the GCV of methane in MJ per kg)

This reaction assumes complete and exact or stoichiometric combustion; in practice satisfactory combustion does not take place unless excess or oxygen is present. The reasons for this are complex, but it suffices to say that the excess of oxygen takes care of imperfect mixing of fuel and air, variations in burner design, differences in flame shape and temperature, and limitations in the control of air (oxygen) supply.

In all the calculation procedures described in **Sections B7.2 to B7.7** inclusive it is assumed that the air supply will at least reach the minimum required for complete combustion; in practice inadequate air supply can occur and the problems associated with this are referred to in **B7.8**.

Pure oxygen is used for combustion only in very special circumstances and the normal source is air. Oxygen represents 23.2% by weight or 21.00% by volume of atmospheric air at 15 °C.

The nitrogen (together with traces of other inert gases in air) plays no part in the reaction, but becomes a burden because it must be raised to the temperature of the flue gases by the energy from the fuel. The amount of air required for stoichiometric combustion of 1 kg of methane is therefore

$$\frac{4 \times 100}{23.2} = 17.2\,kg$$

Equation (7.2.3) may now be rewritten:

$$1\,kg\,CH_4 + 17.2\,kg\,air \rightarrow 2.75\,kg\,CO_2 + 2.25\,kg\,H_2O + 13.2\,kg\,N_2 + 55.6\,MJ \qquad (7.2.4)$$

The total weight of flue gases is 18.2 kg and the percentage by weight of CO_2 is 15.1%. The mass relationship can be changed to a volume relationship using the ideal gas density values in **A12**:

$$\frac{1}{0.716}\,m^3\,CH_4 + \frac{17.2}{1.293}\,m^3\,air \rightarrow \frac{2.75}{1.964}\,m^3\,CO_2 + \frac{2.25}{0.804}\,m^3\,H_2O$$

$$+ \frac{13.2}{1.257}\,m^3\,N_2 + 55.6\,MJ \qquad (7.25)$$

On a 'per m^3 of fuel' basis and at 0 °C:

$$CH_4 + 9.52\,air \rightarrow CO_2 + 2H_2O + 7.52\,N_2 + 39.8\,MJ \qquad (7.2.6)$$

Volume calculation

For each volume of oxygen supplied there will be (79/21) = 3.76 volumes of nitrogen, hence

$$CH_4\,(1\,volume) + 2O_2\,(2\,volumes) + N_2\,(2 \times 3.76\,volumes) \rightarrow$$
$$CO_2\,(1\,volume) + 2H_2O\,(2\,volumes) + N_2\,(2 \times 3.76\,volumes) \qquad (7.2.7)$$

or

$$CH_4 + 9.52\,air \rightarrow CO_2 + 2H_2O + 7.52\,N_2 \qquad (7.2.8)$$

This is the same equation as that developed from mass considerations and is clearly a much simpler way of dealing with gaseous fuels.

The volumetric proportion of CO_2 in the flue gases may now be calculated on either a 'wet' or 'dry' basis, i.e.

$$\frac{1.0}{1.0 + 2.0 + 7.52} \times 100\% = 9.51\% \text{ ('wet' basis)}$$

$$\frac{1.0}{1.0 + 7.52} \times 100\% = 11.74\% \text{ ('dry' basis)}$$

The procedure used for methane can be applied to other gaseous hydrocarbon fuels using the data in **Tables A12, A13** and **A14**.

All natural gases are mixtures and most contain incombustible substances such as nitrogen and carbon dioxide (properties of natural gases and of liquefied petroleum gases are given in **Tables A5** and **A6**).

B7.3 'Bond' Energy calculations

Values of GCV are given in **Tables A5, A6, A7** and **A8** and for commercial fuels these values are established experimentally by means of calorimeters; however, an understanding of the source of the heat of reaction can be gained by calculating these values using the 'bond energy' concept which derives from the forces holding molecules together. These values have been expressed in kilojoules per mol according to the bonds between the atoms as shown in **Table B7.3**.

Apply these values to the methane reaction:

$$CH_4 + 2O_2 \rightarrow CO_2 + 2H_2O \qquad (7.3.1)$$

$$
\begin{array}{cccc}
\overset{\displaystyle H}{\underset{\displaystyle H}{H-\overset{|}{\underset{|}{C}}-H}} \quad + & \begin{array}{c} O=O \\[6pt] O=O \end{array} & \longrightarrow \quad O=C=O \quad + & \begin{array}{c} H-O-H \\[6pt] H-O{-}H \end{array} \\[18pt]
4 \times 414 & 2 \times 498 & 2 \times 803 & 4 \times 464 \\
= 1656\ kJ & = 996\ kJ & = 1606\ kJ & = 1856\ kJ
\end{array}
$$

The energy released when the bonds in the products are formed thus exceeds the energy to break the bonds in the reactants by

$$1856 + 1606 - 1656 - 996 = 810\ kJ/mol$$

$$50.6\ MJ/kg\ CH_4$$

To this figure must be added the latent heat of vapourization of 2.25 kg of water (*see* 7.2.4), i.e.

$$2.258 \times 2.25 = 5.08\ MJ$$

(2.258 MJ/kg is the latent heat of saturated steam at 1 atmosphere from **A32**).

GCV is therefore
$50.6 + 5.08 = 55.68\ MJ/kg$
$= 55.68 \times 0.716$ (from **A12**) $= 39.9\ MJ/m^3$ (at 0 °C)

B7.3 Bond energies

Bond	Energy (kcal/mol)	Energy (kJ/mol)
C−H (single)	99	414
C−C (single)	83	347
C=O (as in CO_2)	192	803
O=O (as in O_2 molecule)	119	498
O−H (single)	111	464

Applying the 'bond energy' approach to the combustion of propane introduces the C−C bond (83 kcal/mol):

$$C_3H_8 + 5O^2 \rightarrow 3CO_2 + 4H_2O \qquad (7.3.2)$$

$$
\begin{array}{cccc}
\overset{\displaystyle H\ \ H\ \ H}{\underset{\displaystyle H\ \ H\ \ H}{H-\overset{|}{\underset{|}{C}}-\overset{|}{\underset{|}{C}}-\overset{|}{\underset{|}{C}}-H}} \quad + \quad 5(O=O) & \longrightarrow & 3(O=C=O) & + & 4(H-O-H) \\[24pt]
8 \times 414 + 2 \times 347 \quad 5 \times 498 & & 3(2 \times 803) & & 4(2 \times 264) \\
= 4006\ kJ \qquad = 2490\ kJ & & = 4818\ kJ & & = 3712\ kJ
\end{array}
$$

The energy released is thus

$$4818 + 3712 - 4006 - 2490 = 2034\ kJ/mol$$

Converting as before, the relative molecular mass of propane being 44, we obtain the heat of combustion as

$$\frac{2034}{44} = 46.22 \, \text{MJ/kg propane}$$

Adding the latent heat of 1.64 kg of water

$$1.64 \times 2.258 = 3.70 \, \text{MJ}$$

we obtain a total of 49.9 MJ/kg or

$$49.9 \times 1.968 \, (\textit{see} \, \textbf{A11}) = 98.2 \, \text{MJ/m}^3 \, (\text{at } 0 \, °\text{C})$$

B7.4 The 'ultimate analysis' approach

This method has the advantage of being applicable to any fuel and is more suitable for calculations involving solid or liquid fuels. Its basis depends on establishing the proportions of combustible elements in the fuel, i.e. carbon, hydrogen and sulphur and relating these to three simple reactions:

$$C + O_2 \rightarrow CO_2 \text{ (one carbon atom two oxygen)} \tag{7.4.1}$$
$$2H_2 + O_2 \rightarrow 2H_2O \text{ (four hydrogen two oxygen)} \tag{7.4.2}$$
$$S + O_2 \rightarrow SO_2 \text{ (one sulphur two oxygen)} \tag{7.4.3}$$

Account may also be taken of the production of carbon monoxide by partial oxidation of carbon:

$$2C + O_2 \rightarrow 2CO \text{ (two carbon, two oxygen)} \tag{7.4.4}$$

The theoretical quantity of oxygen required to burn 1 kg of any fuel may be stated as

$$\left(\frac{C}{12} + \frac{H}{4} + \frac{S}{32} \right) \times 32 \, \text{kg per kg fuel} \tag{7.4.5}$$

The numerator in each case is the mass of each element as a proportion of 1 kg of fuel; the denominator is the relative molecular mass to one molecule of oxygen in each of reactions (7.4.1), (7.4.2), and (7.4.3); the factor 32 converts molecules of oxygen to kmole. If the fuel has an oxygen content then this must be deducted from the sum of the elements in the bracket in (7.4.5) on the basis O/32 (32 being the relative molecular mass of O^2).

The mass of air required for combustion will then be

$$\left(\frac{C}{12} + \frac{H}{4} + \frac{S}{32} - \frac{O}{32} \right) \times 32 \, \text{kg per kg fuel} \tag{7.4.6}$$

We now apply this approach to methane, which consists ($C = 12$, $H_4 = 4$) of 75% carbon and 25% hydrogen, as shown in **Table B7.4.1**. The content of flue gas can also be assessed, as shown, by noting that in equations 7.4.1, 7.4.2, and 7.4.3 one kmole of product comes from one kmole of fuel element (CO_2 from C, $2H_2O$ from H_2, SO^2 from S).

From the above data the proportions by weight or volume of CO_2 and H_2O in the 'dry' and 'wet' flue gases may be calculated and they will be consistent with those calculated in **B7.2**.

The analysis of solid fuels is expressed in carbon, hydrogen, sulphur, nitrogen, moisture and ash; the procedure for calculating flue gas content is exactly as for methane as described above, noting that nitrogen, moisture and ash are incombustible and will have no effect on the requirement for air. The same approach can be used for liquid or gaseous fuels where ultimate analysis detail is available; **Table B7.4.2** below uses the volume analysis figures for North Sea gas from **A5**.

B7.4.1 Ultimate analysis calculation for combustion of methane

Methane

Constituent	*Mass* (kg/kg fuel)	*Oxygen needed* (kmol/kg fuel)
C	0.75	$0.75/12 = 0.0625$
H_2	0.25	$0.25/4 = 0.0625$
		Total 0.1250

Combustion gases

Gas	*Mass* (kg/kg) or (kmol × rmm)	*Volume* (m^3/kg) (kmol × 22.41)$_a$	*Volume* (m^3/m^3) (kmol × 22.41 × 0.716)^b
Oxygen	$0.1250 \times 32 = 4$	$0.1250 \times 22.41 = 2.80$	$2.80 \times 0.716 = 2.00$
Nitrogen^c	$0.1250 \times \frac{79}{21} \times 8 = 13.17$	$0.1250 \times \frac{79}{21} \times 22.41 = 10.54$	$10.54 \times 0.716 = 7.54$
Air	$4 + 13.17 = 17.17$	$2.80 + 10.54 = 13.34$	$2.00 + 7.54 = 9.54$

Products

Constituent	*Mass* (kg/kg fuel)	*Product*	kmol *product* *per* kg *fuel*
C	0.75	CO_2	$0.75/12 = 0.0625$
H_2	0.25	H_2O	$0.25/2 = 0.1250$

Flue gases

Gas	*Mass* (kg/kg)	*Volume* (m^3/kg)	*Volume* (m^3/m^3)
CO_2	$0.0625 \times 44 = 2.75$	$0.0625 \times 22.41 = 1.40$	$1.4 \times 0.716 = 1.00$
H_2O	$0.1250 \times 18 = 2.75$	$0.1250 \times 22.41 = 2.80$	$2.8 \times 0.716 = 2.00$

^{a}22.41 is the volume (m^3) of 1 kmol at 0 °C.
^{b}0.716 is the ideal gas density (kg/m^3) of methane (from **A12**).
^{c}79/21 is the ratio of nitrogen to oxygen in air.

B7.4.2 North Sea Gas

Constituent	*Volume* *analysis* (m^3/m^3 NSG)	*Ideal gas* *density* (kg/m^3)	*Mass of* *constituent* (kg/m^3 NSG)	*Ultimate analysis* (kg/m^3 NSG) N_2	C	H_2	O_2
CH_4	0.944	0.716	0.676		0.507	0.169	
C_2H_6	0.030	1.342	0.040		0.033	0.007	
N_2	0.015	1.250	0.019				
C_3H_8	0.005	1.968	0.010		0.008	0.002	
CO_2	0.002	1.964	0.004		0.001		0.003
C_3H_{10}	0.002	2.593	0.005		0.004	0.001	
C_5H_{12}	0.001	3.220	0.003		0.002	0.001	
C_6H_{14} etc.	0.001	5.0 (avg)	0.005		0.004	0.00	
Totals			0.762	0.019	0.559	0.181	0.003

The ultimate analysis figures are calculated in ratio to the relative atomic masses of the elements in the compound. On conversion to a 'per kg' basis, the analysis becomes

N_2 $\dfrac{0.019}{0.762} = 0.0249$

C $\dfrac{0.559}{0.762} = 0.7336$

H_2 $\dfrac{0.181}{0.762} = 0.2375$

O_2 $\dfrac{0.003}{0.762} = 0.0039$

The contribution to combustion of the oxygen content of North Sea gas is almost negligible, but for clarity it is allowed for. Thus

$$O_2 \text{ needed } = \frac{0.7336}{12} + \frac{0.2375}{4} - \frac{0.0039}{32} = 0.1207 \text{ kmol}$$

Mass of air per kg of fuel

$$= \frac{0.1207 \times 32}{0.232} = 16.65 \text{ kg}$$

Volume of air at 0 °C

$$= \frac{16.65}{1.293} = 12.87 \text{ m}^3$$

The composition of the flue gases is calculated as for methane, and the results are shown in **Table B7.4.3**.

B7.4.3 Flue gases from combustion of North Sea Gas

Constituent	Mass (kg/kg NSG)	Product	kmol *product per* kg NSG
C	0.734	CO_2	0.734/12 = 0.0611
H_2	0.238	H_2O	0.238/2 = 0.1190
N_2	0.025	N_2	0.025/28 = 0.0009

Flue gas	Mass (kg/kg NSG)	Volume (m³/kg NSG)	Volume (m³/m³ NSG)
CO_2	$0.0611 \times 44 = 2.68$	$0.0611 \times 22.41 = 1.37$	$1.37 \times 0.762 = 1.04$
H_2O	$0.1190 \times 18 = 2.14$	$0.1190 \times 22.41 = 2.67$	$2.67 \times 0.762 = 2.03$
N_2	$\left(\dfrac{0.1207 \times 32 \times 76.8}{23.2}\right) +$ $0.0009 \times 28 = 12.81$	$\left(\dfrac{0.1207 \times 22.41 \times 79}{21}\right) +$ $0.0009 \times 22.41 = 10.20$	$10.20 \times 0.762 = 7.77$

CO_2 content of 'dry' flue gas: $\dfrac{1.04 \times 100}{1.04 + 7.77} = 11.8\%$ by volume

$$\frac{2.68 \times 100}{2.68 + 12.81} = 17.3\% \text{ by weight}$$

B7.5 Carbon dioxide content of flue gases

Based on chemical methods of flue gas analysis the carbon dioxide content has long been established as the main indicator of the effectiveness of combustion of hydrocarbon fuels and of the extent to which excess air is being supplied to the burners. With the availability of the zirconium oxide cell as a means of continuous detection of the proportion of oxygen in flue gases, instruments have been developed which rely on oxygen measurement only (*see* **B33**). Similarly techniques are available for accurate monitoring of carbon monoxide as an indicator of incomplete combustion.

As commercial hydrocarbon fuels burn essentially to carbon dioxide and water (the sulphur content being ignored for the moment) the extent of the presence of CO_2 depends on the ratio of carbon to hydrogen in the ultimate analysis of the fuel; the greater the proportion of carbon in the fuel the greater the percentage of CO_2 in the flue gases.

In **B7.2** the combustion of methane has been examined and methane has the lowest carbon/hydrogen ratio among the hydrocarbon gases, i.e. 3:1 based on the mass ratio of the atoms in CH_4 (C = 12, H = 4). There are fuel gases which have a lower ratio, but these are normally mixtures containing a high proportion of hydrogen. The ultimate is hydrogen itself, which has a carbon/hydrogen ratio of zero.

At the other extreme is pure carbon which has a ratio of infinity; between the two lie the other hydrocarbon gases, the light distillates such as petrol, naphtha and gas oil, the heavier oil fuels, and the range of coals up to anthracite and coke having ratios up to 30:1. Information on ratios for liquid and solid fuels is given in **Tables A7** and **A8**. The combustion of pure carbon will produce the highest proportion of carbon dioxide in its flue gases.

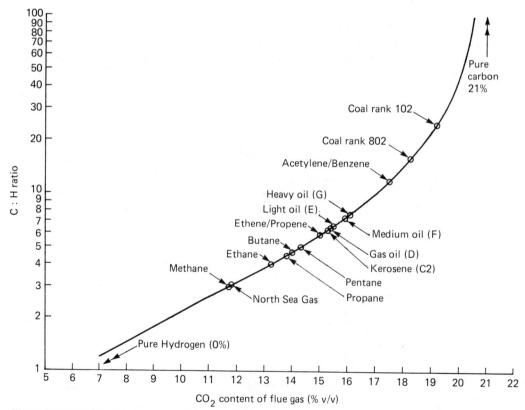

Figure B7.5.1 *Relationship between carbon: hydrogen ratio and carbon dioxide content of flue gases in stoichiometric combustion of hydrocarbon fuels (the presence of sulphur being neglected)*

The equation $C + O_2 \rightarrow CO_2$ states that 12 kg carbon will react with 32 kg oxygen to produce 44 kg carbon dioxide; this on a 'per kg of fuel' basis and including the nitrogen element may be restated:

$$1\,kg\ C + 2.67\,kg\ O_2 + 8.84\,kg\ N_2 \rightarrow 3.67\,kg\ CO_2 + 8.84\,kg\ N_2$$

Alternatively on a volume basis the equation may be stated as in **A13**:

$$0.51\,kg\ C + 1.0\,m^3\ O_2 + 3.76\,m^3\ N_2 \rightarrow 1.0\,m^3\ CO_2 + 3.76\,m^3\ N_2$$

The CO_2 content of flue gas by volume when carbon is burned with the exact requirement of air is therefore

$$\frac{1.0}{1.0 + 3.76} \times 100\% = 21\%$$

This is the maximum CO_2 content and contrasts with 11.74% on a 'dry' basis when methane is burned at the stoichiometric point. Chart **B7.5.1** shows how the CO_2 content relates to the C:H ratio for the complete range of fuels and this chart forms the starting point for calculations of CO_2 and O_2 content with excess of air. The CO_2 content is calculated from

$$\frac{1}{4.76 + (3/z)\,(4.76 - 1)} \times 100\%$$

where
z = C:H ratio

B7.6 Relationship between excess air and the carbon dioxide and oxygen content of flue gases

Reverting to the volume calculation for CH_4 in **7.2**

$$CH_4 + 2O_2 + 7.52N_2 \rightarrow CO_2 + 2H_2O + 7.52N_2$$

If, say 40% excess air is supplied there will be 40% of the oxygen supplied in the flue gas as well as the total volume of nitrogen and the equation becomes

$$CH_4 + 2.8O_2 + 10.53N_2 \rightarrow CO_2 + 2H_2O + 0.8O_2 + 10.53N_2$$

The percentage of CO_2 in the dry flue gas is thus

$$\frac{1}{1 + 0.8 + 10.53} \times 100\% = 8.11\%$$

and the percentage of O_2 is

$$\frac{0.8}{1 + 0.8 + 10.53} \times 100\% = 6.49\%$$

In order to derive general equations to evaluate CO_2 and O_2 content for any hydrocarbon and for any proportion of excess air, the starting point is the equation

$$C_xH_y + \left(x + \frac{y}{4}\right)O_2 = xCO_2 + \frac{y}{2}H_2O$$

(This equation enables all the reactions in **Table A13** to be derived.) The equation can be modified to take count of the C:H mass ratio by substituting $(12x/z)$ for y, where z is the mass ratio and 12 is the ratio of the relative atomic masses of carbon and hydrogen. Thus

$$O_2\ requirement = x + \frac{3x}{z}$$

$$CO_2\ content\ of\ flue\ gas = x$$

$$\text{N}_2 \text{ content of flue gas} = 3.76 \left(x + \frac{3x}{z} \right)$$

$$\text{H}_2\text{O content of flue gas} = \frac{6x}{z}$$

If excess air is supplied in ratio R to the minimum requirement the N_2 content will rise by the same ratio and there will be surplus oxygen in the flue gas because the amount supplied will be greater than the amount required for combustion.

This leads to the general formulae

$$\text{CO}_2 \text{ content} = x$$

$$\text{O}_2 \text{ content} = \left(x + \frac{3x}{z} \right) (R - 1)$$

$$\text{N}_2 \text{ content} = \left(x + \frac{3x}{z} \right) \times 3.76 \times R$$

$$\text{H}_2\text{O content} = \frac{6x}{z}$$

The CO_2 percentage by volume of dry flue gas is

$$\frac{x}{x + \left(x + \frac{3x}{z} \right) (R - 1) + \left(x + \frac{3x}{z} \right) \times 3.76 \times \text{R}} \times 100\%$$

x can now be eliminated and the formula simplified to

$$\frac{1}{4.76\,R + \frac{3}{z}\,(4.76\,R - 1)} \times 100\%$$

The formula for O_2 content can also be simplified to

$$\frac{\left(1 + \frac{3}{z} \right) (R - 1)}{4.76\,R + \frac{3}{z}\,(4.76\,R - 1)}$$

From the above two equations the percentages of CO_2 or O_2 in the flue gas can be calculated for any hydrocarbon when the C:H ratio is known and for any ratio of total air supplied to stoichiometric air requirement.

This approach ignores flue gases derived from the sulphur or oxygen content of the fuel; it also ignores the dilution of flue gas by the nitrogen content; the pattern of CO_2 and O_2 content for various fuels and various ratios of excess air is shown in **Charts B7.6.1** and **B7.6.2**.

From **Chart B7.6.1** it can be seen that gaseous fuels have the lowest C:H ratios and the lowest CO_2 content, and solid fuels the highest, with liquid fuels falling in between.

Fuels which are mixtures, or for which ultimate analysis details are available should be assessed as described in **B7.4**.

Variation in C:H ratio has a pronounced effect on CO_2 content, but chart **B7.6.2** shows that O_2 content varies little for the range of hydrocarbon fuels. The percentage of oxygen in flue gas with 100% excess of air is 11.09% for methane, 10.57% for heavy fuel oil, while the theoretical minimum (for pure carbon) is 10.50%.

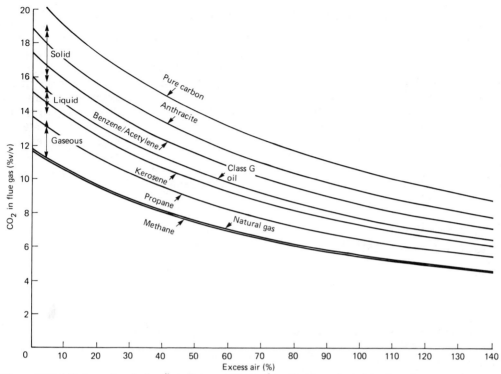

Figure B7.6.1 *Hydrocarbon fuels: effect of excess air on carbon dioxide content of dry flue gas*

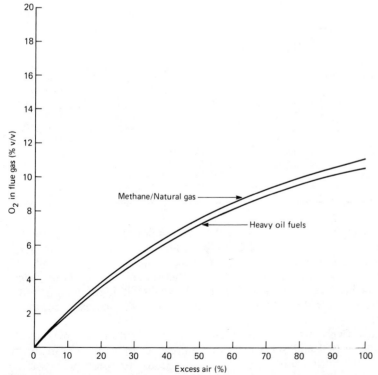

Figure B7.6.2 *Relationship between oxygen in dry flue gas and excess air for hydrocarbon fuels*

B7.7 The effect of flue gas temperature on combustion losses

The energy loss in flue gases may be calculated from the mass of each gas in the flue, the specific heat capacity of the gas and the temperature rise above the temperature of the air supplied to the burners. Because the specific heat capacity of gases varies with temperature account must be taken of this variation. **Chart A15.2** shows this relationship for the gases relevant to combustion calculations. It is also necessary to assess the components of the gases in the flue using volume or ultimate analysis as appropriate to the fuel, to separate the 'wet' loss and the 'dry' loss and to relate the losses to gross or nett CV as appropriate.

Methane can again be used to illustrate the procedure. From **B7.3** the GCV of methane can be taken as 55.9 MJ/kg and the NCV as 50.8 MJ/kg. The full calculation of dry loss as related to temperature of flue gas is tabulated in **B7.7.1**. Note that ambient temperature is taken at 15 °C; note also that the CO_2 loss is calculated from

$$2.75 \text{ kg} \times S \times (F - 15)$$

and the N_2 loss from

$$13.23 \text{ kg} \times S \times (F - 15)$$

where
S is specific heat capacity (kJ/kg)
F is flue gas temperature (°C)

B7.7.1 Effect of flue gas temperature on combustion loss

Flue gas temp	CO^2		N_2		Total dry loss	Loss
	Sp. heat capacity	Loss	Sp. heat capacity	Loss		
(°C)	(kJ/kg)	(MJ)	(kJ/kg)	(MJ)	(MJ/kg fuel)	(% of NCV)
100	0.914	0.21	1.030	1.16	1.37	2.7
200	1.000	0.51	1.051	2.57	3.08	6.1
300	1.066	0.84	1.085	4.09	4.93	9.7
400	1.118	1.18	1.113	5.67	6.85	13.5
500	1.164	1.55	1.137	7.30	8.85	17.4
600	1.201	1.93	1.163	9.00	10.93	21.5
700	1.231	2.32	1.186	10.75	13.07	25.7
800	1.258	2.72	1.204	12.50	15.22	30.0
900	1.280	3.12	1.216	14.24	17.36	34.2
1000	1.300	3.52	1.227	15.99	19.51	38.4
1200	1.326	4.32	1.245	19.52	23.84	46.9
1500	1.357	5.54	1.284	25.23	30.77	60.6
2000	1.400	7.64	1.304	34.25	41.89	82.5

The loss due to the specific enthalpy (total heat) of the water vapour produced during the combustion process may be taken direct from steam tables, and converted into 'per kg of fuel' by multiplying by 2.25 (*see* equation 7.2.3), as shown in **Table B7.7.2**.

The figures for temperatures beyond 1000 °C are theoretical; in practice such temperatures tend to be unachievable without recourse to special techniques such as pure oxygen firing to reduce the nitrogen loss. However, the extent of the losses in furnaces operating at high temperatures is clear.

An important side effect of flue gas temperature is that, at temperatures below 180 °C, the sulphur bearing fuels release SO_2 and SO_3 which, in the presence of water vapour, form and deposit sulphurous and sulphuric acids. This can cause serious corrosion problems.

It should be noted that **Table B7.7.1** is based on stoichiometric air supply and that the losses will be greater with excess air because of the higher volume of nitrogen in the flue

B7.7.2 Loss due to specific enthalpy of water vapour

Temp. (°C)	Total heat (MJ/kg)	Loss (MJ/kg fuel)	Loss (% of GCV)
100	2.68	6.03	10.8
200	2.88	6.48	11.6
300	3.08	6.93	12.4
400	3.28	7.38	13.2
500	3.49	7.85	14.0
600	3.71	8.35	14.9
700	3.90	8.78	15.7
800	4.11	9.25	16.6
900	4.33	7.72	17.4
1000	4.53	10.19	18.2

gases together with the excess oxygen present. This is illustrated in **Chart B7.7.1** showing the theoretical losses in the combustion of methane and carbon as a percentage of NCV. The performance of North Sea Gas will be close to that of methane and the loss with oil fuels will fall between the methane line and the carbon line. **Chart B7.7.2** gives the same information in a more open scale. The calculation of flue gas losses can be greatly simplified by using average values for specific heat capacity of the flue gas and working on the total mass of gases. **Chart A15.1** gives average figures which are sufficiently accurate for most normal calculations.

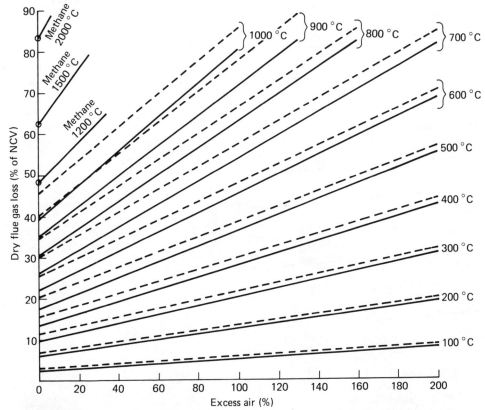

Figure B7.7.1 *Theoretical effect of flue gas temperature and excess air on combustion losses*
Full lines: methane   Broken lines: carbon
Flue gas temperatures are shown; ambient temperature 15 °C

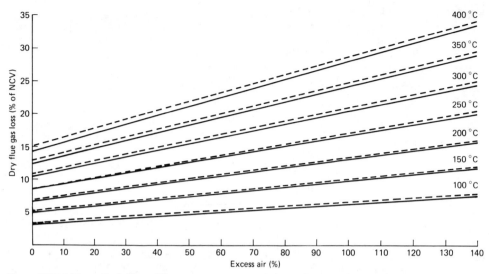

Figure B7.7.2 *Theoretical effect of flue gas temperature and excess air on combutstion losses*
Full lines: methane   Broken lines: carbon
Temperature rise is shown on the curves

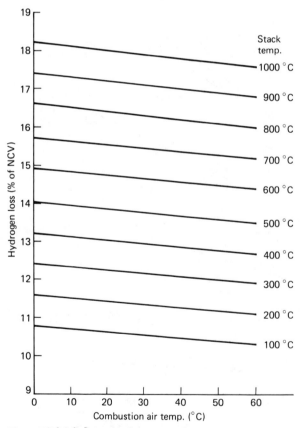

Figure B7.7.3 *Relationship between stack temperature, combustion air temperature, and loss in flue gases due to water vapour*

Chart B7.7.3 shows the loss resulting from the water vapour content of the flue gas and this loss is expressed as a percentage of GCV; the chart also illustrates the reduction in loss which results from increasing the combustion air temperature, i.e. reducing the temperature rise of the flue gases. The chart has been drawn on the basis of methane, but will vary very little for North Sea Gas. Preparation of a table similar to the above, but showing dry loss related to temperature for North Sea Gas instead of methane would involve slightly different mass values, i.e. CO_2, 2.68 kg; N_2, 12.7 kg; H_2O, 2.15 kg (all per kg of fuel). These values have been established in **B7.4**.

B7.8 The effects of incomplete combustion

Partial oxidation will take place according to the general equation

$$C_x H_y + \left(\frac{x}{2} + \frac{y}{4} \right) O_2 \rightarrow xCO + \frac{y}{2}H_2O$$

Substitution of the C:H ratio z (= 12 x/y) as in B7.6 leads to

$$O_2 \text{ required} = \frac{x}{2} + \frac{3x}{z}$$

$$N_2 \text{ in flue gases} = 3.76 \left(\frac{x}{2} + \frac{3x}{z} \right)$$

$$CO \text{ in flue gases} = x$$

If now we consider an oxygen deficiency of Q, where Q is the fraction of stoichiometric requirement, part of the combustion will be as in **B7.6** and part as above with Q being the full oxidation proportion and $(1 - Q)$ the partial oxidation proportion.
Thus

$$\text{Total } O_2 \text{ required} = Q \left(x + \frac{3x}{z} \right) + (1 - Q) \left(\frac{x}{2} + \frac{3x}{z} \right)$$

N_2 in flue gases $= 3.76 \times O_2$ required
CO_2 in flue gases $= Qx$
CO in flue gases $= (1 - Q) x$

The volume percentages in the flue gases are thus

$$CO = \frac{1 - Q}{\left\{ \left(\frac{Q}{2} + \frac{1}{2} + \frac{3}{z} \right) \times 3.76 \right\} + 1} \times 100\%$$

$$CO_2 = \frac{Q}{\left\{ \left(\frac{Q}{2} + \frac{1}{2} + \frac{3}{z} \right) \times 3.76 \right\} + 1} \times 100\%$$

Use of these formulae enables the charts at **B7.8.1** to be prepared showing the effects of oxygen deficiency on flue gas content. Measurement of CO content of flue gases enables a check to be made that there is an oxygen deficiency, but it also performs the vital function of detecting the presence of potentially poisonous fumes. It must be emphasized that the charts at **B7.8.1** are theoretical and that CO can be present in flue gases even though there is excess air supplied to the burner; the partial oxidation occurs for other reasons connected with the complex nature of the combustion process. As an approximation every 0.25% of carbon monoxide in the flue gases will cause a loss in efficiency of 1%.

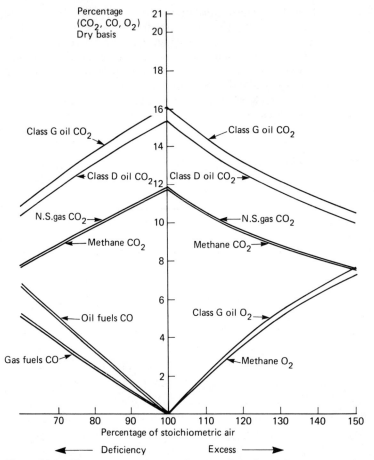

Figure B7.8.1 *Flue gas content diagram showing effect of air deficiency on combustion of oil and gas*

B8 Boiler and burner efficiencies

It is important to distinguish between the various efficiencies as quoted from time to time.

$$\text{Boiler thermal efficiency} = \frac{\text{Useful output}}{\text{Gross energy in fuel}} \times 100\%$$

This takes count of the gross calorific value of the fuel and of all the losses up to the point of delivery of the output from the boiler; these losses are defined in BS 845 and referred to also in BS 2885, the latter giving test procedures for gas-fired equipment not covered by BS 845. The complete range of losses and the approach to their measurement is as follows:

L_G Sensible heat in dry flue gas which is normally assessed from the CO_2 or O_2 content.

L_W Enthalpy of water vapour in flue gas; this is determined by the moisture content of the flue gas and the latent and sensible heat in the moisture. For temperatures up to 300 °C this value may be taken as 2.75 MJ/kg. However, the British Standard sets arbitrary values for this loss, i.e.
Coal 4%   Oil 6%   Natural gas 11%

L_U Unburnt gas loss (presence of CO in flue gases). An approximate rule is that 0.25% CO represents 1% drop in efficiency.

L_A Combustibles in ash. This is either estimated or determined by analysis.
L_B Blow-down loss (steam boilers only).
L_R Radiation, convection and miscellaneous. The British Standard lays down the percentages shown in **Table B8** related to the maximum continuous rating of the plant.

$$\text{Combustion efficiency} = \frac{\text{Input} - \text{Flue gas losses}}{\text{Input}} \times 100\%$$

With the chemical method of flue gas analysis the enthalpy of the water vapour is excluded because the sample is removed from the flue and the vapour condenses; for this reason the efficiency is on a 'dry' basis and expressed as a percentage of the nett calorific

B8 Percentage losses due to L_R for various boilers

Boiler	Loss (%)
Modern package boiler	3.0
Water tube boiler	4.0
Economic wet back boiler	4.0
Economic dry back or sectional boiler	5.0
Cornish or Lancashire vertical boiler	6.5
Sectional with no heat transfer under surface	8.5

value of the fuel. Where measurements are taken directly in the flue gases notably when O_2 is measured by the zirconium oxide sensor the percentage is related to the total volume of flue gas and in these circumstances the combustion efficiency is normally expressed using the total or 'wet' loss in the flue gases. Care must be taken to ascertain which measurement is being used and which efficiency is being quoted.

Where it is possible to measure the output from the boiler as is frequently the case with steam systems the efficiency may be expressed as

Efficiency = Output/Input

With a steam boiler the output is the enthalpy of the steam produced less the heat content of the feed water; the input is the gross calorific value of the fuel.

The examples which follow are from actual case studies and serve to emphasize the tolerances which are inherent in these measurements and the possible pitfalls in placing too much reliance on a single monitoring signal. Combustion is a highly complex process and one should be wary of the glib, easy answer to maintaining high operating efficiencies.

B8.1 Low pressure hot water boilers fired by natural gas

The information in **Chart B8.1** was recorded using the chemical method of analysis of flue gases. The excess air based on the CO_2 and O_2 measurements can be taken from **Charts B7.6** and **B7.6.2** respectively. The dry loss as a percentage of nett calorific value can then be

B8.1.1 Boilers fired by natural gas

Boiler		Flue gas content (%)			Flue gas temp. (%)	Ambient temp. (°C)
No.	Output (kW)	CO_2	O_2	CO		
1	129	7	10	0.2	154	10
2	175	4.5	11.5	0.2	154	10
3	234	6	11	minimal	149	10
4	234	5	11	minimal	141	10

taken from **Chart B7.7.2** using the temperature *rise*, i.e. (flue gas temperature) – (ambient temperature). This leads to the summary in **Table B8.1.2**. The inconsistencies between CO_2 and O_2 are a reminder of the need to allow a tolerance of a minimum of $\pm \frac{1}{2}\%$ in combustion efficiency measurements). The proportion of CO recorded for numbers 1 and 2 represents a loss of efficiency of approximately 0.8%.

The gross calorific value of North Sea Gas is 53.4 MJ/kg (**A5**), the enthalpy of water vapour in flue gas at 150 °C is 2.8 MJ/kg (**A33**), the mass of water vapour in flue gas is 2.14 kg/kg fuel (**B7.4**); hence 'wet' loss is $2.8 \times 2.14 = 6.0$ MJ/kg, i.e. 11.2% of GCV.

B8.1.2 Excess air and percentage loss

Boiler	Based on CO_2		Based on O_2	
	Excess air (%)	Loss (%)	Excess air (%)	Loss (%)
1	60	6.7	81	7.8
2	145	10.7	110	9.1
3	83	7.8	96	8.6
4	120	8.8	96	8.0

B8.1.3 Dry loss, wet loss correction, and efficiency

Boiler	Dry loss (%)	After reduction and correction (%)	Efficiency on GCV (%)
1	7.3 + 0.8 = 8.1	7.2 + 11.2 = 18.4	81.6
2	9.9 + 0.8 = 10.7	9.5 + 11.2 = 20.7	79.3
3	8.2	7.3 + 11.2 = 18.5	81.5
4	8.4	7.5 + 11.2 = 18.7	81.3

The dry loss, based on the average of the $O_2 + CO_2$ readings, with allowance of 0.8% for the CO content (boilers 1 and 2) is shown in the second column of **Table B8.1.3**. Reduction of these figures in the ratio NCV: GCV, i.e. 47.6:53.4 and addition of 11.2% for the wet loss gives the figures in the third column, and the efficiencies based on GCV are given in the fourth column.

By trimming excess air and ensuring that burners are set correctly target efficiencies of 83% are reasonable.

B8.2 3870 steam boiler fuelled by Class G oil

For data, *see* **Table B8.2**.

The C:H ratio is 7.5:1 which indicates a maximum CO_2 content of 16% (**B7.5.1**); 11% CO_2 indicates 46% excess air (**B7.6.1**), 3% O_2 indicates 16% excess air. At 305 °C rise the indicated loss (**B7.7.2**) is 15% based on CO_2 and 12.4% based on O_2, on average 13.7% of NCV.

B8.2 Fuel and flue gas of 3870 kW boiler

Fuel		Flue gas	
Carbon	85.4%	Temperature	332 °C
Hydrogen	11.4%	CO_2 content (dry)	11%
Sulphur	2.8%	O_2 content	3%
Other	0.4%		
GCV	42.9 MJ/kg	Ambient temp.	27 °C

The ultimate analysis procedure as described in **B7.4** may be used to assess air requirement and flue gas make up, but in practice the main need is to establish the enthalpy of the H_2O content.

The mass of H_2 per kg of fuel is 0.114 kg and this will yield $0.114/2 = 0.057$ kmole of H_2O per kg of fuel; this represents a mass of $0.057 \times 18 = 1.03$ kg.

The enthalpy at 300 °C is 3.08 MJ/kg (**A33**) and the loss is thus $3.08 \times 1.03 = 3.17$ MJ/kg fuel, i.e. 7.39% of GCV. The flue gas loss based on GCV is

$$13.7 \times \frac{(42.9 - 3.08)}{42.9} \times 100\% = 12.7\%$$

The losses can be summarized:

L_G = 12.7%
L_W = 7.4%
L_U = 1.0% (estimate to allow for unburnt gases)
L_R = 4.0% (economic wet back boiler)
L_B = 2% (estimated)

The total loss is 27.1% and the boiler efficiency 72.9%.

B8.3 Two steam boilers supplying heating and process load

For data, *see* **Table B8.3.1**.

The C:H ratio is 5.2:1 indicating a maximum CO_2 content of 14.4% (**B7.5.1**); 12% CO_2 indicates 20% excess air; 6% CO_2 indicates 130% excess (**B7.6.1**). This leads, from **B7.7.2**, to a dry loss of 17.8% for No. 1 and 23.8% for No. 2, based on NCV.

The mass of H_2 per kg of fuel being 0.156 kg, the yield of water vapour is $0.156/2 = 0.078$ kmol per kg fuel, and the mass of water vapour is 0.078×1.4 kg per kg of fuel.

B8.3.1 Data for two steam boilers

Boiler	Maker	Steam (lb/hr)	Fuel		Flue gas	
1	Lumby	1000	C	81%	Temp. (No. 1)	440 °C
2	Marshall	1500	H	15.6%	Temp. (No. 2)	330 °C
			Other	3.4%	CO_2 content (No. 1)	12%
			GCV	42.6 MJ/kg	CO_2 content (No. 2)	6%
					Ambient temp.	27 °C

B8.3.2 Summary of losses in the two boilers

Loss	Boiler 1 (%)	Boiler 2 (%)
L_G	15.8	21.4
L_W	11.0	10.3
L_U	0.7[a]	
L_R	10.0[b]	7.5[b]
L_B	0.5[c]	0.5[c]
Total loss	38.0	39.7
Efficiency	62.0	60.3

[a] Trace of smoke.
[b] Very high because of age and condition.
[c] Blow down well controlled.

The enthalpy of water vapour is

No. 1 $1.4 \times 3.35 = 4.69 \, \text{kJ/kg}$
No. 2 $1.4 \times 3.13 = 4.38 \, \text{kJ/kg}$

i.e. 11.0% and 10.3% of GCV respectively.
   Flue gas losses based on GCV are therefore

No. 1 $17.8 \times \dfrac{42.6 - 4.69}{42.6} = 15.8\%$

No. 2 $23.8 \times \dfrac{42.6 - 4.38}{42.6} = 21.4\%$

The losses are summarized in **Table B8.3.2**.

B9 Energy balance in furnaces

High temperature fossil fuel fired furnaces inevitably have high flue gas losses for the reasons referred to in B7.7; however, the overall energy performance of furnaces may be improved by better insulation and the development of ceramic fibre high temperature insulation has had a major effect on these improvements. In spite of its higher cost, electricity can be justified as a fuel for well insulated furnaces because the flue gas losses of fossil fuels are avoided. Melting furnaces for non-ferrous metals or holding furnaces with good capping facilities and good control of metal to the casting process are examples where electricity can be advantageous. Section **B32** covers electrical heating in general and **B32.3** outlines some of the applications.
   Detailed information on the characteristics and properties of furnace insulation materials is best obtained from the material manufacturers or from furnace insulation specialists; reference to section **C3/C16** should enable suitable contact to be made.
   Some of the materials used for the manufacture of furnace insulation are fireclays, China clay (kaolin), ball clay, calcined clay, bauxite and various aluminosilicate materials (either occurring naturally or synthesized). Other synthetic materials used are silicon carbide and fibres synthesized from alumina or aluminosilicate. High purity refined zirconium silicate is also used for specialized application.
   Manufacturers normally publish chemical analysis of their products and the dominant materials are alumina (Al_2O_3), silica (SiO_2) lime (CaO) in the so-called castable materials, but refractories generally contain much less CaO.
   Iron oxide (Fe_2O_3) can constitute up to 10% but for the high temperature materials the Fe_2O_3 content tends to be very much less. Other constituents are titanium oxide (TiO_2), magnesium oxide (MgO) and small quantities of alkalis and other impurities.
   Some very broad statements may be made on the performance of refractories and castables:

(1)   Thermal conductivity almost always increases with increase in operating temperature; a particular exception is refractory material containing 95% or more alumina where the reverse applies.
(2)   Increase in bulk density results in higher crushing strength but with higher thermal conductivity.
(3)   Thermal conductivity tends to increase with higher classification temperature.
(4)   High silica content is usually associated with lower classification temperature and lower thermal conductivity whereas the reverse applies with high alumina content.

   The tables at **A22** which are abridged from information published by Morgan Refractories, Morganite Ceramic Fibres and MPK Insulation, give the chemical constituents and main properties of typical materials; however, it is emphasized that an extremely wide range of products is available and manufacturers' advice should be sought on specific application.

It is usual to distinguish between refractories, which are manufactured building blocks for high temperature insulation, castables, which are supplied as a dry blend for hydration at the time of installation and mouldables which are prepared in a plastic condition and supplied at the proper consistency for application.

Castables need to be cured through the hydration stage and during this process lose some of their crushing strength; there is a continued loss in strength up to 900–1000 °C but thereafter crushing strength rises owing to vitrification. Mouldables have a low initial crushing strength but, being heat setting, their strength rises as they are exposed to high temperature.

Ceramic fibre insulation is best described by reference to the very well established trade names of the principal manufacturers concerned:

SAFFIL fibres (an Imperial Chemical Industries product) are microcrystalline and of high purity containing at least 95% alumina; such fibres exhibit less shrinkage at high temperature when compared with aluminosilicate fibres and consequently achieve higher classification temperatures (*see* **A22.7**).

TRITON KAOWOOL which is a blown aluminosilicate fibre made from blends of high purity alumnina and silica (triton is the trade name of the Morgan Refractories product and Kaowool the trade name of the Babcock and Wilcox product).

These basic fibre materials can be supplied as bulk fibre or made up into blanket, paper, board, sheet, felt, rope or pre-formed shapes. Their application to finished insulation is handled by specialists using suitable cements, anchoring systems and furnace construction techniques. The impact which ceramic fibre can have on furnace insulation rests on their relatively low thermal conductivity associated with low density and hence low heat capacity; this is illustrated in the thermal profile calculations at **B9.1**. Ceramic fibre sheet is made up into modules of varying thickness by the process of 'stack bonding' using an organic binder; details of heat loss and heat storage for a range of such modules are given in **A22.3**. Performance of high temperature furnaces can be further improved if the emissivity of the hot face can be raised (refer to **B24.3**). Raising the emissivity towards unity will cause re-radiation to take place within the furnace chamber with consequent reduction in losses. The charts at **A22.8** (reproduced by courtesy of MPK Insulation) illustrate the range of emissivity values for furnace materials and show the superior performance of ENE-COAT which is a silicon carbide based coating material.

The rules of conductive heat transfer are defined in **B24.1** and these rules may be applied to furnace insulation calculations, thus:

$$\text{Heat flow rate } (Q) \text{ (W/m}^2) = \frac{\text{Temp. differential (K)} \times \text{thermal conductivity (W/m K)}}{\text{Thickness (m)}}$$

Alternatively

$$Q = \frac{\text{Temperature differential (K)}}{\text{Thermal resistance (m}^2 \text{ K/W)}}$$

The other properties commonly defined in relation to furnace work are:

Hot face and cold face temperature (°C) (in the case of multilayer insulation intermediate temperatures are also defined).
Heat loss normally stated in W/m^2 but may also be stated as MJ/m^2 hr (1 MJ/m^2 hr = 277 W/m^2).
Heat storage (MJ/m^2).

Three examples of calculation procedures follow; **B9.1** covers thermal profile calculations and **B9.2** and **B9.3** cover case histories showing the overall energy balance in furnaces.

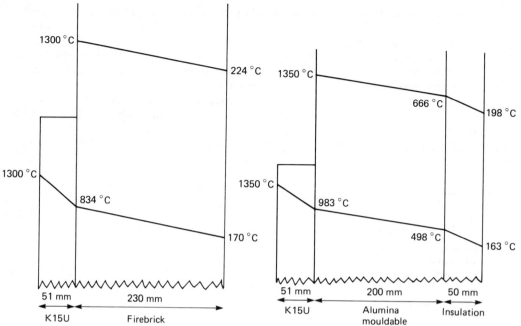

Figure B9.1(a) *Thermal profile of a typical forge furnace roof*

Figure B9.1(b) *Thermal profile of a typical walking beam steel billet reheating furnace roof*

B9.1 Thermal profile calculations for furnace linings

The calculations which are based on case histories have been adapted from information published by Morganite Ceramic Fibres Ltd.

Forge furnace roof

For data, *see* **Table B9.1.1**.

Heat flow rate Q in firebrick alone is

$$\frac{1300 - 224}{0.23} \times 1.00 = 4680 \text{ W/m}^2$$

The resistance to heat flow after adding the ceramic fibre will be the sum of the resistances of the two elements, i.e.

$$\frac{0.23}{1.00} + \frac{0.051}{0.31} = 0.395 \text{ K m}^2/\text{W}$$

B9.1.1 Data for forge furnace roof

Material	Average temperature	Thermal conductivity (w/m K)	Density (kg/m³)
Firebrick lining at	762 °C	1.00	
Firebrick lining at	509 °C	0.98	
Ceramic fibre at	1054 °C	0.31	
Firebrick			1100
Ceramic fibre			120

Thus heat flow rate with fibre added is

$$\frac{1300 - 170}{0.395} = 2860 \text{ W/m}^2$$

The intermediate temperature may be checked because the flow rate will be the same across both layers, i.e.

Fibre $\quad \dfrac{(1300 - 834) \times 0.31}{0.051} = 2830 \text{ W/m}^2$

Firebrick $\quad \dfrac{(834 - 170) \times 1.00}{0.23} = 2890 \text{ W/m}^2$

Reduction in heat flow rate from addition of fibre lining is

$$4680 - 2860 = 1820 \text{ W/m}^2 \ (39\%)$$

Using specific heat capacity values from **A22.4** and **A22.5**:

Heat storage per m^2 surface area firebrick only:

$$1100 \ (\text{kg/m}^3) \times 0.23 \ (\text{m}) \times 1.13 \ (\text{kJ/kg K}) \times 1300 \ (\text{k}) = 371\ 700 \text{ kJ/m}^2$$

Heat storage with fibre added:

$$(1100 \times 0.23 \times 1.12 \times 834) + (120 \times 0.051 \times 1.00 \times 1300) = 244\ 700 \text{ kJ/m}^2$$

Reduction from addition of fibre 127 400 kJ/m^2 (34%)

Walking beam reheat furnace

For data, see **Table B9.1.2**.

B9.1.2 Data for walking beam reheat furnace

Material		Thermal conductivity (w/m K)	Density (kg/m³)
60% Alumina mouldable at	1008 °C	1.1	1750
60% Alumina mouldable at	741 °C	1.05	
50 mm Insulation at	432 °C	0.40	800
50 mm Insulation at	330 °C	0.375	
Ceramic fibre at	1167 °C	0.35	120

Resistance to heat flow before adding fibre

$$\frac{0.05}{0.40} = \frac{0.20}{1.1} = 0.307 \text{ K m}^2/\text{W}$$

Check for intermediate temperature:

Hence heat flow rate $\quad \dfrac{(1350 - 198)}{0.307} = 3750 \text{ W/m}^2$

60% Alumina $\quad \dfrac{1350 - 660}{0.20} \times 1.1 = 3760 \text{ W/m}^2$

Insulation $\quad \dfrac{666 - 198}{0.05} \times 0.4 = 3740 \text{ W/m}^2$

Combined resistance to heat flow *after* adding fibre:

$$\frac{0.051}{0.35} + \frac{0.2}{1.05} + \frac{0.05}{0.375} = 0.469 \text{ K m}^2/\text{W}$$

New heat flow rate

$$\frac{1350 - 163}{0.469} = 2530 \text{ W/m}^2$$

Check for intermediate temperatures:

Fibre $\qquad \dfrac{1350 - 983}{0.051} \times 0.35 = 2520 \text{ W/m}^2$

60% Alumina $\quad \dfrac{983 - 498}{0.20} \times 1.05 = 2550 \text{ W/m}^2$

Insulation $\qquad \dfrac{498 - 163}{0.051} \times 0.375 = 2500 \text{ W/m}^2$

Reduction from addition of fibre

$$3750 - 2530 = 1220 = \text{W/m}^2 \ (33\%)$$

Heat storage before adding fibre:

60% Alumina $\qquad 1750 \times 0.2 \times 1.12 \times 1350 = 529\,000$
50 mm Insulation $\quad 800 \times 0.05 \times 1.06 \times \ 666 = \ \ 28\,000$
$\qquad\qquad\qquad\qquad\qquad\qquad\qquad$ Total $\ \ 557\,000 \text{ kJ/m}^2$

Heat storage after adding fibre:

60% Alumina $\qquad 1750 \times 0.2 \times 1.12 \times 983 \quad = 385\,000$
50 mm Insulation $\quad 800 \times 0.05 \times 1.05 \times 498 \quad = \ \ 28\,000$
51 mm Fibre $\qquad 120 \times 0.051 \times 1.05 \times 1350 = \ \ \ 9\,000$
$\qquad\qquad\qquad\qquad\qquad\qquad\qquad$ Total $\ \ 415\,000 \text{ kJ/m}^2$

Reduction from addition of fibre

$142\,000 \text{ kJ/m}^2 \ (25\%)$

B9.2 Heat treatment of steel laminations for electric motor manufacture

Packs of laminations for stators treated at 800 °C
Laminated rotors with die cast aluminium cages treated at 600 °C
Fuel: Propane GCV 50 GJ per tonne (92 MJ per m^3)

(1) *Energy consumed by burners*
$\qquad$ Stators $\quad$ 2 Furnaces with 8 burners, each 35 kW input
$\qquad\qquad\qquad$ 1 Furnace with 8 burners each 26 kW input
$\qquad\qquad\qquad$ 1 Furnace with 8 burners each 22 kW input
$\qquad$ Rotors $\quad$ 1 Furnace with 8 burners each 35 kW input
$\qquad\qquad\qquad$ 1 Furnace with 8 burners each 26 kW input
$\qquad$ Duty $\qquad$ 3 cycles per day, 5 days per week. Analysis of records indicated the fuel
$\qquad\qquad\qquad$ consumption could be related to the equivalent of 6¾ hr at full fire each day.
$\qquad$ Total weekly design input to burners
$\qquad$ Stators $\quad 8 \times [(2 \times 35) + 26 + 22] \text{ kW} \times 6.75 \text{ hr} \times 5 \text{ days} = 31\,860 \text{ kWhr}$
$\qquad$ Rotors $\quad 8 \times (35 + 26) \text{ kW} \times 6.75 \text{ hrs} \times 5 \text{ days} = 16\,470 \text{ kWhr}$
$\qquad$ Total $\qquad (31\,860 + 16\,470) \times 3.6 \text{ (MJ per kWhr)} = 174 \text{ GJ per week}$

(2) *Energy stored in work*
60 batches per week   840 kg stators   500 kg rotors
Specific heats: Steel 0.13   Aluminium 0.23   (Rotors based on 90% steel and 10% aluminium) : $(0.9 \times 0.13) + (0.1 \times 0.23) = 0.14$ kcal/kg K

Weekly stored energy (assuming 20 °C ambient)
Stators $60 \times 0.13 \times 840$ kg $\times (800 - 20)$ °C $= 5\ 110\ 560$ kcal
Rotors $60 \times 0.14 \times 500$ kg $\times (600 - 20)$ °C $= 2\ 436\ 000$ kcal

Total $\dfrac{(5\ 110\ 560 + 2\ 436\ 000) \times 4.184 \text{ (kJ per kcal)}}{10^6 \text{ (kJ to GJ)}} = 31.6$ GJ per week

(3) *Energy stored in furnaces*
Inside surface area (all furnaces): Walls $4.4\,\text{m}^2$ Roof $1.2\,\text{m}^2$
Storage capacity of furnace linings: Walls $121.5\,\text{MJ/m}^2$ Roof $138.6\,\text{MJ/m}^2$
Stored energy at full temperature (all six furnaces);

$6 \times [(4.4 \times 121.5) + (1.2 \times 138.6)] = 4206$ MJ

Add 100% to allow for cold start each day and to provide for cooling between batches. Weekly Total

$\dfrac{4206 \times 2.0 \times 5 \text{ (days)}}{1000 \text{ (MJ to GL)}} = 42.0$ GJ

(4) *Losses in exhaust gases*
Temperature of gasses leaving furnace: Stators 610 °C   Rotors 400 °C
Measured volume flow of exhaust gases: Stators $1.14\,\text{m}^3/\text{s}$ Rotors $0.51\,\text{m}^3/\text{s}$
A check on volume flow may be made by applying the combustion theory (**B7**) to propane and from **Table A14** it can be seen that five volumes of oxygen are needed for combustion of each volume of pure propane (C_3H_8). Hence the air requirement is

$\dfrac{5 \times 100}{21 \text{ (\% by volume O}_2 \text{ in air)}} = 23.8$ volumes

This is confirmed by **Table A6** which shows that commercial propane requires 23 volumes of air per volume of fuel.
From **B9.2(1)** above the total weekly energy input is 174 GJ, and splitting this in ratio to burner input gives 100 GJ for stators and 74 GJ for rotors. Converting these figures to MJ per second:

Stators $\dfrac{100 \times 1000 \text{(GJ to MJ)}}{5 \text{ (day)} \times 6.75 \text{ (hr)} \times 3600 \text{ (second)}} = 0.823$ MJ/s

Rotors $\dfrac{74 \times 1000}{5 \times 6.75 \times 3600} = 0.60$ MJ/s

Based on a GCV of 92 MJ/m^3 the volume input in m^3/s is

Stators $\dfrac{0.823}{92} = 0.0089$

Rotors $\dfrac{0.609}{92} = 0.0066$

The flue gas volume (m^3 per second) for stiochiometric combustion will be $(23 + 1) \times$ fuel volume, thus

Stators $0.0089 \times 24 = 0.214$

Rotors $0.0066 \times 24 = 0.159$

Correcting these figures for flue gas temperature

$$\text{Stators} \quad 0.214 \times \frac{273 + 610}{273 + 20} = 0.64 \text{ m}^3/\text{s}$$

$$\text{Rotors} \quad 0.159 \times \frac{273 + 400}{273 + 20} = 0.31 \text{ m}^3/\text{s}$$

These values indicate approximately 85% excess air to stators and approximately 60% to rotors. Following the procedure in **B7.7** and using the stoichiometric values in **A13**, we obtain the flue gas contents (kg/kg fuel):

$$
\begin{array}{llll}
\text{Stators} & N_2 & 12.1 \times 1.85 & = 22.4 \\
& CO_2 & & 3.0 \\
& O_2 & 3.64 \times 0.85 & = 3.1 \\
\text{Rotors} & N_2 & 12.1 \times 1.6 & = 19.4 \\
& CO_2 & & 3.0 \\
& O_2 & 3.64 \times 0.60 & = 2.2
\end{array}
$$

Taking ambient temperature as 20 °C and specific heat capacities from **Table B7.7**, the energy in dry flue gases per kg of fuel is

$$\text{Stators} \quad \frac{\begin{array}{c}(22.4 \times 1.163 \times 590) + (3.0 \times 1.201 \times 590) \\ + (3.1 \times 1.07 \times 590)\end{array}}{1000} = 19.4 \text{ MJ}$$

$$\text{Rotors} \quad \frac{\begin{array}{c}(19.4 \times 1.11 \times 480) + (3.0 \times 1.118 \times 480) \\ + (2.2 \times 1.01 \times 480)\end{array}}{1000} = 13.5 \text{ MJ}$$

The loss due to water vapour in flue gases may be taken at 2.7 MJ per kg of H_2O for both stators and rotors, i.e. $2.7 \times 1.64 = 4.4$ MJ

The weekly input of fuel is:

$$\text{Stators} \quad \frac{1000 \text{ GJ}}{50 \text{ (MJ/kg)}} \times 1000 \text{ kg} = 2000 \text{ kg}$$

$$\text{Rotors} \quad \frac{74 \text{ GJ}}{50} \times 1000 \text{ kg} = 1480 \text{ kg}$$

The annual flue gas losses is therefore:

$$\text{Stators} \quad (19.4 + 4.4) \times \frac{2000}{1000} = 47.6 \text{ GJ}$$

$$\text{Rotors} \quad (13.5 + 4.4) \times \frac{1480}{1000} = 26.4 \text{ GJ}$$

Total 74 GJ

(5) *Energy balance*

Flue gas losses	74 GJ
Stored in furnaces	42 GJ
Stored in work	32 GJ
Radiation and unaccounted	26 GJ
	174 GJ

From **A6**, propane yields 544 m^3 of gaseous fuel per tonne and based on gross CV of 93 MJ/m^3 the weekly tonnage of fuel is

$$\frac{174 \times 1000}{544 \times 92} = 3.47 \text{ tons per week}$$

With fuel at say £200 per tonne, the weekly fuel cost is £694 and the cost per GJ input £4. The average fuel cost per tonne of material processed is

$$\frac{694}{60 \text{ (batches)} \times (0.84 + 0.50) \text{ tonnes per batch}} = £8.6$$

The above study puts values to the losses in the heat treatment process and enables evaluation of heat recovery systems to be made, i.e.:

The use of recuperative burners.
Recovery of energy by heat exchangers in flues.
Reduction of stored energy by use of ceramic fibre furnace linings.
Recovery of energy released when work is cooling.
As assessment may be made of the effect of using ceramic fibre insulation using the data contained in **Table A22.3** and assuming a cold face temperature of 52 °C for the rotor furnaces and 68 °C for the stator furnace; also assuming that the storage capacity of the roof will be the same as the walls.
The stored energy for the six furnaces will become

$$(3 \times 3.53 \text{ (MJ/m}^2) \times 5.6 \text{(m}^2) + (3 \times 4.41 \times 5.6) = 164 \text{ MJ}$$

The allowance for cold start and for cooling between batches will be much lower, say 40%.
The weekly stored energy thus becomes

$$\frac{164 \times 1.4 \times 5}{1000} = 1.15 \text{ GJ}$$

This would have the effect of reducing the fuel bill from £694 to £530 per week; however, very careful evaluation of the effect on the process and of the mechanical suitability of ceramic fibre would be needed.

B9.3 Heat treatment of dies for aluminium extrusion

B9.3.1 Funnace No. 1

Nominal operating temperature 1050 °C (based on analysis of thermograph charts) temperature of gases leaving furnace 420 °C; continuous operation with three complete cycles during each 24 hr; fired by North Sea Gas and consuming on average of 7.2 kg gas per hour for 18 of the 24 hr; 7 days per week; 46 weeks per year.
Annual energy input:

$$\frac{7.2 \text{ (kg)} \times 53.4 \text{ (MJ per kg from A5)} \times 18 \text{ (hr)} \times 7 \text{ (day)} \times 46 \text{ (wk)}}{1000 \text{ (MJ to GJ)}} = 2228 \text{ GJ}$$

From **Table A13** and calculations at **B7.4**, with the assumption of 50% excess air, the approximate masses of flue gasses are, in kg/kg fuel:

N_2 19.5
CO_2 2.75
H_2O 2.25
O_2 2.0

The total weight of 'dry' flue gas is thus 24.25 kg per kg fuel. For simplicity use average flue gas as chart **A15.2** which indicates 1.13 kJ/kg K at 420 °C.
The total energy content of the flue gases is

$$\frac{7.2 \times 24.25 \times 1.13 \times (420 - 30) \times 18 \times 7 \times 46}{10\,000\,000} = 446 \text{ GJ per year}$$

Estimated energy stored in work based on 500 kg of material per 24 hr = 50 GJ.

The energy stored in the furnace and lost by radiation is thus

$$2228 - (446 + 50) = 1736\,\text{GJ}$$

The rates of energy flow are

Fuel input $\dfrac{7.2 \times 53.4 \times 1000\ (\text{MJ to KJ})}{3600\ (\text{hr to s})} = 107\,\text{KW}$

Flue loss $\dfrac{7.2 \times 24.25 \times 1.13 \times 390}{3600} = 21\,\text{kW}$

B9.3.2 Furnace No. 2

Nominal operating temperature 600 °C (based on analysis of thermograph charts) temperature of flue gases 450 °C; ambient temperatures 30 °C; continuous operation 4 to 5 cycles each 24 hr with average consumption 5.1 kg gas per hour for 22 of the 24 hr; 7 days a week; 46 weeks per year.

Annual energy input

$$\frac{5.1 \times 53.4 \times 22 \times 7 \times 46}{1000} = 1929\,\text{GJ}$$

Specific heat capacity of average flue gas at 450 °C = 1.135 kJ/kg K.
Energy of content of flue gases (assuming 50% excess air)

$$\frac{5.1 \times 24.25 \times 1.135 \times (450 - 30) \times 18 \times 7 \times 46}{1\,000\,000} = 342\,\text{kJ/yr}$$

Annual energy stored in work, say 50 GJ
Energy stored in furnace and lost by radiation

$$1929 - (342 + 50) = 1537\,\text{GJ per year}$$

Rates of energy flow:

Fuel input $\dfrac{5.1 \times 53.4 \times 1000}{3600} = 76\,\text{kW}$

Fuel loss $\dfrac{5.1 \times 24.25 \times 1.135 \times 420}{3600} = 163\text{kW}$

The total of 37 kW flue gas loss can be recovered at an efficiency of around 80% and used for space and water heating in adjacent parts of the factory and offices.

The efficiency of conversion can be further improved by utilizing the already warmed ambient air in the furnace room as input to an air-to-air heat exchanger.

In theory the surplus energy is of value only during the working day in the heating season, but in practice the problem of maintaining comfort conditions in adjacent areas is much reduced if these areas are heated overnight during winter. The alternatives are air-to-air heat exchanger to provide warm air heating or air to water to supply a low pressure hot water system.

B10 Environmental temperature and thermal comfort

The concept of environmental temperature is used principally to compensate for radiation from internal surfaces in a building where the temperature of these surfaces is lower than the air temperature within the building. Clearly these lower surface temperatures are likely to occur on external walls particularly those with a high degree of exposure.

The environmental temperature is calculated by taking two-thirds of the mean radiant temperature and adding it to one-third of the air temperature. To illustrate the calculation

B10.1 Data for environmental temperature calculation

Region	Temperature (°C)	Area (m²)
Floor	17	690
External walls	18	210
Wall glazing	10	25
Roof	12	600
Roof glazing	10	230
Internal air	19	—

the data quoted in **B11.7** and summarized in **Table B10.1**, covering a small machine shop may be used.

Mean radiant temperature:

$$\frac{(690 \times 17) + (210 \times 18) + (25 \times 10) + (600 \times 12) + (230 \times 10)}{690 + 210 + 25 + 600 + 230} = 14.39\,°C$$

Environmental temperature:

$$(\tfrac{2}{3} \times 14.39) + (\tfrac{1}{3} \times 19.00) = 15.9\,°C$$

Work by Professor P. O. Fanger in the Laboratory of Heating and Air Conditioning at the University of Denmark has introduced further concepts into the study of temperature and comfort conditions. This work co-ordinates six factors affecting both physiological and psychological reactions to comfort conditions; these factors are

air temperature, mean radiant temperature, air velocity, water vapour pressure (humidity), metabolic rate of the individual and the thermal insulation of his clothing.

Professor Fangar's approach is to demonstrate that when a person feels comfortable in his environment these six factors are in equilibrium. The first four can be measured by well established techniques and Professor Fangar* has postulated scales of measurement for the last two, i.e. MET factor and CLO factor.

Table A4 gives empirical values for the metabolic heat generated by people varying for an average person from 100 w at rest to 400 w when doing heavy work. Corresponding MET factors are 1.0 to 4.0.

CLO factors run from zero (no clothing) to 3.0 (full winter outdoor clothing).

A further approach to comfort conditions uses the Effective Temperature which is intended to compensate for the degree of humidity; this assumes that people will feel comfortable provided conditions fall short of the point where they are perspiring continuously (conversely it must also ensure that they feel warm enough!). This approach results in an effective temperature scale and a zone of comfort conditions for a person engaged in sedentary activity.

In practice a relative humidity of between 40% and 70% satisifies most people and this is the normal criterion for air conditioning calculation; in winter time people will tolerate a dry bulb temperature two or three °C below the summer level.

There is no ideal: people vary, activity varies and clothing worn varies.

B11 Calculation of energy loss through building structures

To calculate the heat loss through a building structure it is first necessary to establish the thermal transmittance or 'U' value of the main structural elements (floor, walls, roof, glazing, doors, etc).

*P. O. Fangar. 'Thermal comfort', McGraw Hill, New York (1972)

In general there are three distinct steps in establishing the 'U' value:

(1) Determine the thermal conductivity of the layers which make up the element, i.e. in the case of a conventional wall the outer leaf, the inner leaf and the plaster.
(2) Establish the external and internal surface resistance.
(3) Allow for the thermal resistance of any ventilated or unventilated air space, e.g. the cavity in a wall or the air space in a hollow building block.

Some references for comprehensive data and guidance are:

British Gypsum White Book, British Gypsum, Ruddington, Nottingham
CIBS Guide A3,   Chartered Institution of Building Services, London
Eurisol – UK 'U' Values,   The Association of British Manufacturers of Mineral Fibres, London

The notes which follow and the associated tables and diagrams are intended to present a shortened practical guide which should enable calculations to be made covering the majority of commercial and industrial building structures.

B11.1 Thermal conductivity

Values for commonly encountered building materials are given in **Table A20.1**. Values for concrete and brickwork vary considerably with the density of the material used and for these reasons the values are portrayed in chart form at **A20.2**. The definition of thermal conductivity is given in **Table A2.3** and it is expressed in watts per metre thickness per degree kelvin or W/m K.

Thermal resistivity is the reciprocal of thermal conductivity and is expressed as m K/W and defined as the resistance per unit thickness.

B.11.2 Thermal resistance

Insofar as thermal resistivity is the resistance per unit thickness, the thermal resistance of an element will be the resistivity times the element thickness in metres, i.e. m^2K/W.

The significance of reciprocals in thermal calculations is best emphasized at this point by stating the universal law for resistances in series (whether they be electrical, thermal or friction), namely that the addition of resistances in series brings about a reduction in the rate of flow whether the flow be electric current, heat or fluid.

Thus for heat transfer:

$$\text{Thermal transmittance ('U' value)} = \frac{1}{R_1 + R_2 + R_3 + \text{etc}}$$

Where R_1, R_2, R_3 etc. are the thermal resistances of the elements of the structure.

'U' value is hence the reciprocal of the total resistance and is expressed in watts per metre square kelvin or W/m^2K.

B11.3 External and internal surface resistance

These values are also part of the build up of the total resistance; however they are special cases because they vary with the emissivity of the surface concerned, with convection loss which is influenced by the direction of heat flow (horizontal, vertically upward or downward) and, so far as external surfaces are concerned, with the degree of exposure.

(1) The emissivity factor of a surface varies from approximately 0.9 for common building materials to 0.05 for polished metal surfaces; the factor indicates the rate at which the surface will lose heat by radiation to adjacent surfaces (*see* **Table A21.8**).
(2) The convective heat transfer coefficient is highest for vertical upward airflow and lowest for vertical downward flow.

(3) The degree of exposure of an external surface is classified as sheltered, normal or severe, and these degrees are defined by CIBS as

Sheltered: Up to third floor buildings in city centres.
Normal: Most suburban and rural buildings; fourth to eighth floor in city centres.
Severe: Buildings on hill or coastal sites; floors above fifth in suburban or rural districts; floors above ninth in city centres.

Values of external resistance according to exposure are given in **A21.2**. Note that glazing is particularly affected by exposure, whereas the effect of exposure on well insulated structures is small.

Values for internal surfaces are given in **A21.1**.

B11.4 Thermal resistance of air spaces

Air space resistance is affected by surface emissivity, the dimensions of the air space, the direction of heat flow, whether the space is ventilated and, to a minor degree, by the temperature difference across the space. A special case is where a corrugated sheet is in contact with a flat sheet (as in certain uninsulated roof structures) leaving a series of narrow air spaces between trough and crest of the corrugations. It has been established that up to 25 mm width of air space resistance increases; thereafter it remains almost constant.

The resistance is reduced where there is a series of vertical subdivisions (as in a hollow block with two or more such spaces) and the extent of this reduction depends on the width of the space. These effects have been taken count of in standard values covered by the CIBS guide and these values are shown at **A21.3** and **A21.4**. The advantage of cavity wall filling as a means of improving insulation have been proved beyond doubt; however, care is necessary to avoid secondary problems arising from moisture migration across the filled cavity.

B11.5 Thermal transmittance of floors

The heat loss from concrete or suspended floors depends heavily on the loss to the surroundings from the edges of the floor and is consequently high for small independent structures and low for large industrial or commercial buildings. The CIBS guide gives a comprehensive range of standard '*U*' values for floors, but the condensed table at **A21.5** should cover normal industrial and commercial situations.

B11.6 Thermal transmittance of glazing

Standard '*U*' values are given at **A21.6** and the values quoted are exclusive of frames. If frames are included in the area of glazing, wooden frames will reduce the overall '*U*' value and metal frames will increase it; the increase with metal frames will be less if there is a thermal barrier or break between the glass and the frame (see **A21.7**). Apart from double or triple glazing there are a number of other approaches to reduce thermal transmittance:

(1) DURATHERM offer a twin wall polycarbonate with stiffening membranes giving semi-rigid panels of overall thickness 20 mm. This is claimed to be easier to fix than secondary glazing, with a '*U*' value improvement from 5.75 to 2.17.
(2) ACOUSTICABS offer ISOFLEX which is a sandwich of three transverse layers of transparent corrugated cellulose. This is a flexible material with an improvement of '*U*' value from 6.0 to 2.0 claimed.
(3) BONWYKE distribute the MADICO range of flims which are made from polyester forming a 0.025 mm sandwich with a layer of aluminium of molecular thickness. These materials are primarily to reduce solar gain, but are claimed to reduce heat loss from the building by internal re-radiation.
(4) DICK de LEEUW use the sunblind principle to improve '*U*' values and also to reduce solar gain. The material is grained aluminium foil bonded to polyester film with an

overall thickness of approximately 0.025 mm. The material is raised or lowered across a window as a normal sunblind. 'U' value with single glazing is claimed as 2.6 when unventilated and 3.5 when lightly ventilated.

All the above notes are based on 'steady state' conditions, i.e. with temperature remaining constant. In practice, temperatures are invariably either rising or falling and the assessment of the dynamic performance of a structure under changing temperature conditions is rather more complex, involving the use of admittance values (expressed in W/m^{2}K) and certain factors which allow for the dynamic thermal performance. Detailed guidance is given in CIBS guide A3.

B11.7 Calculation procedure for a factory building

The example concerns a small machine shop in a group of factory buildings built some 70 years ago and consisting of the following elements:

Floor.   'L' shaped consisting of solid concrete 57.5 m × 12 m, area 690 m^2; for calculation purposes floor may be considered as having two exposed edges with 'U' value from **Table A21.5** of 0.24 W/m^{2}K.

Exposed Walls.   Total area 210 m^2 solid 9 inch brick with piers. The summation of the R values is:

    (1) Surface resistance based on normal exposure and high emissivity from **Table A21.2**   0.06 m^{2}K/W

    (2) Outer leaf of brickwork 115 mm; Thermal conductivity based on density 1500 kg/m^3 (from chart **A20.2** 'exposed' curve)   0.65 W/mK Resistance will be reciprocal of conductivity times thickness of brick, i.e.

$$\frac{1}{0.65} \times \frac{115}{1000} = 0.18 \text{m}^2 \text{K/W}$$

    (3) Inner leaf of brickwork 115 mm; thermal conductivity based on density 1500 kg/m^3 (from 'protected' curve **A20.2**)   0.47 W/mK

$$\text{Resistance} = \frac{1}{0.47} \times \frac{115}{1000} = 0.25 \text{ m}^2\text{K/W}$$

    (4) Inside surface resistance (**Table A21.1** 'high emissivity') = 0.12 m^2 K/W

    The Total of (1) + (2) +(3) + (4) = 0.61 m^{2}K/W
    Hence

$$\text{'}U\text{' value} = \frac{1}{0.61} = 1.64 \text{ W/m}^2\text{K}$$

Wall Glazing.   Area 25 m^2: 'U' value from **A21.6**   5.6 W/m^{2}K. As the windows were set in steel frames and divided into small panes, increase 'U' value by 10% to 6.16 W/m^{2}K.

Roof.   Pitched roof 600 m^2 (exclusive of glazing and allowing for pitch). Construction: timber trusses carrying 10 mm tiles with 2 mm sarking felt supported by 10 mm roof boarding. Summation of 'R' values:

    (1) Outside surface resistance (**A21.2**)   0.050 m^{2}K/W
    (2) 10 mm tile, from **A20.1** conductivity = 0.85 W/mK
    Hence

$$\text{Resistance} = \frac{1}{0.85} \times \frac{10}{1000} = 0.012 \text{ m}^2\text{K/W}$$

(3) 2 mm sarking felt, from **A20.1** conductivity = 0.20 W/mK
Hence

$$\text{Resistance} = \frac{1}{0.20} \times \frac{2}{1000} = 0.010 \text{ m}^2\text{K/W}$$

(4) 10 mm roof boarding, from **A20.1** conductivity = 0.13
Hence

$$\text{Resistance} = \frac{1}{0.13} \times \frac{10}{1000} = 0.080 \text{ m}^2\text{K/W}$$

(5) Inside surface resistance (**A21.1**) = 0.10 m^{2}K/W

Total (1) + (2) + (3) + (4) + (5) = 0.252 m^{2}K/W

Hence

$$\text{`}U\text{' value} = \frac{1}{0.252} = 3.97 \text{ W/m}^2\text{K}$$

Roof Glazing. Area 230 m^2 'U' value from **A21.6** 6.6 W/m^{2}K
Large panels in metal frames allow, say, 6.7 W/m^{2}K

Normal procedure is to establish the average 'U' value for the complete building and this is done by adding together the total losses through the elements and dividing the result by the total exposed area, i.e.

Floor (690 × 0.24) + Walls (210 × 1.64) + Wall Glazing (25 × 6.16) + Roof (600 × 3.97) + Roof Glazing (230 × 6.7)

Total loss = 166 + 344 + 154 + 2382 + 1541 = 4587 W/K

Note the loss is in watts per degree Kelvin temperature difference between outside and inside the building. Note also the major effect on losses of the roof glazing and of the poorly insulated roof itself.
The average 'U' value is thus

$$\frac{4587}{690 + 210 + 25 + 600 + 230} = \frac{458}{1755} = 2.6 \text{ W/m}^2\text{K}$$

To assess the heating requirement to compensate for structure losses, it is normal in the United Kingdom to assume a minimum external temperature of zero centigrade and to work to the desired environmental temperature (**B10**).
Based on 18 °C inside the heating requirement is

$$\frac{2.6 \times 1755 \times 18}{1000 \text{ (W to kW)}} = 82 \text{ kW}$$

The moves which can be made to improve the thermal performance of the building are

(1) Reduce roof glazing to 5% of roof area, i.e. 0.05 × 830 = 42 m^2.
Total loss from roof now becomes

42 × 6.7 + 788 × 3.97 = 3410 W/K

compared with

2382 + 1541 = 3923 W/K

The reduction in heating requirement will be

$$\frac{3923 - 3410}{1000} \times 18 = 9.2 \text{ kW}$$

i.e. from 82 kW to 73 kW

(2) Retaining 5% roof glazing, reduce the 'U' value of the roof by introducing below the sarking 50 mm glass fibre supported by 10 mm plaster board secured to rafters. The addition to 'R' value will be

Glass fibre (from **A20.1**)

$$\frac{1}{0.040} \times \frac{50}{1000} = 1.25 \, \text{m}^2\text{K/W}$$

Gypsum plaster board (from **A20.1**)

$$\frac{1}{0.18} \times \frac{10}{1000} = 0.06 \, \text{m}^2\text{K/W}$$

Total addition to R 1.31 m²K/W

New total 'U' value of unglazed part of roof

$$\frac{1}{0.252 + 1.31} = 0.63 \, \text{W/m}^2\text{K}$$

New average 'U' value

$$\frac{(690 \times 0.24) + (210 \times 1.64) + (25 \times 6.16) + (788 \times 0.65) + (42 \times 6.7)}{1755}$$

$$= \frac{1457}{1755} = 0.83 \, \text{W/m}^2\text{K}$$

New heating requirement

$$\frac{0.83 \times 1755 \times 18}{1000} = 26 \, \text{kW}$$

This dramatic improvement resulting from a better insulated roof is apparent. Similar evaluations may be made of, for example, a false ceiling using the values for loft space in **A21.4** or for external treatment of the roof by urea–formaldehyde foam.

A comparison may be made with a modern building complying fully with Part F (1.4.82) of the 1976 Building Regulations. This stipulates maximum 'U' values for roof of 0.35 and walls 0.60. Under these conditions, assuming similar wall glazing and a maximum of 5% roof glazing, the average 'U' value becomes

$$\frac{(690 \times 0.24) + (210 \times 0.6) + (25 \times 6.16) + (788 \times 0.35) + (42 \times 6.7)}{1755}$$

$$= 0.57 \, \text{W/m}^2\text{K}$$

The heating requirement is

$$\frac{1755 \times 0.57 \times 18}{1000} = 18 \, \text{kW}$$

B12 Heating and cooling requirements of buildings

In general winter heating of buildings is required to compensate for losses which take place through the fabric (**B11**) and losses from air changes (**B23**); however, the heating requirements must take account of gains as well as losses, and in summer time the gains can predominate, necessitating some form of cooling. A balance must therefore be struck after examining the following sources of heat gain:

(1) *Process heat* This is by far the most important consideration for energy conservation. Process heat must always be studied for its potential for recovery, recuperation or reduction of losses by better insulation or control. **Section B9** covers surplus heat from furnaces and **Sections B22** (heat pumps) and **B25** (heat exchangers) are also relevant. A common source of loss which affects many industrial processes is the evaporation loss from the surface of hot water, and **Table A32.5** gives some useful data.

(2) *Metabolic heat* This can be a significant factor in factories with a high population density or in public rooms, particularly ballrooms or discos, where activity rates are high. **Table A4** gives data on watts loss per person at various activity rates.

(3) *Electric lighting* The watts loss from electric lamps can be a significant factor particularly in areas where a high standard of illumination is provided. Techniques are available to recover this heat particularly where air conditioning is provided (**B20**) and **Table A39** gives the data necessary to assess the losses from light sources.

(4) *Electric motors* Whether motors are industrial or for commercial or domestic machinery it must be remembered that *all* the energy supplied is dissipated as heat either in losses in the motors or in friction in driven equipment.

(5) *Solar gains* It is not part of this manual to examine the potential merits of solar energy as a means of ultimately solving the energy needs of this planet; suffice to say that the main concern is to *reduce* the impact of solar heat on buildings in summer time and thus reduce the need for cooling. In energy conscious buildings the reduced area of glazing and the use of reflective films goes far to alleviate the discomfort from solar gains in working areas. However, in heating and air conditioning design calculations solar gains are a major factor, together with building aspect, prevailing winds and degree of exposure. Whether the final balance dictates the need for air conditioning (**B20**) or adequate ventilation (**B10**) can be judged only against the full facts of a particular situation.

B13 Selection and sizing of heating equipment

B13.1 Boilers

In recent years development effort towards more efficient use of fuel has tended to concentrate on four main objectives, i.e. better control systems, higher efficiency boilers, effective burning of difficult fuels and the modular approach enabling boilers to run at close to their design load and therefore at high efficiency.

(1) *Control systems* Monitoring and control of boiler performance by feed-back from flue gas analysis, either O_2 or CO_2, is firmly established technology for all but the smallest installations and a wide range of systems is now available (*see* **B33.1**).

(2) *Higher efficiency boilers* Detailed attention to heat exchanger design can yield improvements in efficiency (but at an inevitable penalty on first cost). A major step forward can be made, particularly with gas-fired boilers, if the latent heat of vaporization of water vapour in the flue gases can be recovered. Condensing boilers with flue exit temperatures as low as 50 °C have been designed, but the penalty is an expensive stainless steel secondary heat exchanger to recover the condensate and avoid the problems of corrosion. Efficiencies on GCV as high as 90% are claimed for condensing boilers; in France and Germany considerable service experience has been gained but adoption of the principle in the UK has been relatively slower (early 1984).

Some improvement to efficiency can be achieved by avoiding the loss to the flue which takes place when a boiler cuts out under thermostatic control; designs are available which maintain circulation from the boiler casing to the system, thus reducing the loss.

(3) *Difficult fuels* Utilization of the lower grades of solid fuel and a wide range of industrial and domestic waste has been the subject of study by many manufacturers

and proven designs are now available to utilize energy from these sources (*see* **A8, A10** and **A11**).

(4) *Modular installation.* Use of the modular approach enabling a group of three, four or more boilers to be sized in such a manner that load fluctuations can be catered for by shutting down unwanted capacity and allowing remaining modules to operate at their rated load and consequently with maximul efficiency.

Other developments are steam boilers with a much increased turn-down ratio enabling them to operate at high efficiency on relatively low steam demand; also the many boilers designed specifically for hot water production. This enables optimum performance to be achieved in relation to hot water demand and storage requirement in industrial and commercial installations. By separating the hot water requirement from the space heating need both services can be provided more efficiently. The sealed heating system, which avoids the problems of header tanks, water damage through failure of a mains connected system and evaporation and corrosion problems from open systems, is beginning to gain some ground in the UK after many years of successful application in Europe and North America. In Summary, there is an extremely wide choice of equipment for generating steam and hot water and a very careful evaluation of alternatives is essential if the best compromise between plant requirements and economy is to be achieved.

B13.2 Low pressure hot water systems

The heating requirement of the building or room must first be assessed as outlined in **B12** and a figure of total kW output required established for the environmental temperature selected, and assuming a minimum external temperature (normally 0 °C). Selection of radiator types and sizes can then be made using the following criteria:

(1) Total output from radiators must have a margin of at least 10% over the calculated kW loss to provide for exceptional conditions.
(2) Disposition of radiators must be such as to give an even distribution of heat.
(3) Positions of doors, windows and permanent fixtures in the building will affect radiator positions and sizes, compromises are therefore invariably necessary.
(4) The use of radiators with fins to improve rate of heat output is in general to be preferred where the slight extra projection from walls can be accepted.
(5) Double radiators have less than double output, but can be used with advantage where wall space for radiator fixing is restricted.

The table at **A37.1** enables radiator selection to be made and also gives correction factors for variation of mean water temperature and different values of air temperature in the building. Normal practice is to design the system on the basis of BS 3528 (1974) which stipulates a mean water temperature of 76.7 °C and a room temperature of 21.1 °C.

The simple abridged table at **A37.3** gives the watts output per nominal m^2 of standard steel radiators and enables a rapid estimate to be made of radiator sizes to suit a given situation.

Alternative types of radiator are gaining in popularity notably continuous low level 'skirting' types, radiators designed for maximum convective heat transfer, and fan assisted type for rapid warm-up.

B13.3 Gas fired radiant heating

In the days of cheap fuel the principal source of energy for radiant heaters was electricity (hot water 'radiators' distribute heat much more by convection than by radiation). At today's fuel costs, electric radiant heaters are uneconomic except at special off-peak rates or in circumstances where their use is occasional for top-up purposes.

The gas-fired ceramic plaque type heater with surface temperature 850–900 °C is well known; modern designs on this principle have inputs up to 30 kW with overall length under

2 metres. The alternative is the 'black' radiant heater where combustion takes place in a steel tube approximately 100 mm diameter; recent development has concentrated on the U tube configuration involving two parallel 100 mm diameter tubes each some 5 metres long. At one end the pipes are connected together by a U tube end and at the other end a gas burner is connected to one pipe and an exhaust fan to the other. The peak value of surface temperature is around 450 °C near the end of the U tube. It can be shown that over the major part of its length the heat released by radiation is well in excess of the heat due to convection.

Heat input of standard units is normally between 22 kW and 35 kW, but smaller sizes have been developed. The points to be borne in mind in applying these 'black' heaters in industrial or commercial premises are:

(1) A mounting height of 5 m or more is usually called for, but some smaller units can function satisfactorily at 3.0–3.5 m.
(2) The heaters are relatively bulky so that overhead mounting, particularly in engineering shops with large machine tools and overhead cranes or services, can cause problems.
(3) Because the energy in the flue gases is very largely dissipated in the tube itself, combustion efficiency is high. Where heaters are ganged together into a common flue the loss is very small resulting in overall conversion efficiencies in excess of 90%.
(4) Because of their infrared radiant quality they provide almost immediate benefit and feeling of comfort to those coming within their range, whereas none of the radiant heat is transmitted to the intervening air.
(5) The problems of heat stratification in high building are very much less than with warm air systems.

Catalytic heaters, sometimes referred to as 'flameless' utilize a catalyst to enable butane gas to release its enthalpy at around 450 °C; such appliances are available in domestic sizes and offer safety advantages because of the absence of a high temperature flame.

Calculation procedure for radiant heating systems follows the pattern of **B23.2** and **B11**, noting that air change loss is assessed on the difference between *air* temperature and external temperature whereas structure loss is based on *environmental* temperature (**B10**). Details of this approach can be obtained from the CIBS guide.

From the point of view of good distribution, the higher radiant heaters are mounted, the better, but some compensation must be made if the height is above 5 m. For every 0.3 m above 5 m, 1% should be added to the calculated heat loss.

The total kW output dictates the number of heaters required and the layout of heaters should take account of:

(1) Assuming that heaters are in a number of lines according to the size of the building, the distance between lines should be twice the height from floor level to the radiant tube.
(2) The number in each line will depend on the total number calculated as above.
(3) Heaters in the lines adjacent to outside walls should be disposed so as to achieve radiation to a height of two metres on the walls.
(4) There should, as far as possible, be a greater concentration of heaters near to areas of high loss such as doors, large glazed areas and exposed outside walls.
(5) There will inevitably be a need to find the best compromise between the layout of heaters and the layout of general lighting.
(6) Care must be taken where overhead services, cranes, large machines tools or other obstructions may be in the line of radiation. Such obstructions can reduce the effectiveness of the heaters and can also cause problems because of overheating of plant, machinery or services within the range of the radiant tubes. In these circumstances satisfactory results can often be achieved by using reflector shields to prevent overheating.

Dr John K. Maund of the University of Aston in Birmingham has written a paper, Radiant heating for industry (March, 1984), which is the outcome of research work into improved reflector design and the maximization of radiant heat as a proportion of the total energy released.

B13.4 Warm-air heating systems

The free-standing gas or oil-fired warm-air heater is a well tried and proven system for heating industrial and commercial premises and is generally available in units with from 50 kW to 300 kW output. Because such heaters are flued the conversion efficiency cannot be as high as for radiant systems and starting with a cold building they certainly take longer to create comfortable conditions for the occupants. They occupy some floor space, but this can be overcome by using suspended units with down blast fans. Care is necessary in siting these units to avoid the creation of an unnecessarily high rate of air change resulting from the power of the heater's fans. These heaters have certain clearly defined advantages, i.e.:

(1) They can be installed quickly and easily.
(2) Moving to a new location requires gas or oil pipework and electricity services, together with a simple flue through an outside wall or roof and is likely to be less costly than changing the location of a radiant system.
(3) They can be used for air circulation either by re-circulation in conjunction with fresh air supply to reduce contamination, or as ventilators by using the heater fans for circulation during summer.

Calculation procedure, as for other systems, involves estimation of loss due to air changes and loss from the fabric of the building; care must be taken to allow for any additional ventilation caused by the heater fans and a rather greater margin is desirable in assessing the heater output to ensure that it can achieve a rapid warm-up when starting from winter temperatures.

A recent development work has focused on the direct gas fired 'flueless' heater where the products of combustion are released into the heated space. This approach offers much higher conversion efficiencies as the flue loss is virtually eliminated. However very careful control of combustion is necessary to ensure that carbon monoxide content of flue gas is such that threshold limit values are not exceeded (A24.2).

B13.5 Calorifiers

Calorifiers fall into two categories, non-storage and storage, and they are both further subdivided according to the primary fluid, i.e. steam, high pressure hot water and, for storage type, also low pressure hot water. Non-storage calorifiers are used for providing process hot water where the demand is constant; alternatively, where steam or high pressure hot water is a mains service in an industrial complex or in a district heating scheme; in such cases they are used so as to draw energy from the mains for local distribution to a low pressure hot water heating system or for supply of domestic hot water where demand may be classed as reasonably constant.

Storage type calorifiers, as their name implies, are designed to deal with widely fluctuating requirements and their selection is governed by the total load required to provide the service needed and by the time taken to recover temperature after periods of peak demand.

The tables at B13.5.1 illustrate the selection procedure with the total storage determined by the size of the shell and the heater battery selected according to recovery time, i.e. 1, 2 or 3 hours.

The primary fluid in all cases flows in a tube 'bundle' in the form of multiple U tubes; in non-storage types the 'bundle' is distributed through the length of the calorifier which is normally horizontal but can be vertically mounted; in storage types it is normally concentrated at the bottom of a vertical configuration with, of course, a much greater relative volume of secondary water.

The design of the heating bundle is normally generous in relation to required capacity and manufactures offer standard ranges from which a selection may be made according to predicted demand.

Chart B13.5.2 illustrates the process of selecting a non-storage calorifier according to output required and pressure of steam available; the curves are based on 82 °C return in the secondary water and the output available must be adjusted for other secondary temperatures. The charts enable a size of calorifier to be selected and this must be the size above the intersection of output and steam pressure.

Chart B13.5.3 illustrates size selection where the primary flow is water; in this case the arithmetic mean temperature difference must be assessed, i. e.

$$\text{AMTD} = \frac{T_1 + T_2}{2} - \frac{t_1 + t_2}{2}$$

where
T_1 is primary inlet temperature to calorifier
T_2 is primary outlet temperature from calorifier
t_1 is secondary intlet temperature
t_2 is secondary outlet temperature

Having established the calorifier size from the chart it is then necessary to check the primary water flow rate against the manufacturers' specification to establish the number of passes of primary fluid necessary to effect adequate heat transfer (normal designs are either two pass or four pass).

B13.5 Calorifier size and flow rates

Size	Flow rate (l/s) Max.	Min.
1	1.90	0.95
2	1.90	0.95
3	2.65	1.33
4	2.65	1.33
5	5.88	2.94
6	5.88	2.94
7	11.36	5.68
8	11.36	5.68
9	15.78	7.89
10	15.78	7.89
11	20.20	10.10

Table **B13.5** gives primary flow details for the eleven sizes illustrated in Chart **B13.5.3**. This can be best illustrated by an example:

Output required 800 kW, primary water 160 °C, inlet 110 °C, outlet, secondary 81 °C, inlet 71 °C outlet

$$\text{AMTD} = \frac{160 + 110}{2} - \frac{81 + 71}{2} = 49 \text{ °C}$$

From the chart this is just within the capacity of size 6.

The output of 800 kW or 800 kJ/s must equate with the total heat released by the primary water in falling from 160 °C to 110 °C. Thus:

$$800 \text{ kJ/s} = \frac{l/s}{1000} \times 988 \text{ (kg/m}^3) \times 4.175 \text{ (kJ/kg K)} \times 50(\text{K})$$

(1000 converts litres to m^3, 988 and 4.175 are taken from **A18**). Thus

litre/s = 3.88

The lower flow rate of size 6 is 2.94 and if the required figure had been less than this a four-pass design would have been needed; if greater than the higher flow rate (5.88) a larger calorifier would be needed. For 3.88 litres per second the two-pass unit is satisfactory.

B13.5.1 Selection of shell and heater battery for storage type calorifier

For outlines, *see* **Figure B13.5.1**

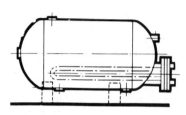 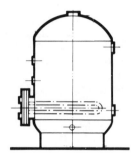

Figure B13.5.1 *Calorifier outlines*

B13.5.1 Range and output of heater units

Size	1	2	3	4	5	6	7	8	9	10	11	12	13	14	15	16	17
Output (kW)	12.2	14.7	17.1	18.3	19.6	22.0	24.2	25.7	29.3	34.2	36.6	39.1	44.0	48.8	51.3	58.7	66.0

Size	18	19	20	21	22	23	24	25	26	27	28	29	30	31	32	33
Output (kW)	68	73	78	88	98	103	110	117	132	147	176	205	220	235	264	293

Storage capacity of shells and heater unit details

Shell storage capacity (l)	Heater unit for recovery time		
	1 hour	2 hour	3 hour
570	11	4	1
570	13	6	2
800	15	8	3
900	16	9	5
1140	19	11	7
1360	21	13	9
1590	23	15	10
1820	25	16	12
2050	26	17	13
2270	27	19	14
2730	28	21	16
3180	29	23	18
3400	30	24	19
3640	31	25	20
4090	32	26	21
4540	33	27	22

Data adapted from catalogue information supplied by Crosse Engineering Ltd.

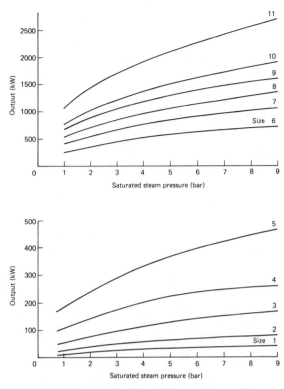

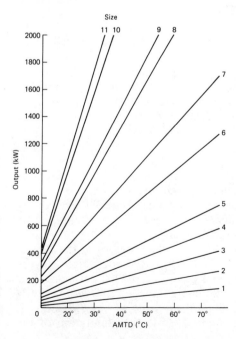

B13.5.2 *Output curves for range of steam-water non-storage type calorifiers; hot water outputs based on 82 °C return*

(Adapted from catalogue information by Heat Transfer Ltd.)

Figure B13.5.3 *Output curves for range of water-water non-storage type calorifiers: hot water output based on 82 °C flow, 71 °C return*

B14 Degree days

B14.1 Background to method

The 'degree day' has been used as a means of assessing the severity of winter weather by the meterological office for almost a century and in 1966 N.S. Billington proposed a straightforward calculation procedure to enable the degree day to be used to establish the heating requirements of buildings. It is now generally accepted that for the latter purpose it has serious limitations but as a means of monitoring energy consumption according to the relative severity of heating seasons it can be useful.

The first essential is to establish a base temperature for external conditions and all published information on degree days is related to 15.5 °C. For accurate assessment of energy requirements in particular situations a base other than 15.5 °C may well be needed, but the outline of calculation procedure which follows uses the 15.5 °C figure.

Experiment and test with particular reference to housing has shown that 15.5 °C external may be related directly to 18.5 °C internally; this implies that heat gains within the building are such that this differential will normally exist. The further basic assumption implicit in the degree day method is that provided the external temperature is maintained at 15.5 °C *no heating will be required in the building*.

In order to calculate degree days three arbitrary approximations have been made:

(1) If during a 24 hour period the external temperature never rises above 15.5 °C the average number of degrees by which temperature falls below base is taken as the difference between 15.5 °C and the simple average of maximum and minimum recorded temperature. Thus if the minimum temperature is 1 °C and the maximum 14 °C the difference is

$$15.5 - \frac{14 + 1}{2} = 8 \,°C$$

and this is classified as 8 Degree Days.

(2) If during a 24 hour period the temperature is below base more that it is above, the degree day figure is calculated from the difference between one half the extent to which it falls below base and one quarter the extent to which it rises above. Thus, if the maximum is 19.5 °C and the minimum is 3.5 °C the figure becomes

$$\frac{15.5 - 3.5}{2} - \frac{19.5 - 15.5}{4} = 5 \text{ Degree Days}$$

(3) If during the 24 hour period the temperature is above base more than it is below, degree days are calculated from one quarter the extent to which the minimum temperature falls below base. Thus for a maximum of 20 °C or above and a minimum of 11.5 °C the figure becomes

$$\frac{15.5 - 11.5}{4} = 1 \text{ Degree day}$$

Clearly an assessment based on maximum and minimum readings only gives an approximate value. To obtain accuracy, continuous temperature readings would be necessary with degree days calculated from logarithmic mean values. With the availability of recording instruments and the readiness with which such records can be computed by microprocessors, the way is open to provide more precise data.

Degree day figures for each month of the year and for seventeen regions of the British Isles are published regularly by the Department of Energy. Recorded figures vary according to region and according to severity of winter from 1700 to 2900 per year. This annual figure may be used to make an assessment of the energy needed for *continuous 24 hour heating* of a given structure; however, as in the great majority of cases, heating is intermittent, various correction factors need to be applied. A further limitation of the method is that it takes no account of exposure or wind velocity and direction, all of which have a significant effect on fuel used. Billingtons equations for *continuous heating* are as follows:

(a) Required base temp. (° C)

$$= \text{inside temperature (° C)} - \frac{\text{Average heat gain (kW)}}{\text{Building heat requirement (kW/° C)}}$$

(b) Equivalent hours of full load plant operation

$$= \frac{24 \times \text{degree days in heating season}}{\text{design temp. difference (° C)}}$$

(c) Energy consumption (kWhr)

$$= \frac{\text{Equivalent hours} \times \text{maximum output of heating plant (kW)}}{\text{Heating plant efficiency}}$$

A recent development from the Shirley Institute (Feb 1984) uses a microprocessor to produce daily degree day information and to cater for working periods less than 24 hours. The device is being manufactured by JEL and should overcome some of the disadvantages of published degree day information.

B14.2 Calculation procedure

Applying the approach to the building referred to in **B11.7**, the following data may be tabulated:

Inside temperature   18 °C
External temperature   0 °C
Design temperature difference   18 °C
Heating requirement to cover structure losses   82 kW

Air change loss based on two changes per hour and a building volume of 3450 m^3 may be calculated using the procedure in **B9**, i.e.

$$\frac{2 \times 3450 \text{ (m}^3\text{/s)} \times 1.22 \text{ (kg/m}^3\text{)} \times 0.989 \text{ (spec. ht. cap.)} \times 18 \text{ (°C)}}{3600 \times 1000} = 42 \text{ kW}$$

Heat gain will come from electric motors, lighting and the occupants (the position of the building is such that winter solar gains may be neglected)

Total, say 15 kW. Hence

$$\text{Required base temperature} = 18 - \frac{15}{(82 + 42) \div 18} = 14.6 \text{ °C}$$

It has been shown that the correction factor to degree day figures follows an approximately straight line law for base temperatures 2.5 °C either side of 15.5 °C and that $\pm 10\%$ on base temperature varies the correction factor (unity at 15.5 °C) by + 20%. Thus for 14.6 °C the percentage decrease is

$$\frac{15.5 - 14.6}{15.5} \times 100\% = 6\%$$

and the correction factor will be 12% below base, i.e. 0.88

Taking degree days for the Severn Valley at 2235 (20 year average)

$$\text{Equivalent hours} = \frac{24 \times 2235 \times 0.88}{18} = 2622$$

$$\text{Energy consumption} = \frac{2622 \times (82 + 42)}{0.75} = 433\,000 \text{ hWhr} = 1560 \text{ GJ}$$

This is clearly a very high figure and reflects the fact that continuous heating has been calculated.

A more practical figure may be arrived at by assuming a plant utilization factor of say 80% and evaluating the energy requirement as in **B4**, i.e.

$$\frac{124\ (kW) \times 40\ (hr/week) \times 30\ (week) \times 0.80}{0.75\ (plant\ efficiency)} = 159\,000\ kWhr = 570\ GJ$$

To reconcile these two very different values of energy requirement it is necessary to reduce the degree day calculation to allow for 5 day week, and to allow for factory hours as opposed to continuous heating.

Assuming 0.75 as the 5 day week factor we have $1560 \times 0.75 = 1170\ GJ$.

To reconcile this with the **B4** calculation, the factor to be applied for factory hours is 0.5.

It is of course a fact that maintenance of continuous comfort conditions in winter time requires more energy per hour during the night than during daytime so that in practical terms the 0.5 factor is probably not far from the truth. A great deal depends on the 'weight' or thermal inertia of the building; light structures will consume relatively more than heavy structures when they are heated intermittently.

What this calculation demonstrates above all is that for other than continuous heating, the degree day method is fraught with many problems. For those interested in pursuing the approach, reference may be made to 'Energy Demand and System Sizing' No. 2 in the British Gas series 'Studies in Energy Efficiency in Buildings'. Reference may also be made to CIBS Guide B18, 'Owning and Operating Costs', which tabulates a range of factors and equivalent hours for intermittent operation and various building structures.

B14.3 Use of degree days for monitoring energy conservation

The chart at **B14.4** is based on an actual case history and shows the relationship between total energy consumption and degree days for the same factory complex as that referred to

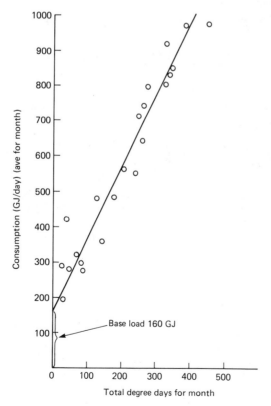

Figure B14.4 *Relationship between energy consumption and degree days; the slope of the line is 2 GJ/degree day*

in **B2.4**. There is clearly an underlying straight line relationship between energy consumption and degree days for a particular factory complex and a particular rate of activity. Using a chart of this type enables the 'better' months to be identified when the actual consumption falls below the degree day line. It must of course be established that the 'better' month is as a result of energy conservation measures and not because of changes in plant or in rate of factory activity.

Note that the chart enables the 'base load' to be identified, i.e. the daily rate of energy consumption when no shop heating is called for. Finally the slope of the curve determines the additional energy consumed for each degree day increase. In summary with a slope of 2 GJ per degree day, a base load of 160 GJ per day and an annual degree day figure of say 2000 the total annual consumption, assuming 240 working days per year and 20 working days in each month, is

$$(160 \times 240) + (2 \times 2000 \times 20) = 118\,400\,GJ$$

B15 The psychrometric chart: its basis and application (refer A36)

Because the moisture content of air is a dominant factor in comfort conditions an understanding of calculation procedures is essential. The chart at **B15.1** is based on the steam tables at **A33** and **A34** and shows the relationship between temperature and absolute pressure up to the critical point for steam (375.7 °C). **B15.2** is plotted on a more open scale and covers temperatures up to 120 °C. On this chart the pressure scale is the vapour pressure of the air–moisture mixture, but the relationship between temperature and pressure is the same as for saturated steam.

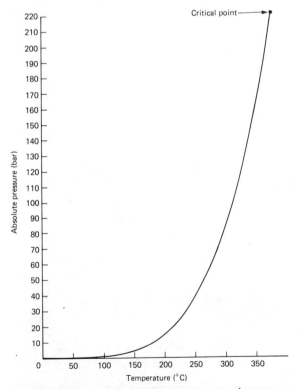

Figure B15.1 *Relationship between temperature and pressure for saturated or superheated steam*

The concept of moisture content in air rests on the proposition of partial pressure or vapour pressure exerted by the molecules of water which are free as vapour in the air–moisture mixture. This vapour pressure is independent of the pressure exerted by the oxygen, nitrogen and other gases in the air. Thus, in the case of an air space above water contained in a vessel and where molecules of moisture are leaving the water surface to form vapour in the air, there will be a particular vapour pressure associated with a particular air temperature with a relationship as in **Chart B15.2**. If the space above the water level is enclosed and if the pressure is raised to a point known as the saturation vapour pressure then condensation will take place. (It should be noted that condensation will also occur if the temperature is lowered to the level well known as the dew point.)

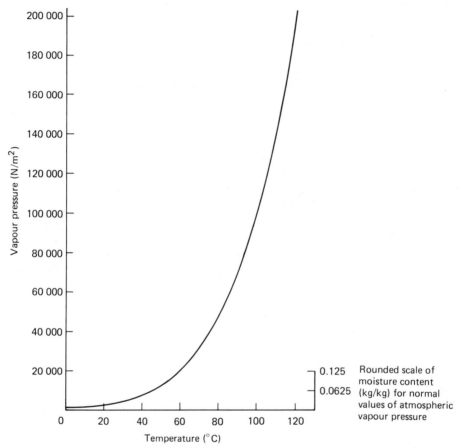

Figure B15.2 *Relationship between temperature of saturated air and vapour pressure (or moisture content)*
The right hand scale is a rounded scale of moisture content (kg/kg) for normal values of atmospheric vapour pressure

For low values of moisture content there is almost a direct relationship between vapour pressure and the absolute humidity of the air as measured in kg/kg (kg of moisture present per kg of dry air). The development of the chart can therefore be illustrated by substituting a scale of moisture content for the scale of vapour pressure in the ratio $10\ 000\,\text{N/m}^2 = 0.0625\,\text{kg/kg}$. The moisture content scale is shown on the right of **Chart B15.2** and applies to pressure below $20\ 000\,\text{N/m}^2$.

The wet and dry bulb thermometer measures the moisture content in the atmosphere. Saturation (100% relative humidity) is indicated by identical wet and dry bulb temperatures, and the extent of the depression of the wet bulb reading indicates the extent to which relative humidity falls below 100%. For accurate measurement the wet bulb must be subjected to a rate of air movement not less than five metres per second and this can be achieved with a 'sling' thermometer.

The next stage in charting the information is to redraw the diagram for a further restricted range of temperatures and to introduce a complete family of relative humidity curves from 100% down to 10%. This is illustrated in **B15.3**. Also shown on **B15.3** are the dry and wet bulb temperature lines; for temperatures up to 25–30 °C the dry bulb lines are nearly vertical, but as temperatures increase – these lines show an increasing slope towards the right at the top of the chart. The use of the wet bulb temperature as a method of

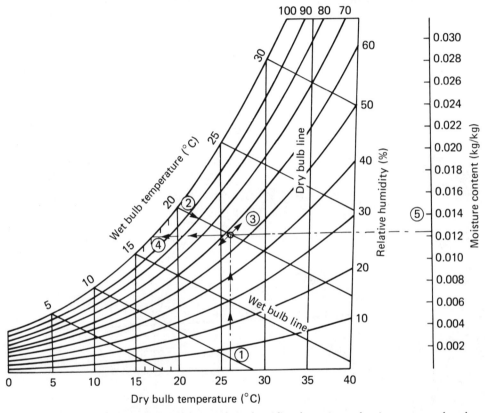

Figure B15.3 *The psychrometric chart showing relative humidity, dew point, and moisture content, based on barometric pressure 101.325 kPa*

measuring humidity is arbitrary in basis but it does enable the essential relationships in moisture laden air to be established. The slope of the wet bulb lines has been accurately determined by experiment and from theoretical considerations; as can be seen these lines slope from left to right on the diagram and are in fact parallel to each other. **Chart B15.3** can therefore be used to illustrate a number of important relationships, i.e.

(1) The intersection of the dry bulb and wet bulb lines gives the relative humidity and as shown in the example on the diagram 26 °C dry bulb with 20 °C wet bulb indicates 58% R.H.

(2) The dew point or the temperature to which the mixture must be cooled in order to induce condensation is determined by the horizontal line from the intersection point to the 100% humidity curve and as can be seen in the example it is 17.1 °C. Reaching the

130

dew point in this manner is known as sensible cooling, in other words, the removal of sensible heat but not latent heat.

(3) The moisture content at 26 °C–20 °C is indicated by the horizontal line to the right and is in fact 0.0124 kg/kg (kg of moisture per kg of dry air).

(4) Saturation of the original mixture can also be brought about by adding further moisture without temperature change and this is known as isothermal humidification. It is clear from **B15.3** that if the 26 °C dry bulb line is continued to meet the 100% humidity curve it will indicate 26 °C as the wet bulb temperature and condensation will start to take place.

(5) The opposites of (2) and (4) are known respectively as sensible heating (which reduces relative humidity without any change of total moisture content) and isothermal dehumidification (which reduces both relative humidity *and* total moisture content).

(6) The 100% R.H. line can of course be reached by a combination of cooling and humidification, and this is known as adiabatic humidification.

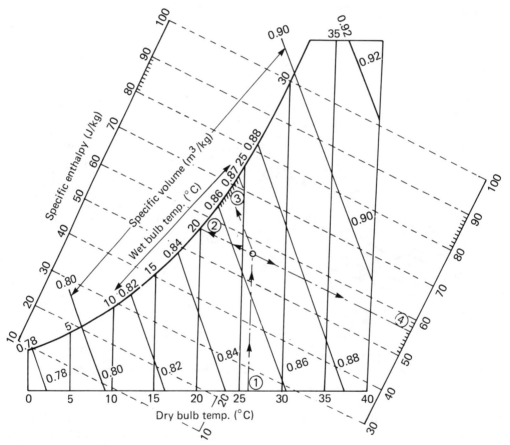

Figure B15.4 *The psychrometric chart showing specific volume and specific enthalpy, based on barometric pressure 101.325 kPa*

The final stage in the development of the complete chart is illustrated in **B15.4**; this shows the lines of specific volume and the value is 0.865 m³/kg for the 26 °C–20 °C condition. It also shows the lines of specific enthalpy which are dotted across the chart and for 26 °C–20 °C the value is 50.67 kJ/kg. Usual practice on comprehensive charts is to show the start and finish of the specific enthalpy lines external to the main chart so that a straight edge may be used to read off the figure for a particular wet–dry bulb condition.

Because of the variety of lines on a comprehensive chart, care is always necessary in interpretation.

It should be noted that charts are drawn for a particular barometric pressure (101.325 kPa for normal calculations). A typical chart (as available from the Chartered Institution of Building Services) is shown at **A36**; other charts are available for different temperature ranges; charts can, of course, also be produced for alternative conditions of pressure.

The theoretical basis of the relationship between vapour pressure and moisture content is dependent on the relative molecular mass of air and water (namely 28.96 and 18.02) and on the difference between partial and total pressure in accordance with the formulae

$$\frac{18.02}{28.96} \times \frac{\text{partial pressure of vapour}}{(\text{Total pressure} - \text{partial pressure})}$$

Thus, if partial pressure is very small compared with total pressure (as is the case in normal humid air calculations) this is a direct relationship as stated above.

The more precise values calculated from the full formula are

Partial pressure (N/m^2)	*Moisture content* (kg/kg)
1000	0.0062
5000	0.0322
20 000	0.1526

and it is these values which are used in constructing the working charts.

The theoretical basis of the relative humidity percentage rests on the ratio of partial pressure (saturation vapour pressure) of the water vapour to the vapour pressure of liquid phase water at the same temperature.

Extending this to the example on diagrams **B15.3** and **B15.4**:

Moisture content = 0.0124 kg/kg
Total pressure   = 101 325 N/m^2

Partial pressure (ρ) may be arrived at from the formula

$$\text{Moisture content } 0.0124 = \frac{18}{29} \times \frac{\rho}{(101\ 325 - \rho)}$$

Hence

$$\rho = 1986\,N/m^2$$

From **Table A18** the partial pressure of water at 26 °C is 3360 N/m^2

Hence

$$\text{R.H.} = \frac{1986}{3360} = 59\%$$

Alternatively R.H. may be stated as the ratio of actual moisture content to the moisture content required to produce isothermal saturation. From the chart at **A36** moisture content at 26 °C, wet bulb = 0.0214

Hence

$$\text{R.H.} = \frac{0.0124}{0.0214} = 58\%$$

B16 Theoretical and empirical approaches to fluid flow

Some appreciation of fluid flow theory is essential to the understanding of practical calculations; **B24.1**, **B24.2** and **B24.3** refer in general terms to some aspects of fluid flow and it is suggested that **B24** be read in conjucntion with the notes which follow. The main formulae are relevant to **B17** and **B18** covering water and steam services and compressed air; there is further overlap with **B19.4** which refers to air flow in ventilation ducts.

Consistency in units used is vital in making these calculations and there are differences in approach in the many works which are available covering fluid flow. The reader is commended particularly to the CIBS guide C4 which gives much information in tabular form and which uses the SI system throughout.

The basic formulae and equations are as follows:

B16.1 The Bernoulli theorem

There are three components which make up the total energy in fluid flow, i.e. the POTENTIAL energy arising from the height of the particles of the fluid above a datum level, the PRESSURE energy arising from the 'head' of fluid in the system and the VELOCITY energy arising from the velocity of fluid flow.

The equation stated in terms of energy per unit mass may be written

$$\frac{P}{\rho} + \frac{V^2}{2} + gZ = \text{constant} \tag{16.1.1}$$

where
P is pressure (Pa)
ρ is density of fluid (kg/m^3)
V is mean velocity of flow (m/s)
g is acceleration due to gravity (9.81 m/s^2)
Z is height above datum (m)

Restating the equation in terms of energy per unit weight and introducing the concept of friction loss in the system together with energy added by a pump we have

$$\frac{P_1}{\varpi} \times \frac{V_1^2}{2g} + Z_1 + H = \frac{P_2}{\varpi} + \frac{V_2^2}{2g} + Z_2 + h_f \tag{16.1.2}$$

where
subscripts 1 and 2 refer to the two points in the system
H is mechanical energy per unit of weight added by the pump (kJ/kg)
h_f is mechanical energy per unit weight lost by friction (kJ/kg)
$\varpi = $ Specific weight of fluid $=$ Density $(\rho) \times g$

Finally the equation may be stated in pressure terms, i.e.

$$P_1 + \tfrac{1}{2}\rho V_1^2 + \rho g Z_1 + \rho g H = P_2 + \tfrac{1}{2}\rho V_2^2 + \rho g Z + \rho g h_f \tag{16.1.3}$$

The expressions in equation (16.1.3) (all in Pa) are defined as

Velocity pressure $= \tfrac{1}{2}\rho V^2$
Pump pressure $= \rho g H$
Pressure due to potential $= \rho g Z$
Pressure loss due to friction $= \rho g h_f$

All these terms can be evaluated from basic information except the friction loss which for a closed circuit will be balanced by the pump pressure. The value of the friction loss (h_f) is obtained from the Darcy equation.

B16.2 The Darcy equation

$$h_f = \frac{4 \times f \times L \times V^2}{2 \times g \times D} \tag{16.2.1}$$

where
L and D are the length and diameter of the pipe (m)
f is the friction factor

The equation can be expressed in terms of volume flow instead of velocity by substitution:

Average volume flow, Q (m³/s) = velocity (m/s) × Cross section area of pipe (m²) i.e.

$$Q = V \times \frac{\pi}{4} \times D^2$$

Alternatively

$$V = \frac{Q}{D^2} \times \frac{4}{\pi}$$

Thus

$$h_f = \frac{4 f L}{2 g D} \times \left(\frac{4Q}{\pi D^2} \right)^2$$

$$= \frac{4 f L Q^2}{D^5} \times \frac{4^2}{2g \times \pi^2}$$

$$= 0.331 \frac{f L Q^2}{D^5} \tag{16.2.2}$$

The important points about this equation are that for a given type of pipe

   (a) The friction loss varies directly as the length of pipe.
   (b) The friction loss varies directly as the square of the volume of flow.
   (c) The friction loss varies inversely as the fifth power of the internal diameter of the
        pipe.

B16.3 The friction factor

To determine the friction loss by means of the Darcy equation a value must be assigned to
the friction factor (f); this in turn necessitates evaluating the Reynolds Number (Re) and
the relative roughness of the pipe (ks/d).

The Reynolds Number, which is referred to further in **B24**, is the product of pipe
diameter and velocity of flow divided by the kinematic viscosity of the fluid, i.e.

$$\frac{D \text{ (m)} \times V \text{ (m/s)}}{\psi \text{ (m}^2\text{/s)}}$$

As can be seen, the units of DV are cancelled out by the units of ψ and the Reynolds
Number is dimensionless.

The relative roughness is the roughness of the pipe (ks) divided by the pipe internal
diameter (d).

The theory behind the calculation of friction factor is quite complex, but the C.I.B.S.
guide C4 gives full information to cover most practical situations. Absolute roughness (ks)
varies from 0.003 for plastic pipes to 2.5 for rusted steel.

Table C.4.5 of the guide gives relative roughness values for mild steel, cast iron, UPVC
and copper piping over the full range of standard sizes; these values vary from 0.014 for
10 mm nominal heavy galvanized steel pipe to 0.0000096 for 159 mm nominal diameter
copper. Finally the guide gives a range of charts relating friction factor to Reynolds Number
for relative roughness values from 0.01 to 0.00001.

B24 outlines the concept of laminar and turbulent flow, but it suffices to say that for
normal fluids in commercial pipes the flow will be turbulent and the Reynolds Number
therefore well in excess of 3000 (it can be as high as 10 000 000). The friction factor for most

practical situations will be between 0.002 and 0.013 and there are empirical values which enable the friction loss to be calculated from the Darcy equation, i.e.:

Gas in ordinary mains   0.005
Flue gases   0.010
Compressed air from 0.0035 for clean pipes to 0.005 for old pipes
Water from 0.005 for ordinary pipes to 0.010 for encrusted pipes

A direct method of assessing pressure drop in compressible (gaseous) fluids is covered at some length in Crane Ltd technical paper M410 and this paper is the source of information in **Tables A27** and **A28**.

The method described in Crane Ltd paper evaluates Pressure Drop (kPa/m) from

$$\frac{C_1 \times C_2}{\text{Density}}$$

where

$C_1 = W^2/10^8 = (W = \text{flow in kg/h})$
C_2 is a factor obtained from **Tables A27.2** and **A28.2**.

Table A27.1 gives wall thickness and inside diameter for the range of nominal inch size pipes to BS 1600 and **Table A27.2** gives values of C_2 for this range.

Table A28.1 covers pipes to BS 3600 having the same external diameter as those in **A27.1**, but with a wider range of wall thicknesses. **A28.2** gives the corresponding values of C_2.

The method employing the factors C_1 and C_2 is illustrated in **B17.4.4** covering steam flow calculations.

Yet a further method of assessing pressure drop is to use velocity as a criterion and to relate this to tables of pressure drop in various pipe sizes; this method is applied to steam in **B17.4.4** and to compressed air in **B18.2** but it has limitations.

B17 Water and steam services

B17.1 Cold water services

These are in a separate category and the detailed study of these is concerned primarily with ensuring that the pressure loss in the system is held within the limits which ensure adequate pressure at the outlet point. A well established technique is based on allocating 'demand units' to various sanitary fittings and the type of use to which they are put, and basing pipe sizes on the flow resulting from these calculations. In general, cold water supplies are of limited significance in terms of energy usage, but where water is used for cooling in industrial processes an entirely different picture emerges.

Industrial consumers pay for water on a metered basis and costs can be substantial; saving water is not direct energy saving, but it can be money saving; looked at in the National context it is also energy saving as the power consumed in recovering, purifying and re-distributing water is very substantial.

Any water which runs to drain after use as a coolant needs close examination; establishing its temperature and mass flow rate enables the energy content to be estimated and this must then be balanced against the cost of heat recovery equipment coupled possibly with filtering and re-use of this water. The external cooler avoids running process water to drain and certainly reduces industrial water costs, but the cooler itself discharges energy to atmosphere and the question to ask is whether this energy can be diverted and utilized.

Table A30 gives an indication of flow rates and pressure losses for water at 15 °C and is a useful basis for preliminary pipe sizing; however, the influence of bends, fittings and valves on pressure loss can be a significant or even the major part of pressure loss in a system. The assessment of pressure loss from fittings by theoretical considerations is quite complex and it is wise to rely on published empirical data which normally expresses loss from fittings in terms of the 'equivalent length' of the appropriate pipe diameter. An indication of equivalent lengths of some common pipe fittings can be obtained from **Table A31.2**.

B17.2 Hot water services

These are of course, of greater significance in terms of energy use and it is important to ensure that temperature is held close to the most economical level and that the pipework is adequately insulated. As far as friction loss is concerned the lower viscosity of hot water **(Table A18)** tends to reduce the loss particularly at low flow rates in the smaller pipe sizes. **section B8** covers boiler efficiencies, and **Section B13.5** covers calorifiers, both of which are relevant to effective hot water systems.

B17.3 High pressure hot water

Whereas hot water systems will normally operate at boiler pressures between 50 kPa and 300–400 kPa, high pressure hot water systems are used for industrial applications with pressures up to 2000–2500 kPa. The boiler construction is as for steam boilers, but the water line is held higher and the pipe system is sealed, with circulation by pumps. Precautions need to be taken against the occurence of flash steam and the whole of the plant including heaters and calorifiers must be designed to withstand the higher pressures.

Pipe sizing calculations must make due allowance for the lower viscosity of the water. Because of the much higher temperatures of the circulating water the requirement of adequate insulation assumes much greater significance. All parts of the system except those directly involved in process work or space heating must be insulated.

B17.4 Steam services

This work cannot attempt to cover the technology involved in steam and steam services, but the steam tables in SI Units at **A33** give the data necessary for calculations and the paragraphs which follow summarize those aspects which are particularly relevant to energy and energy calculations.

B17.4.1 The advantages of steam as a means of distributing energy

(1) The pressure and temperature of saturated steam are inter-related and this enables close control of process temperature to be maintained by regulating steam pressure; the heat released at constant pressure comes from the latent content enabling the temperature to remain unchanged.

(2) The heat content of steam can be up to 25 times that carried at the same temperature by air or flue gases thus enabling relatively small pipes to handle large amounts of energy.

(3) Superheated steam, by virtue of its pressure can release a substantial proportion of its energy to do mechanical work and still retain a heat content for process work or space heating.

(4) If used at progressively lower pressures its energy content can be used at several stages, and finally the condensate resulting from the energy released can be used as an energy source or returned to the boiler system.

B17.4.2 The essential requirements for the efficient utilization of steam

(1) Understand fully the energy requirements and operating temperatures of the process plant and heating equipment and relate these accurately to the capacity of the boiler plant after making due allowances for loss of energy and pressure in the distribution system.

(2) Select boiler sizes in such a manner that, as far as possible, individual boilers may operate at close to full capacity. The examples quoted in **B2** illustrate the need to match the output of one boiler to the summer (process) load with a second or third unit available to meet peak winter demand.

(3) Maintain the necessary 'quality' of steam providing for:

Purging of air from system when starting from cold.

Maintenance of adequate 'dryness' of saturated steam by removing particles of moisture entrained with the steam.

Feed water treatment to ensure that the solids content is within the limits which can be tolerated by the boiler and the distribution system.

Condensate recovery, not just to save energy, but also to ensure that pipework remains free to carry steam efficiently.

(4) Design and maintain a distribution system which is adequate to handle the volume of steam needed, will keep steam velocity at acceptable levels and will ensure that the pressure appropriate to the process is available where needed. In achieving these objectives the losses in pipework must be balanced against the pressure requirements at different parts of the distribution system.

B17.4.3 Calculation procedures for piping sizing

The velocity method

This is referred to in **B.16.3** and can be useful provided its limitations are recognized. Various formulae, factors and tables are available relating velocity and mass flow to the nominal sizes of pipes but it is very simple to calculate if the *bore* of the pipe is known together with either volume flow or mass flow since:

$$\text{Velocity (m/s)} = \frac{\text{Mass flow rate (kg/s)}}{\text{Density (kg/m}^3) \times \text{area of pipe bore (m}^2)}$$

or

$$\text{Velocity} = \frac{\text{Mass flow rate (kg/s)} \times \text{specific volume (m}^3\text{/kg)}}{\text{Area of pipe (m}^2)}$$

or

$$\text{Velocity} = \frac{\text{Volume flow rate (m}^3\text{/s)}}{\text{Area of pipe (m}^2)}$$

The area of the pipe is $\pi/4 \times (\text{diameter})^2$

Table A35.1 gives a range of values for various gauge pressures, velocities and pipe sizes.

The simplified flow formula

This is also referred to in **B16.3** and the tables at **A27.2** and **A28.2** enable pressure drop to be calculated.

The CIBS Guide C4

Comprehensive tables are published by the Chartered Institute of Building Services relating absolute pressure, initial and final pressure conditions, mass flow rate and equivalent length of pipe to a range of nominal pipe sizes fro 10 mm to 300 mm. For all these tables the absolute pressure quoted corresponds to a steam velocity of 30 m/s.

The somewhat complicated formula used for derivation of the tables is

$$\text{Pressure drop due to pipe friction} = P_1^{1.929} - P_2^{1.929}$$

$$= \frac{0.003\,032\, M^{1.889}\, l}{d^{5.027}}$$

Where

P_1 and P_2 are initial and final gauge pressures in the pipe (kPa)

M is mass flow rate (kg/s)

l is the length of the pipe (m), with allowance for pipe fittings

d is the internal diameter of the pipe

The similarity of the formula to the Darcy equation is apparent with an almost exact fifth power inverse relationship with pipe diameter and rather less tham a direct square law relationship with mass flow.

The pressure factor method

This has the attractions of simplicity and is covered fully in SPIRAX SARCO information book 'Steam Distribution'. In essence this starts with a pressure factor (P) which varies with the absolute pressure of saturated steam and which for the sake of clarity and easy access is tabulated in **A33** along with many other properties of saturated steam.

Table **A35.2** relates the pressure drop factor (F) to a range of pipe sizes and gives corresponding values of pipe capacity (kg/s). The formula used is:

$$F = \frac{\text{Initial pressure factor } (P_1) - \text{final pressure factor } (P_2)}{\text{Equivalent length of pipe}}$$

Two simple examples illustrate all four methods.

B17.4.4 Examples of pressure drop calculations for steam pipes

Example 1

Initial pressure	700 kPa (absolute)
Final pressure	600 kPa (absolute)
Pipe length including allowance for fittings	300 m
Steam flow	0.25 kg/s (900 kg/hr)
Density of saturated steam at 700 kPa (A32)	3.67 kg/m^3

Velocity method based on 30 m/s

$$\text{Velocity (m/s)} = \frac{\text{Mass flow rate (kg/s)}}{\text{Density (kg/m}^3) \times \text{Area of pipe (m}^2)}$$

Hence

$$\text{Area of pipe} = \frac{0.25}{30 \times 3.67} = 0.00227 \, \text{m}^2$$

$$\text{Pipe diameter} = \sqrt{\frac{0.00227 \times 4}{\pi}} = 0.054 \, \text{m} \, (54 \, \text{mm})$$

From **A27.1** a schedule 80 pipe, external diameter 73 mm, has a bore of 59 mm and should therefore be satisfactory.

Simplified flow formula

$$\frac{(\text{kg/hr})^2}{10^8} = \frac{(900)^2}{10^8} = 0.0081$$

$$\text{Permissible pressure drop} = \frac{700 - 600}{300} = 0.33 \, \text{kPa/m}$$

But,   Pressure drop (kPa/m) $= \dfrac{C_1 \times C_2}{\text{density}}$

Hence, $C_2 = \dfrac{0.33 \times 3.67}{0.0081} = 150$

From **Table A27.2** for schedule 80 73 mm external diameter pipe, $C_2 = 162$ and this is satisfactory.

CIBS Guide C4 method

For the pipe to be satisfactory using this approach, $P_1^{1.929} - P_2^{1.929}$ should be greater than

$$\frac{0.003\ 032 \times M^{1.889} \times l}{d^{5.027}}$$

Now

$$700^{1.929} - 600^{1.929} = 307\ 747 - 22\ 588 = 79\ 000 \text{ (approx)}$$

and

$$0.003\ 032 \times 0.25^{1.889} \times 300 = 0.0663$$

Hence $d^{5.027}$ must be greater than

$$\frac{0.0663}{79\ 000} = 8.39 \times 10\_^7$$

If d is taken as 59 mm,

$$d^{5.027} = (0.059)^{5.027} = 6.62 \times 10^{-7}$$

Schedule 40 pipe has 62.7 mm bore, hence

$$(0.0627)^{5.027} = 8.99 \times 10^{-7}$$

which should be satisfactory.

The pressure factor method

Pressure used for this method are gauge, not absolute, i.e. 600 kPa and 500 kPa. From **A33** the pressure factors are 43.54 and 32.32 respectively.
Hence

$$\text{Pressure drop factor } F = \frac{43.54 - 32.32}{300 \text{ (m)}} = 0.0377$$

From **A35.2** the values of capacity (kg/s) are

$F = 0.040$ 0.30 for 65 mm bore pipe
$F = 0.030$ 0.0255 for 65 mm bore pipe

65 mm bore pipe is therefore satisfactory.

It is clear from the alternative calculations that around 63 mm bore should produce the required final pressure and this enables a check on actual velocity to be made, i.e. taking a pipe from **Table A28.1**, 76.1 external diameter 63.5 bore and 6.3 mm wall thickness

$$\text{Velocity} = \frac{0.25 (\text{kg/s})}{3.67\ (\text{kg/m}^3) \times \dfrac{\pi \times (0.063\ 5)^2}{4}\ (\text{m}^2\ \text{area})} = 21.5 \text{ m/s}$$

This calculation exposes the weakness of the velocity method; clearly 30 m/s is too high.

Example 2

Pressure required at process load   750 kPa (absolute)
Steam flow   0.08 kg/s (288 kg/hr)
Pipe length   200 m

Assume that fittings increase equivalent length of pipe to 220 m. To determine pressure required at supply end of pipe an assessment must be made of pipe losses and this can be checked later against actual pipe size selected. A reasonable assumption is 2½% loss over 100 m, i.e. $2.2 \times 2\frac{1}{2} = 5.5\%$.

This equates closely with a supply pressure of 800 kPa
Density of saturated steam at 800 kPa (A15.2)   4.16 kg/m^3

Velocity method based on 30 m/s

$$\text{Area of pipe} = \frac{0.08}{30 \times 4.16} = 0.000\,641\,\text{m}^2$$

$$\text{Diameter} = \sqrt{\frac{0.000\,641 \times 4}{\pi}} = 0.0286\,\text{m}$$

From A28.1 choice lies between 48.3 mm external diameter (28.3 mm bore) and 42.4 mm external diameter (29.8 mm bore)

Simplified flow formula

$$C_1 = \frac{(288)^2}{108} = 0.000\,83$$

$$\text{Required pressure drop per m} = \frac{50\,\text{kPa}}{220\,\text{m}} = 0.227\,\text{kPa/m}$$

Hence

$$0.227 = \frac{0.000\,83 \times C_2}{4.16}$$

giving a desirable value for C_2 of 1129
The choice lies between schedule 80 (38.1 mm bore) (see A27.1 and A27.2) or 39.3 mm bore (external diameter 48.3 mm) (see A28.1 and A28.2). These have values of C_2 of 1590 and 1350 respectively.

CIBS Guide C4 method

$$\frac{P_1^{1.929} - P_2^{1.929}}{0.003\,032 \times 0.08^{1.889} \times 220} = \frac{800^{1.929} - 700^{1.929}}{= 0.005\,65} = 46\,600$$

Hence $d^{5.027}$ must be greater than

$$\frac{0.000\,65}{46\,600} = 1.21 \times 10^{-7}$$

$d^{5.027}$ can now be evaluated for 39.3 or 42.7 mm bore, giving

$$(0.0393)^{5.027} = 0.87 \times 10^{-7}$$

or

$$(0.0427)^{5.027} = 1.30 \times 10^{-7}$$

This approach indicates 42.7 mm bore

The pressure factor method

From **A33** pressure factors are 56.38 and 49.76 for 700 and 650 kPa gauge pressures. Hence

$$\text{Pressure drop factor } F = \frac{56.38 - 49.76}{220} = 0.030$$

From **A35.2**, 40 mm bore will carry 0.065 kg/s at a value of $F = 0.030$ and this confirms that 42.7 mm is about right. the velocity with 42.7 mm bore pipe is

$$\frac{0.08}{4.16 \times \left(\dfrac{\pi \times (0.0427)^2}{4} \right)} = 13.4 \text{ m/s}$$

which once again exposes the weakness of the velocity method unless it is applied to large diameter steam mains

The above calculations are intended to illustrate pipe-sizing procedures and it would be foolish to pretend that the subject is a simple one; the use of these procedures enables straightforward installations to be checked, but in a multi-circuit system the question of balancing becomes of paramount importance.

For any major installation it is good practice, before commissioning, to use flow measuring devices and install balancing valves so as to ensure satisfactory operation of the system.

'Systems Balancing for Heating and Chilled Water Circuits' by Crane Ltd givers general guide lines, together with comprehensive information on the characteristics of measuring devices and balancing valves.

B18 Compressors and compressed air services

B18.1 Compressors

The true cost of compressed air as a form of energy is seldom understood or effectively monitored. The vast majority of compressors are electric motor driven and chart **B18.1.1** depicts a possible starting point for monitoring compressed air costs; this shows the

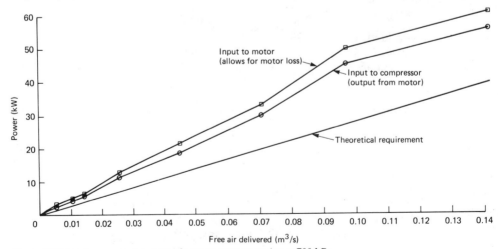

Figure B18.1.1 *Power requirements of compressors operating at 700 kPa*

theoretical power requirement for compressed air, together with the requirement of a standard range of compressors* showing both the power input to the compressor and the power *input* to the motor (after allowing for electric motor efficiency). The theoretical requirement comes from **B18.1.2** which shows the power needed to bring about adiabatic compression at pressures up to 1400 kPa of 0.10 m³/s of air supplied. Two curves are shown; one depicts single stage and the other two stage compression. The two stage is more efficient in energy terms and also results in a rather lower air temperature rise (around 100 °C) compared with single stage (200 °C or more). Three stage would be more efficient still, but the major gain is from single to two.

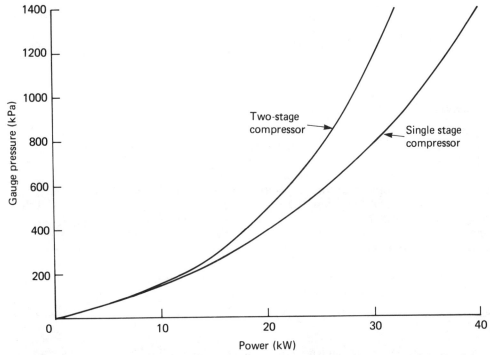

Figure B18.1.2 *Theoretical relationship between pressure and power input for compression of air, based on adiabatic compression for 0.10 m³/s*

Returning now to **B18.1.1** the theoretical requirement can be seen to be only a little over half the energy input to the motor; other points to note are:

(1) The curve of motor output is not a smooth one because motor sizes are selected by the compressor manufacturer from British Standard outputs. The precise power consumption under load would need to come from test bed information.
(2) The theoretical line is based on the normal industrial pressure of 700 kPa.
(3) Because the operating pressure is gauge pressure (i.e. above atmosphere) the compression ratio at 700 kPa will be 7.91 approximately. This means that the volume of air supplied to the compressor will be reduced in the ratio 1:7.91 in reaching the operating pressure. At the same time its temperature will be raised to 120 °C or higher.
(4) Based on the largest compressor in the range, which handles 0.142 m³/s of supply air and consumes 61 kW, the volume per hour is

$$0.142 \times 3600 = 511 \, m^3$$

*Based on the Hydrovane PU/PUA range

and the cost per hour based on, say 4p per kWhr is

$$\pounds\,\frac{61 \times 4}{100} = \pounds 2.44$$

The cost of air compressed is thus

$$\frac{2.44 \times 100}{511} = 0.48\,\text{p/m}^3$$

For a 48 hr week, a 70% utilization factor and a 47 week year, the annual running cost is

$$2.44 \times 48 \times 0.70 \times 47 = \pounds 3850$$

It is, of course, perfectly normal for industrial compressors to be operating throughout working hours; wear and tear may be reduced by having multiple units with sequence control.

B18.2 Compressed air reservoirs and distribution

Pipe friction in a wrongly designed system can absorb a significant proportion of the pressure delivered by the compressor; apart from the nuisance of inadequate pressure on production lines there is a very real waste of energy. It is sound practice for normal industrial applications to be generous in pipe sizes as the volume of the pipe runs can form a useful reservoir to supplement the capacity of the air receivers. There is inevitably fluctuation in the demand for air and the compressed air held in the receivers and pipe system helps to smooth the load on the compressors themselves. The simple way of assessing pipe sizes is to use velocity as a guide but it is first necessary to convert the air delivered to the compressor to its new volume after compression: for 700 kPa the ratio as already stated is 7.91:1 **Table A29** gives the capacity of a wide range of pipe sizes at velocities between 3 and 9 m/s, and good practice is not to exceed 6 m/s.

For the largest compressor on **Chart B18.1** the volume of *compressed* air will be

$$\frac{0.142}{7.91} = 0.018\,\text{m/s or } 1.08\,\text{m}^3/\text{minute}$$

From **Table A29** it can be seen that a 62.7 mm bore pipe will result in a velocity of 5.5 m/s for this volume of compressed air.

Referring now to **Table A30** the pressure drop at 1.138 m^3/min in a 62.7 mm pipe is 0.058 kPa per metre which appears to be quite satisfactory.

The alternative method is to use the 'compressible fluid approach' referred to in **B16.3**:

Flow (kg/hr) = 0.142×1.293 (air density) $\times 3600 = 660$ kg/hr

Hence

$$C_1 = \frac{(660)^2}{10^8}\,0.004\,37$$

C_2 for 62.7 mm bore schedule 40 pipe is 117 (from **Table A27.1**)
Hence pressure drop per 100 metre is

$$\frac{0.004\,37 \times 117}{1.293} = 0.40 \text{ bar guage}$$

or

0.4 kPa/m

As a check, C_2 for 77.9 mm bore pipe is 37.7 and pressure drop is

$$\frac{0.004\ 37 \times 37.7}{1.293} = 0.13\,\text{kPa/m}$$

The safer choice is to use the larger size, i.e. 77.9 mm.

To form some assessment of the cost of wasting compressed air, reference may be made to **Chart B18.2.1** which shows that free discharge through a 10 mm orifice will at 700 kPa use up $0.129\,\text{m}^3$/s of air which is almost the full output of the largest compressor in the range. This loss will vary as the square of the diameter of the orifice so that 5 mm will lose $0.017\,\text{m}^3$/s which is still very significant in relation to the outputs of the smaller units in the range. The following notes are relevant to **Chart B18.2.1**:

(1) The coefficient of discharge used is 1.0; it could fall to 0.7 for a sharp edged orifice with consequent reduction in discharge.
(2) The curve as plotted is for a 10 mm orifice. For other sizes, discharge will vary as the square of the diameter of the orifice. Thus at 700 kPa if the loss is $0.129\,\text{m}^3$/s, for 1 mm it will be $0.001\ 29\,\text{m}^3$/s.
(3) Between 250 and 1000 kPa the curve is approximately a straight line with (for 10 mm diameter) an increase in discharge of $0.0156\,\text{m}^3$ for every 100 kPa increase in pressure.

In the process of compressing air, work is done which manifests itself in heat in the compressor body, in the driving motor and in the compressed air itself. The motor is normally fan cooled and the heat from this source will warm the air in the compressor house.

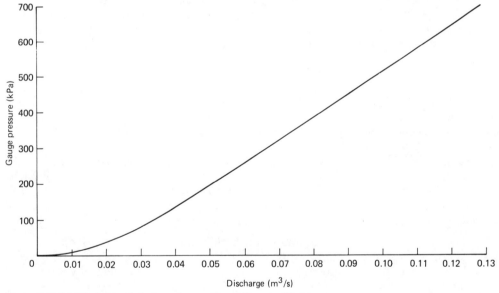

Figure B18.2.1 *Discharge of air through an orifice of diameter 10 mm*

If the compressor is air cooled this also will add to ambient temperature; however, for larger compressors water cooling is normal either by radiators or external cooling towers. For these reasons compressor installations should always be examined with a view to utilizing the surplus heat in winter. This can frequently be achieved by means of ducting and dampers so that in winter the warmed air is used and in summer it is discharged to atmosphere. Either way it is important to keep the ambient temperature of the compressor house under control to avoid malfunction of the compressor or even burn out in the motor.

If the heat added to the compressed air is retained by it and the air delivered hot to the pneumatic plant, then in general terms the moisture present in the supply air will be retained as a vapour in the compressed air; however, if the compressed air is allowed to cool in the pipe and reservoir system then water will appear and cause problems with malfunction of tools, blockage and rusting. For these reasons it is normal to 'dry' the compressed air before distribution and this can be achieved by cooling it at the compressor house. For two stage compressors an intercooler is used and for single stage an after cooler.

Depending on the quality of air required cooling can be by circulating water or by refrigeration, but either way means must be provided to drain away the moisture which is condensed out. Unless elaborate drying takes place at the compressor, means must be provided to drain water from pipework and receivers as well as from coolers.

Cooling plant, particularly if based on refrigeration, is itself energy consuming so that it is important to relate the 'dryness' requirement to the needs of the production process.

B19 Ventilation, fans and filtration

The objective of ventilation is to create comfort conditions for normal respiration and reduce the possibility of inhaling noxious fumes and particles; it is also needed to provide air for combustion and industrial processes. The subject breaks down into five headings.

B19.1 Ventilation requirements

Table A23.1 is derived from CIBS guide B2 and shows the minimum ventilation rates recommended according to the air space occupied per person. These rates are intended to provide adequate fresh air to remove water vapour resulting from perspiration, to reduce body odours to a level where they are not noticeable and to deal with the smoking pollution problem. If there are other domestic contaminants such as cooking, toilets and bathrooms the rates need to be increased; in industrial situations artificial contaminants from process work must be dealt with according to their severity.

Draughts are a special problem and the CIBS guide B2 publishes a chart of maximum air speeds according to temperature and this information is reproduced at **A23.2**.

CIBS guide B2 also deals at some length with the ventilation requirements of Animal Farming, Car Parks, Garages, Tunnels, Food Processing and Kitchens; reference is also made to the requirements of some industrial processes. In any particular industry specialized knowledge is essential if ventilation requirements are to be dealt with effectively and *specialization is particularly relevant when toxic substances are involved*. However, there are general guide lines which can be applied and these are illustrated in tables at **A23** which are derived from the CIBS guide and at **A24** which are derived from 'Threshold Limit Values for Chemical Substances and Physical Agents in the Work Environment with Intended Changes for 1983–84' published by The American Conference of Governmental Industrial Hygienists. These tables cover:

A23.3	Empirical values of ventilation rates for buildings with natural ventilation.
A23.4	Air change rates for natural ventilation in factories.
A23.5	A summary of recommended air change rates for mechanically ventilated buildings.
A23.6	Some recommended values of outdoor air supply for air conditioned buildings.
A24.1	Tables of threshold limit values (TLV) for a range of gases and vapours with an indication of the effectiveness of activate carbon filters; also TLVs for a range of toxic fumes and dusts.
A24.2	An indication of pollution values from engine exhausts which gives a dramatic illustration of carbon monoxide contamination from petrol engines.
A24.3/ A24.4	Some limits for mineral dusts including the so called 'nuisance dusts' which have a long history of little adverse effect on the lungs and have no toxic effect when exposures are kept under reasonable control.
A24.5	The 'simple asphyxiants' which have no other significant physiological effects. The limiting factor for these is a minimum 18% by volume of oxygen.

The TLVs quoted give the maximum concentration in parts per million of substances which have been shown to be poisonous, hazards to health, irritants, narcotics, nuisances or a cause of stress. TLVs are normally on a time weighted average basis related to an 8 hour day, 40 hour week, but some values are ceiling levels which should not be exceeded. The correct interpretation of TLVs is a specialized field and the reader is invited to study the ACGIH publication.

B19.2 Natural ventilation

CIBS guide A4 deals with air infiltration at length and scientific approaches have been devised to deal with a method of ventilation which is unavoidably haphazard in nature. Natural ventilation is affected by tastes of individuals in terms of window opening, variations of wind strength and direction, exposure and height of the building and temperature variations inside and outside.

Two mechanisms have been defined in an effort to evaluate the effectiveness of natural ventilation, these being wind effect which is of course subject to unending variation except in very stable climates, and stack effect which arises from differences between inside and outside temperatures; the latter is affected by the height of the building and the extent to which the buoyancy of warmed air can influence the upper storeys of a building.

Natural ventilation in summer time should come in on the windward side of the building near the floor and out on the opposite side at high level (this of course pre-supposes that the windward side can be defined with any certainty!).

By far the most important consideration in energy conservation is to restrict accidental and unplanned natural ventilation taking place during the heating season. The very fact that a building is heated causes external cold air to be induced into the building through whatever openings are available. One aspect of natural ventilation of particular relevance to industrial premises is fire ventilation; there are of course proprietary designs of roof top ventilators having automatic louvres to avoid unnecessary heat loss under normal conditions. In the event of fire the effect of such ventilators is to remove smoke, reduce the spread of heat at high level, reduce the explosion risk by dilution of unburnt gases and to confine the fire by inducing colder air from nearby areas. They also avoid unnecessary starting of sprinklers except in the immediate area of the fire and they give a visible external warning that a fire has started.

B19.3 Mechanical ventilation

The most common and cheapest form of mechanical ventilation is by means of extractor fans which pull out contaminated air and allow replacement fresh air to find its way in by whatever route is available. Propeller type fans are commonly used for this work and are best sited close to the source of contamination. XPELAIR has become synonymous with domestic and light commercial extractors; there is a wide choice both of manufacturers and of specialized designs to suit kitchens, bathrooms and toilets. For industrial appliations where contamination results from spray painting, fettling, welding, grinding, general machining, foundry work, timber and textile processes much more powerful extract systems are needed and they are frequently associated with ductwork; to deal effectively with these sources of contamination it is essential for the fan unit or ducting from it to be adjacent to the source.

During the heating season the loss of warmed air in these extract systems can be very high and they are all worthy of close study if energy efficiency is to be achieved. In many cases the use of filters with recirculation of filtered air may be used but where there is a risk of the returned air being contaminated, then some form of heat exchanger must be introduced. Supply ventilation (as opposed to extract) is based on bringing fresh air into the building to meet the needs of the occupants or the activities in the building; in industrial situations it is frequently associated with warm air heating systems which can be designed to provide up to 100% recirculation or up to 100% fresh air. Clearly maximum winter energy economy comes from 100% recirculation but in many situations this is not possible without the return air becoming unduly vitiated. In summer these heaters can be used to introduce fresh air with the burners off.

Balanced ventilation, as its name suggests, is a combination of extract and supply and is designed to supply accurately and without discomfort the total ventilation needs of the building. The location and velocity of air intake points is critical to avoid discomfort or draughts and to meet the criteria in **A23.2** regarding acceptable air velocity.

The natural extension of balanced ventilation leads to air conditioning where temperature and humidity are controlled as well as volume and velocity. This subject is dealt with in **Section B20**.

B19.4 Air flow and resistance to air flow

A simple propeller fan mounted in a hole in a wall will move air through the wall and thereby extract noxious or unpleasant fumes, but this is a crude and wasteful form of ventilation. The more controlled the ventilation system becomes the greater is the resistance to flow that the fans must overcome. Canopies, diffusers, ducting and filters all introduce resistance to air flow and the art of fan selection and application lies in the calculation of this resistance and the balancing of fan performance against the volume flow required and the resistance imposed on the fan.

Calculations of the flow of air in ducts may be considered as a special case of Bernoulli's equation for the flow of fluids in pipes. In its origins Bernoulli's equation stems from Newton's second law of motion and the law of the conservation of energy.

If there are no external influences there are three components which go to make up the total energy in fluid flow, i.e.:

(1) The POTENTIAL energy arising from the height of the particles of the fluid above a datum. Being gravity dependent this energy will remain constant for level flow; furthermore because air has a low density its effects can be ignored for normal calculations of flow in ducts.
(2) The PRESSURE energy or the energy due to 'head' (*see* **B16.1**). The pressure in the duct will act in all directions and is the pressure which will tend to burst the duct if positive or cause it to collapse if negative.
(3) The kinetic or VELOCITY energy which arizes from the velocity of the air stream and is defined by the fundamental expression $\frac{1}{2} \times$ mass $\times$ (velocity)^2.

The simplification resulting from using Pascal in place of 'head' is best illustrated by considering a column of water one metre high exerting pressure on one m^2. The mass of this column at the point of maximum density (4 °C) will be 1000 kg and it will exert a pressure of 1000 (kg/m^2) $\times$ 9.807 (m/s^2) (i.e. mass $\times$ gravitational acceleration).

The pressure will therefore be 9807 Pa and the dimensions kg/m s^2. Other conversions from 'head' pressure are given in **Table A3**.

The effects of potential energy being ignored, there are three ways of stating the relationship between pressure and flow:

(1) In terms of energy per unit mass:

$$\frac{\text{Pressure}}{\text{Density}} + \frac{(\text{Velocity})^2}{2} = \text{Constant} \tag{19.4.1}$$

(2) In terms of energy per unit weight:

$$\frac{\text{Pressure}}{\text{Specific weight}} + \frac{(\text{Velocity})^2}{2g} = \text{Constant} \tag{19.4.2}$$

(3) In pressure terms:

$$\text{Pressure} + \frac{\text{Density} \times (\text{Velocity})^2}{2} = \text{Constant} \tag{19.4.3}$$

(Note that specific weight $=$ density $\times g$)

The Bernoulli equation is usually stated in terms of energy per unit mass and the full equation is

$$\frac{P_1}{\rho} + \frac{V_1^2}{2} + gZ_1 = \frac{P_2}{\rho} + \frac{V_2^2}{2} + gZ_2 \tag{19.4.4}$$

where
P is pressure
V is velocity
ρ is density
Z is potential head

The same equation in pressure terms, the potential head being ignored is:

$$P_1 + \frac{\rho V_1^2}{2} = P_2 + \frac{\rho V_2^2}{2} \tag{19.4.5}$$

For fan and duct calculations two further pressure values must be included namely the fan pressure Q and the total resistance of the duct system R giving the full equation:

$$P_1 + \frac{\rho V_1^2}{2} + Q = P_2 + \frac{\rho V_2^2}{2} + R \tag{19.4.6}$$

In fan parlance reference is frequently made to total pressure, being the sum of head pressure and the velocity pressure, and the expression static pressure is used to describe the head pressure.

Measurement of the pressure in a moving air stream in a duct is effected by means of the Pitot tube connected to two manometers, one of which registers the total pressure and the other static pressure; the velocity pressure is the difference between the two.

In the equation (19.4.6) the values of P, Q and R will normally be available in pascal or can be calculated in or translated into pascal, so it is necessary only to assign a value to the velocity pressure; this comes quite simply and directly from the density of air which at 20 °C may be taken as 1.2. Thus

$$\text{Velocity pressure} = \frac{1.2 \times V^2}{2} = 0.6V^2$$

Fan pressures (velocity and total) are available from the characteristic curves issued by manufacturers for various types of fan and various fan blade configurations. It therefore remains to evaluate the resistance of the ventilation system in order to balance requirements against the performance capability of the fan.

An assessment of friction loss may be made by adapting the Darcy formula (B16.2) for water flowing in a pipe, i.e.

Loss of head due to friction (m) $=$

$$\frac{4 \times \text{coeff of friction } (f) \text{ length } (l) \times (\text{velocity})^2}{2 \times g \times \text{diameter } (d)} \tag{19.9.7}$$

For practical calculations the coefficient of friction in galvanized sheet metal air ducts may be taken at 0.006. The pressure loss (Pa) is

$$\text{Head loss} \times \text{density} \times g = \frac{4flV^2}{2\,gd} \times \rho g = \frac{4fl}{d} \times \frac{\rho V^2}{2} \tag{19.4.8}$$

148

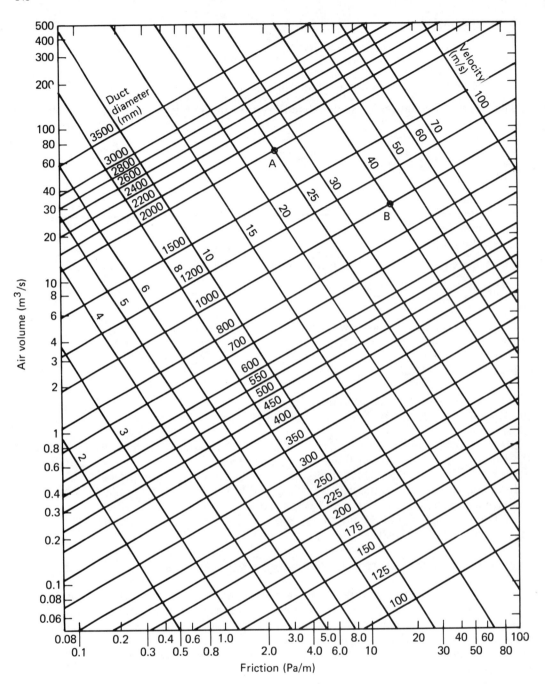

Figure B19.4.1 *Sizing chart for ventilation ducts. The intersection of vertical, horizontal and both oblique lines relates the four values*

A: *2000mm diameter, 25m/s, 2.3Pa/m, 72m³/s*
B; *1000mm diameter, 40m/s, 1.4Pa/m, 31m³/s*

The formula can be further adapted for rectangular ducts with equal velocity of flow by substituting $2ab/(a + b)$ for the duct diameter where a and b are the dimensions of the sides of the duct. If equal flow is being considered the formula becomes

$$\text{Diameter} = 1.265 \times \left[\frac{(ab)^3}{(a + b)}\right]^{0.2}$$

The pressure loss due to duct fittings is calculated from

$$k \times \rho \times \frac{V^2}{2}$$

where
V is velocity at the fitting
k is a factor depending on the fitting and its design.

CIBS Guide C4 gives full information on fittings. As an example, a good right-angle bend has a value of k of 0.10, but a square elbow has $k = 1.25$, equivalent to between 4 and 50 metre of ducting of 1 m diameter.

The above formulae relate to uniform lengths of ducting and corrections are needed for variations in duct size and for bends and branches. Some empirical figures enabling the velocity head loss from fittings to be calculated are given in **Table A31.3**.

The velocity head may be calculated from the expression

$$\text{Factor for fitting} \times 0.6V^2 = \text{pressure (Pa)}$$

The calculation becomes very straightforward if the pascal is used as the unit thus avoiding the translation to mm of water gauge.

Duct sizes may also be estimated by means of a chart relating volume, velocity, friction and duct diameter and an example of such a chart is given at **B19.4.1**.

A typical procedure for calculation of ventilation requirements is given at **B19.7** which shows how system details are related to fan capabilities.

B19.5 Fans and their characteristics

The fans are the energy consumers in a ventilation or heat recovery system and their importance in evaluating either system stems both from the energy they use, which affects the running cost and energy effectiveness of the system, and also from their initial cost, which can be a substantial part of the investment involved.

The three *basic* types of fan available are:
(1) *Propeller fans* which can move large volumes but which generate only low pressure and in general are not suitable for use with ducting.

Chart **B19.5.1** shows the characteristics of a 600 mm diameter fan which will deliver 2.65 m³/s with free air flow, falling to 1.2 m³/s against a pressure of 116 Pa.

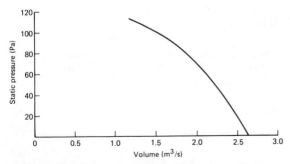

Figure B19.5.1 *Characteristic curve of 600mm propeller fan*

150

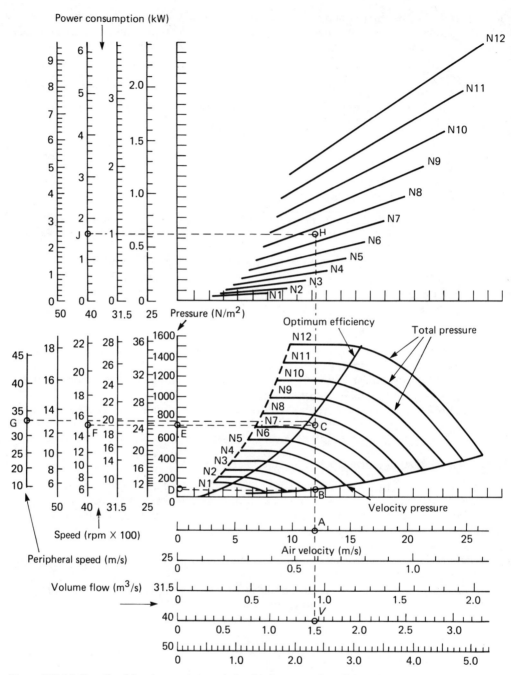

Figure B19.5.2 *Centrifugal fan characteristics: relationship between volume flow and dynamic pressure, total pressure, air velocity, power consumption and rotational and peripheral speed. Power, speed and volume are shown for four standard fans 25, 31.5, 40 and 50*

(Reproduced by courtesy of Myson Brooks Ltd.)

(2) *Centrifugal fans* where the air enters axially and is distributed radially; this configuration causes the axes of the supply and delivery ducts to be at right angles to one another and thus imposes some constraints on the mounting of the fan. However, they are, in general, capable of developing high pressures and by means of variation in blade angle and blade shape can be produced with a wide range of characteristics in terms of volume, pressure and efficiency. With forward curved blades they can overload the drive motor if operated above rated air volume; this tendency can be avoided with backward curved blades. They can be designed for very high pressure, and are commonly used as blowers or boosters for gas supplies.

Roof Units offer a range of 'in-line' fans which are based on the centrifugal principle; they are available for operation against pressures up to 400 Pa with a corresponding volume of 1.4 m³/s and are suitable for duct mounting.

B19.5.2 Pressure ranges for efficient working at various volumes for Myson fans

Pressures (Pa)	Volume (m³/s)
100–1500	up to 20
700–4500	up to 12
500–4000	up to 60
1000–5500	up to 45

Chart B19.5.2 illustrates the characteristics of a group of four fans from a range and it is from charts of this type that fan selection may be based on the volume and pressure called for by the application. This group offers pressures from 200 to 1500 pascals with volumes up to 3 m³/s for high efficiency or up to 4.5 m³/s if a relatively lower efficiency can be tolerated.

The chart is derived from catalogue information issued by Myson fans, and within the complete range the characteristics shown in **Table B19.5** are also available.

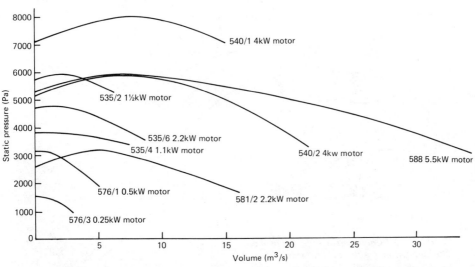

Figure B19.5.3 *Characteristics of a range of belt driven centrifugal gas boosters (natural gas performance based on gas density of 0.6 standard air)*

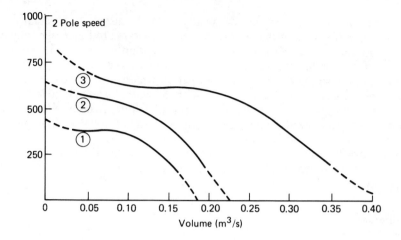

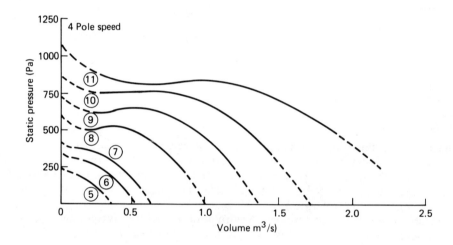

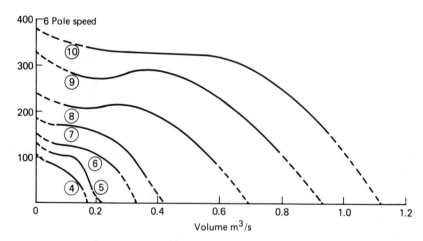

Figure B19.5.4 *Characteristics of a range of centrifugal fans*
The numbers on the curves indicate the fan sizes

(Reproduced by courtesy of Airflow Developments Ltd.)

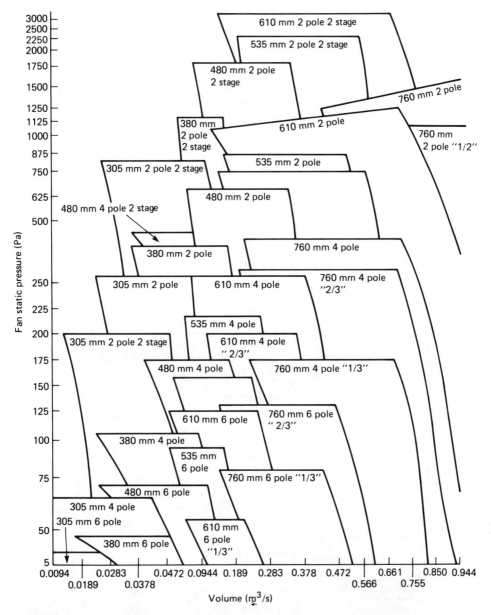

Figure B19.5.5 *Selection chart for axial flow fans of diameter 305-760mm, based on 50Hz motors with nominal speeds: 2 pole, 2900 rev/min; 4 pole, 1440 rev/min; 6 pole, 970 rev/min*

(Reproduced by courtesy of Myson Brooks Ltd.)

The chart is self-explanatory and the example illustrated concerns a volume of 1.5 m³/s (*V*) achieved with the size 40 fan. This gives an air velocity of 12 m/s (*A*), a velocity pressure of 90 Pa (*B*), a total pressure of 720 Pa (*CE*), a speed in r.p.m. of 1520 (*F*) a peripheral speed of 33 m/s (*G*) and a power consumption of 1.6 kW (*HJ*).

As can be seen the fan will operate between the characteristic lines N7 and N8 and will be close to the curve of optimum efficiency.

Chart B19.5.3 reproduced by courtesy of Secomak Air Products Ltd shows the characteristics of a range of blowers for boosting gas supplies and these charts

154

illustrate the higher power needed as the impeller shape and design is varied to give higher volumes.

Chart B19.5.4 reproduced by courtesy of Airflow Development gives details of a range of small centrifugal fans and indicates the need to operate these fans within the scope of their characteristic curves; the fall off in pressure as volume increases is clearly shown and the fan must be operated only on the solid line of the curve.

(3) *Axial flow fans* have a straight through air stream and because of this, and of their pressure and volume characteristics, are very suitable for mounting in ventilating ducts.

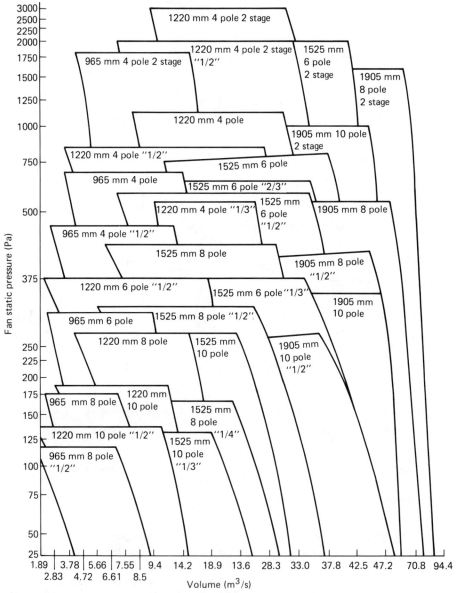

Figure B19.5.6 *Selection chart for axial flow fans of diameter 965-1905mm, based on 50Hz motors with nominal speeds: 4 pole, 1440 rev/min; 6 pole, 970 rev/min; 8 pole, 720 rev/min; 10 pole, 580 rev/min The fraction (e.g. "½") indicates reduced number of blades*

(Reproduced by courtesy of Myson Brooks Ltd.)

In their basic form the delivery air stream itself rotates, thus increasing the delivery side resistance of the system. This tendency can be corrected by the use of guide vanes or by contrarotating impellers within the same fan casing.

The characteristics of axial fans are varied by the diameter of the casing, the angle of the fan blade (*see* **Chart B19.5.7** and **Table B19.5.7**) and by the 'solidity' of the impeller. Mounting the maximum number of blades on the hub gives maximum solidity. Reduced solidity reduces the static pressure and reduces the maximum volume which can be handled but it does reduce power consumption and enable a good efficiency to be achieved.

The scope of a range of axial fans is illustrated by **Charts B19.5.5** and **B19.5.6** reproduced by courtesy of Myson Brooks Ltd. giving pressure-volume relationships for the Myson range from 305 mm to 1905 mm diameter. These give volumes up to 8 m³/s and pressures up to 3000 Pa. Motor speed and power is very much a factor of the specific application of the fan. The speeds are necessarily restricted to the standard

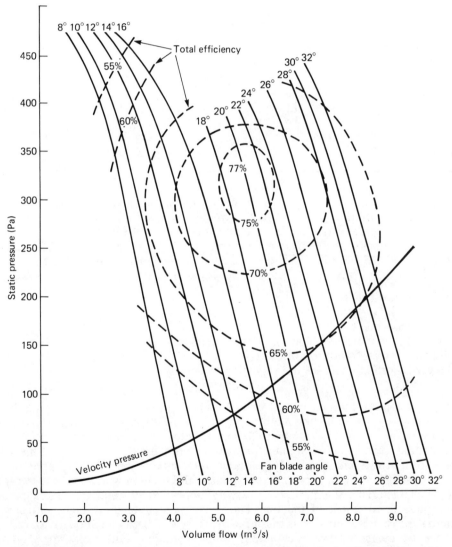

Figure B19.5.7 *Volume-pressure relationships for a range of blade angles for 760mm 4 pole axial fan*

(Reproduced by courtesy of Myson Brooks Ltd.)

range of induction motor speeds, i.e. 2900, 1450, 970 rev/min etc. for 50 Hz and 3500, 1750, 850 rev/min etc. for 60 Hz.

(4) *Mixed flow and crossflow fans* are also available. The former seek to achieve the combined advantages of axial and centrifugal types and the latter, sometimes known as 'tangential', is the type used for small domestic blower heaters and for industrial air curtains. The configuration of the crossflow fan is such that air enters on one side of the casing and is discharged on the other, normally through a slot or a short rectangular length of ducting.

B19.5.7 Fan blade angle and power consumption for a Myson 760 mm axial fan driven by a 4 pole motor

Fan blade angle	Max. power consumption (kW)
8	1.5
10	1.8
12	2.0
14	2.3
16	2.6
18	3.0
20	3.4
22	3.7
24	4.1
26	4.8
28	5.4
30	5.7
32	6.3

(5) *Fan laws:* assuming that the density of the air remains constant there are three basic relationships with the angular velocity of the impeller, i.e. the discharge varies directly, the pressure varies as the square and the power absorbed varies as the cube of the angular velocity. If the air density is changed the flow volume remains constant, but both pressure and the power absorbed vary directly with the change of density.

If the fan speed is N r.p.m. the laws may be restated:

Volume flow (m³/s) $\propto N$
Pressure (kPa) $\propto N^2$
Power absorbed (kW) $\propto N^3$
Volume flow is independent of air (gas) density
Pressure increases in proportion to density
Power absorbed increases in proportion to density

(6) *Variable volume fans:* variation of the air volume delivered according to the needs of the ventilation system is clearly advantageous and can be obtained by varying the fan speed; however, the induction motor is basically a fixed-speed machine and to achieve speed variation adds substantially to first cost. Pole Amplitude Modulation (PAM) is one method which enables an induction motor to run at either of two fixed speeds. A crude form of variable speed may be obtained by varying the volatage applied to the motor terminals without varying the frequency; this method works because the speed torque characteristic of a motor with varying supply voltage approximates to the cube law power requirement of the fan. The cheapest method of voltage control is by external resistance, but this is wasteful of energy; a better method is to vary the voltage by means of 'wave chopping' using thyristors. Full motor-speed control necessitates both voltage and frequency variation and this method can involve control equipment with first cost comparable to that of the fan and motor together.

Other approaches utilize a special drive motor and a good example is the FISCHBACH variable speed fan with 'disc' motor; this enables 100% controlled speed variation to be achieved.

The alternative to speed variation is to vary the mechanics of the fan itself. WOODS of Colchester offer their VAROFOIL range of axial flow fans which are designed to vary the fan blade angle when the fan is running. The variation can be controlled by feed-back from the ventilation system with consequent major savings in energy consumption. These fans are available in diameters from 630 mm to 2800 mm.

B19.6 Air filters and filtration plant

All forms of air filtration must add some resistance to the system and if filters, owing to lack of maintenance, are allowed to become clogged the additional resistance can cause a complete failure of the system. The subject is reviewed under five headings.

B19.6.1 Airborne particles

The chart at **B19.6.1** supplements the information in **A24** in that it indicates the range of particle sizes normally encountered when studying filtration requirements. Various claims are made as to the ability of different filters to handle the range of particle sizes stretching as it does from 0.005 to 500 μm (0.000 005 to 0.5 mm); in general terms high voltage electrostatic units cover from 0.01 to 10 μm, fabric and viscous filters from 0.1 to 50 μm and centrifugal separators from 2 to 500 μm (and higher if need be) (1 μm = 10^{-6}m and has been called 1 micron).

B19.6.2 Activated carbon filters

These are in a special category in that they remove odours and chemical vapours by the process of adsorption and are dependent for their effectiveness on a particular porous form of carbon which presents a large surface area to the air passing through the filter. Reference to **Table A24.2** gives an indication of the effectiveness of these filters against some of the gases subject to threshold limit values.

The chart at **B19.6.2** illustrates the resistance imposed by these filters related to air flow rates. As can be seen this should be within the band of 50–100 Pa for flow rates 0.2 to 0.3 m/s. In assessing their effect on the total resistance of a system it is probable that 200–300 Pa should be allowed to cover the build up of resistance before maintenance or replacement of the units.

B19.6.3 Dry cell filters

The filter medium can be woven cloth, felt or plastic, and these filters can be designed as throw-away units or for on-line cleaning by agitation and/or reverse air flow. Satisfactory arrangements for maintenance and cleaning are vital as many filters in this category can impose a substantial addition to pressure when clogged. Manufacturers should be consulted with regard to the initial resistance and the level expected before cleaning or changing filters. **Chart B19.6.3** shows the characteristics of a range of washable plastic filters with initial resistances up to 100 Pa. as a further example a manufacturer of resin bonded synthetic fibre filters quotes for 1.5 m³/s an initial resistance of 12–60 Pa building up to 125–250 before cleaning. Collection efficiencies are high, up to 99% being claimed when operating under optimum conditions. 95% is more usual but lack of maintenance can depress the efficiency to 50%.

B19.6.4 Viscous filters

These employ water or oil in suspension as the means of removing the dust but they bring the attendant problems of sludge or effluent disposal when used for industrial application.

158

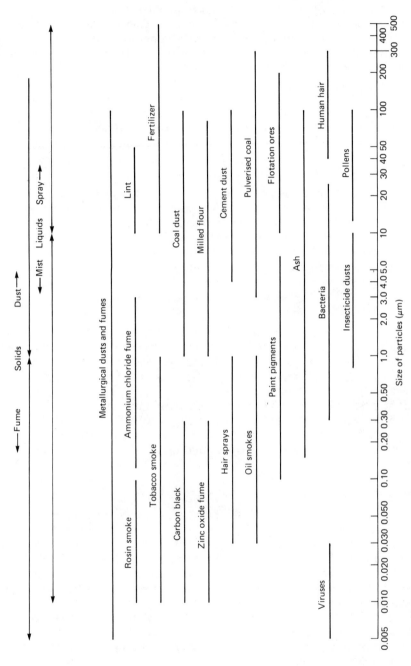

Figure B19.6.1 *Size range of airborne particles*

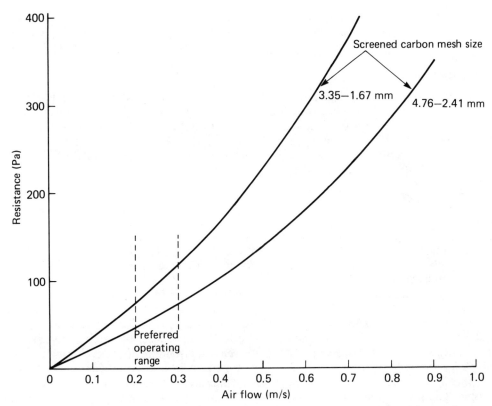

Figure B19.6.2 *Resistance to airflow of activated carbon filters with filter areas from 0.08m² to 0.84m² and handling volumes from 0.02m³/s to 1.9m³/s*

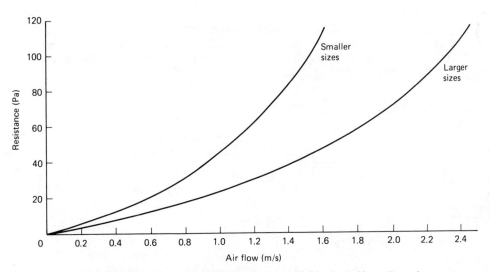

Figure B19.6.3 *Initial resistance to airflow of 50-75mm thick washable plastic filters. Typical curves are shown for filter areas from 0.015m² to 0.50m²*

160

The throw-away petrol engine type filter is familiar to all. This can be used with dust laden air containing up to $2\,g/m^3$ of dust but is not suitable for heavy particles (density $1\,g/cm^3$ or greater). Efficiencies up to 95–98% can be obtained, 85–90% is more usual.

B19.6.5 Cyclones

Traditionally these are used for industrial applications where dust (e.g. from a timber mill) is carried to a central point through ducting with inlet nozzles adjacent to machines. They have relatively low efficiency but can handle heavy concentrations. For effective filtration of air to be returned to the workshop a secondary filter is desirable. The present day trend is to move away from central plant towards unit filters which are self-contained and installed adjacent to the machines producing the dust.

B19.6.6 Electrostatic precipitators

'Low' voltage (approx. $11\,kV$) units are suitable for light dusts with concentrations up to $25\,mg/m^3$ and are valuable for domestic and light industrial applications. They are widely used for dealing locally with welding fumes. For heavy industrial applications 'high' voltage units are used ($30\,kV$) these have high collection efficiencies and are used widely in industries such as cement and steel.

The essential need in considering the energy implications of filter systems is to weigh the energy loss from simple extraction systems against the cost and power consumption of systems which filter and clean the air so that it can be returned to the environment.

Welding, fettling, spray painting and any manufacturing process producing dust are all common sources of energy loss during the heating season. An interesting development from Information Transmission Ltd. senses CO_2 content of air in a building continuously and adjusts, through a control system and relays, the actual rate of ventilation according to need at the time; this suggests worthwhile possibilities for energy saving in public buildings with intermittent use.

B19.7 Calculation procedures for ventilation systems

The assessment of ventilation requirements is reviewed in **B19.1** and due count having been taken of the use and population of the building, together with any special needs arising from heat gains or contamination from process work, a figure of required air changes per hour will be arrived at; this figure forms the basis of calculation procedures which are best illustrated by considering a practical situation.

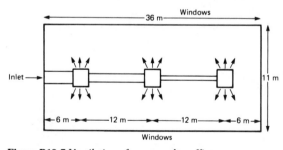

Figure B19.7 *Ventilation of an open plan office*

Consider an open plan office, $36\,m \times 11\,m \times 3.2\,m$ high where ventilation is uneven, particularly in the centre, remote from opening windows. A simple scheme may be considered providing supply ventilation through a single galvanized steel rectangular duct running along the centre line and having three outlets with suitable diffusers, one $6\,m$ from the inlet point, one $18\,m$ and one $30\,m$, as shown in **Figure B19.7**. Duct velocity is to be restricted to $4\,m/s$ and duct size to be reduced after each outlet, so as to maintain this

velocity. The scheme to be based on three air changes per hour, with inlet air at 18 °C. Power is to be from a fan on the inlet side; two bends are required in ducting between fan and inlet to office, with 15 m of ducting between fan and inlet point.
The air volume required is

$$\frac{36 \times 11 \times 3.2 \times 3}{3600} = 1.056\,\text{m}^3/\text{s (say 1.05 m}^3/\text{s)}$$

The density of air at 18 °C is

$$\frac{1.292 \times 273}{291} = 1.212\,\text{kg/m}^3$$

Hence

Mass flow $= 1.212 \times 1.056 = 1.28\,\text{kg/s}$

The size of main supply duct for 4 m/s is

$$\frac{1.056}{4} = 0.264\,\text{m}^2$$

Hence the diameter of a circular duct is

$$\sqrt{\frac{4 \times 0.264}{\pi}} = 0.58\,\text{m}$$

The equivalent diameter of rectangular ducting $0.7 \times 0.5\,\text{m}^2$ is

$$\frac{2 \times 0.7 \times 0.5}{0.7 + 0.5} = 0.583\,\text{m}$$

(The figures 0.7 m and 0.5 m are based on trial and error to match a diameter of 0.58 m)

Velocity pressure $= \dfrac{1.212 \times (4)^2}{2} = 9.7\,\text{Pa}$

(more simply, by use of the value $1.2\,\text{kg/m}^3$ for the density of air, pressure $= 0.6 \times 4^2 = 9.6\,\text{kPa}$)

As the flow in the duct may easily be checked by means of a Pitot tube and manometers it is desirable to express the velocity head in mm water gauge, thus:

Head in m of air $= \dfrac{(4)^2}{2g} = \dfrac{16}{2 \times 9.81} = 0.815$

Head in m of water $= \dfrac{0.815 \times 1.212}{998} = 0.000\,99\,\text{m} = 0.99\,\text{mm}$

From the modified Darcy formula, the pressure loss in pascal per metre run of ducting (from **19.4.8**), is

$$\frac{4 \times f}{d} \times \frac{V^2}{2} = \frac{4 \times 0.006}{0.583} \times \frac{1.212 \times (4)^2}{2} = 0.40\,\text{Pa/m}$$

k value for the bends being assumed at 0.65 the loss in the two bends is

$$2 \times \frac{0.65 \times 1.212 \times (4)^2}{2} = 6.3\,\text{Pa}$$

The total run of ducting from the fan to the first outlet is 15 m (from fan to inlet) + 6 m (from inlet to first outlet) = 21 m.

Total pressure loss to the first outlet is thus

$(21 \times 0.40) + 14.7\,\text{Pa}$

The next step is to establish the size of ducting to maintain velocity between the first and second outlets. We start with the smaller side of the main duct (0.5 m) and by trial and error arrive at the correct size between outlets one and two. Try 0.5×0.4 and 0.5×0.3 and find that 0.5×0.35 is about right, i.e.

$$\frac{2 \times 0.5 \times 0.35}{0.5 \times 0.35} = 0.412\,\text{m equivalent diameter}$$

The volume flow rate will be

$$1.05 - \frac{1.05}{3} = 0.7\,\text{m}^3/\text{s}$$

Thus the velocity flow rate is

$$\frac{0.7}{0.5 \times 0.35} = 4\,\text{m/s}$$

The pressure loss between 1 and 2 will be

$$12 \times \frac{4 \times 0.006}{0.412} \times 1.212 \times \frac{(4)^2}{2} = 6.8\,\text{Pa}$$

$$= 0.57\,\text{Pa/m}$$

By similar methods the size of the final length of ducting can be assessed at $0.35 \times 0.25\,\text{m}^2$ with equivalent diameter

$$\frac{2 \times 0.35 \times 0.25}{0.35 + 0.25} = 0.29\,\text{m}$$

With volume flow

$$0.7 - 0.35 = 0.35\,\text{m}^3/\text{s}$$

the velocity of flow is

$$\frac{0.35}{0.35 \times 0.25} = 4\,\text{m/s}$$

and the pressure loss

$$12 \times \frac{4 \times 0.006}{0.29} \times 1.212 \times \frac{(4)^2}{2} = 9.6\,\text{Pa}$$

$$= 0.80\,\text{Pa/m}$$

The k factor for the six diffusers (two at each outlet) may be taken at 0.7 and the total pressure loss

$$6 \times \frac{0.7 \times 1.212 \times (4)^2}{2} = 40.7\,\text{Pa}$$

The pressure loss in the system is thus

$6.3 + 14.7 + 6.8 + 9.6 + 40.7 = 78.1\,\text{Pa}$

to which must be added the velocity pressure giving 87.8 Pa as the fan total pressure required.

It is interesting to revert to the sizing chart at **B19.4.1** from which the details for 4 m/s flow velocity shown in **Table B19.7** can be taken. The reconciliation is not exact but it is certainly within normal practical limits.

B19.7 Pressure loss calculated from sizing Chart B19.4.1

Volume (m³/s)	Duct diameter (m)	Pressure loss (Pa/m)
1.05	0.6	0.4
0.7	0.47	0.53
0.35	0.33	0.81

Alternative methods of duct sizing which can be used are

(a) based on constant pressure drop per unit length,
(b) based on velocity reduction at each branch,
(c) calculated so that the static pressure at every branch is the same.

These methods will clearly give somewhat different duct sizes but the constant volume method serves to illustrate the procedure.

The final step is to select a fan capable of delivering $1.05\,m^3/s$ against a velocity pressure of 9.7 Pa and a system resistance of 78.1 Pa. In order to do this it is necessary to construct a system characteristic curve based on the square-law relationship between pressure and volume such that

Pressure $=$ (volume)$^2 \times$ constant

for the system described above the constant is

$$\frac{87.8}{(1.05)^2} = 79.6$$

The curve of pressure against volume using this constant is shown at **B19.7.1** and superimposed on this curve are three characteristics for possible axial flow fans, each

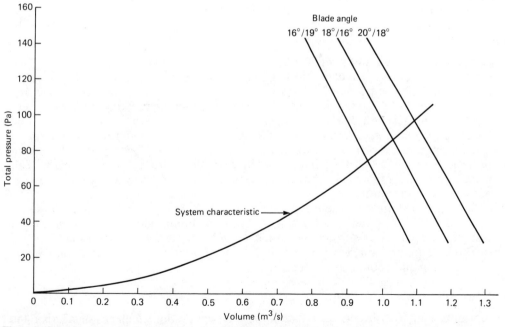

Figure B19.7.1 *Matching ventilation system characteristics against fan data for 480mm 960 rev/min two stage axial fans*

480 mm diameter, two stage units, but with three different blade angles. The 10°/16° angle is the one which most closely matches the system; manufacturer's literature shows this fan to have an efficiency of 72% and a noise level of 79 dBW.

Some indication of power requirement can be obtained from the energy equation (**19.4.1**) which when multiplied by the mass air flow in kg/s gives the energy in joule/s (or watt), thus:

$$\left(\frac{78.1}{1.212} \times 1.28 \right) + \left(\frac{(4)^2}{2} \times 1.28 \right) = 93\,\text{W (approx)}$$

The fan input will be

$$\frac{93\,\text{watt}}{0.72\,\text{(efficiency)}} = 129\,\text{W}$$

and the motor input

$$\frac{129}{0.75\,\text{(motor efficiency)}} = 172\,\text{W}$$

(Actual motor rating should be checked with fan manufacturers.)

B20 Air conditioning

True air conditioning entails complete automatic control of the environment in a building by heating (or cooling), humidifying (or dehumidifying), cleaning, delivering and exhausting the air according to the needs of the occupants or the process. The design of air conditioning systems is highly specialized but many of the calculation procedures involve the basic considerations referred to in **B15, B19** and **B24**. This section therefore will attempt no more than a review of the main types of systems employed, a brief resumé of central plant and an indication of the ways in which energy is used. Energy use in air conditioning should be monitored and controlled just as for other processes and must be part of the audit procedure (*see* **B4.1**).

B20.1 Systems

There are three main types of system.
(1) Central plant with air circulated throughout the building. This is best suited to single large rooms such as banqueting halls, department stores or industrial processes, but has disadvantages if the requirement of various rooms in the building complex varies.
(2) Central plant but with dual ducts enabling hot and cold air to be circulated so that blending may take place according to the varying requirements of different areas.
(3) The induction convector system where hot and/or cold water pipes are used as a means of local adjustment whilst retaining a primary air supply from the central plant. This primary air enters the rooms through nozzles and induces a secondary circulation of air already in the room.

Self-contained units for use in individual rooms can be supplied as a complete single package with the option of a heating as well as a cooling facility, and these can be installed as split systems to minimize space requirements and reduce noise.

The heat pump, which is covered by **B22**, is an important possible option where both heating and cooling are called for.

B20.2 Operation of central plant

In a centralized air conditioning installation there is, in addition to means of heating and cooling, a need for humidifying or dehumidifying the air before it is ducted to the building. The air is first filtered using appropriate filtration systems as described in **B19.6**. The next process in winter is to pass the air over a preheater to raise it to the required temperature. The preheater is normally a coil supplied from steam or hot water heating system. Heating winter air will tend to reduce humidity below acceptable levels and this is corrected by passing the air across sprayed water in a 'washer'; before discharge to ducting, the air is passed across 'eliminator' plates which remove droplets of water in suspension whilst allowing the air to retain its humidity within design tolerance.

In summer time the air after filtration is passed across a cooling coil which may be fed with chilled water or brine from a refrigeration system. After this process it is likely that the air will be saturated and must therefore be cooled in the washer so that a proportion of the moisture is condensed out. The temperature of the air must then be raised again to an acceptable level whilst retaining the humidity content within tolerance. The reader is referred to **B15**, noting that

(a) Isothermal humidification takes place when the winter air after heating passes over the spray.
(b) Sensible cooling takes place when the summer air is passed over the cooling coil.
(c) Sensible heating takes place when the winter air is heated or when the summer air is reheated.
(d) The dehumidification of the summer air is not isothermal because some cooling of the air also takes place; in fact all the changes in temperature and moisture content are to some extent combinations of adiabatic and isothermal.

B20.3 Energy usage

Energy is consumed in air conditioning systems in a number of ways:

(a) The main electric motor driving the compressor in the refrigeration system (**B31** covers the checks necessary to ensure effective use of electrical power).
(b) Supply and exhaust fan motors (*see* **B19**).
(c) Auxiliary motors on boiler systems or driving pumps where water circulation is used. Motors may also be employed to drive fans on external coolers.
(d) Fuel-fired boilers to provide the heat requirements in winter (**B8**).
(e) Electric heater batteries as auxilliaries in the air stream to control temperature and humidity (particular care should be taken to ensure that the main heat energy comes from boilers and not from electric heaters)

Opportunities for heat recovery should be examined as the circulating air can be ducted to pick up heat from any available source which is clean. A good example is ducting the air through fluorescent fittings to pick up the energy dissipated by the tubes.

B21 Refrigeration and cold stores

The refrigeration cycle is referred to in section **B22** and as with air conditioning this brief review will confine itself to creating an awareness of the energy implications of cooling.

A useful starting point is to remind ourselves that cooling consists of extracting heat energy and one kilowatt means the same thing whether heating or refrigerating is involved. **Section B12** outlines the assessment of heating and cooling requirments and the balance of these will determine whether or not cooling is called for.

Wherever refrigeration is at work, whether for normal refrigeration, freezers, refrigerated display or cold rooms, there is a release of heat energy in the system resulting from the extraction of heat from the refrigerator or cabinet. The possibility or utilizing this heat energy must always be examined, and in considering new installations this should be closely studied.

The efficient use of electric motors, both for the main compressor drive and for fans and auxiliaries should always be checked and in large installations manufacturers' data on

166

power *input* to motors should be related to the claimed output in cooling capacity. The coefficient of performance which is now common parlance in relation to heat pumps is not new; it has long been a measure of the effectiveness of refrigeration plant.

Just as buildings need to be insulated and sealed against heat loss so cold rooms are insulated and sealed to retain their lowered temperature. The losses from poor insulation, ill fitting doors and carelessness in closing cold room doors can be a major energy waster.

The design and layout of hotel and commercial kitchens should take count of the relative positions of storage and cooling equipment and ensure as far as possible that heat from kitchens is not adding to the cooling requirement of freezers, refrigerators and cold stores; it is also worth checking that the heat released by refrigeration plant is not adding to extract ventilation requirements of the kitchens themselves.

B22 Heat pumps

B22.1 *Principles of operation and main types available*

The heat pump uses the vapour compression cycle of a refrigeration system in order to extract heat energy from the environment and return the heat to a hot water or space heating system. The heat source can be the atmosphere, the ground, external ponds, streams or rivers or any process fluid. The vapour compression cycle is best illustrated by Figure **B22.1**.

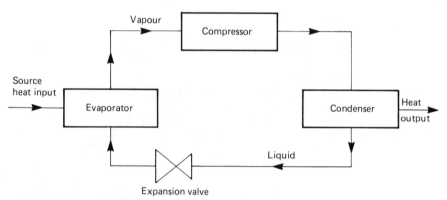

Figure B22.1 *Heat pump cycle*

The mechanism is identical to refrigeration with the heat input source being the cooling stage of the cycle and the heat output taking the place of the rejected energy. The work done in the system is the energy consumed by the compressor which is normally electric motor driven. The heat input may be regarded as 'free' because it is taken from the environment.

Heat pumps can be 'air to air' where the atmosphere is the source and the output is delivered as warmed air to the building; 'air to water' where the output is delivered to a low pressure hot water system or to supply domestic hot water. If the external source is water then 'water to air' or 'water to water' units may be applied.

Because in the cooling mode a heat pump will release humidity when the dew point is reached, there is a well established application for industrial drying and this is illustrated by the chart, table and diagrams at **B22.5** which are reproduced by courtesy of Westair Ltd.

Heat pumps can be supplied to carry out either a heating or cooling duty according to need and in such situations they are likely to be at their most effective. A particular example is in stores or supermarkets where there can be a high heat gain at certain times from lighting and metabolic heat, coupled with a requirement for comfort heating at other times or in other parts of the store. There is also the need for refrigerated display, and in a totally integrated system the heat pump can make radical energy savings.

A special and strongly established application is for indoor swimming pools where there is a need for frequent air changes to control pollution and where the air is at a temperature comparable with the pool; in such applications heat can be extracted from the exhaust air which has a high humid content, and returned either to the water or to the air space above the pool.

When electricity is generated in a fossil fuel fired power station 70% of the GCV of the fuel is lost at source and for this reason considerable development work has been carried out on engine driven compressors for use in the heat pump cycle. This approach enables a major part of the waste heat from the engine to be utilized and thus leads to more efficient use of fossil fuel. The application of a gas engine in this manner is illustrated by the diagrams at **B22.6**, which are reproduced by courtesy of West Midlands Gas (Mr O. B. Gender's paper on Heat Pumps).

The principle of the absorption type refrigerator is well known and although this principle is confined to domestic sizes it does present a special advantage in that the fuel used to maintain circulation in the system can be gas. When used in the heat pump mode the absorption unit therefore uses fuel resources to better advantage than a vapour compression system with electric motor drive.

Standard heat pumps are produced either as packaged systems which are supplied complete (requiring only connections to ductwork and to electrical services) or as split systems; the latter normally have an air handling unit inside the building and the compressor unit outside. Alternatively the air handling unit and compressor may be inside with the condenser only outside; in either case split systems involve connections to refrigerant pipework with the consequent need for a high standard of installation skill.

B22.2 Heat pump performance

Because the heat source is 'free' the efficiency of a heat pump cannot be measured on the same basis as a boiler or heat exchanger; it has therefore become normal practice to state heat pump effectiveness in the manner long associated with cooling systems, based on refrigeration units. These have related the cooling output in tons or kilocalories of cooling to the energy consumed in operating the refrigeration system. With the progressive adoption of the SI system, output either cooling or heating may be stated in kW. Based on the Carnot cycle the maximum or ideal performance of a heat pump may be stated in terms of the input and output absolute temperature (T_1 and T_o) such that the coefficient of performance or ideal COP =

$$\frac{T_o}{T_2 - T_1}$$

The ideal COP can never be achieved in practice, but this relationship shows clearly that the smaller the temperature difference the greater the theoretical COP; this, in turn implies that heat pumps show a radical fall off in performance if output temperatures are relatively high. As a result it is not normally possible to reach the same circulation temperatures as can be achieved with a boiler system.

The normal method of designating the effectiveness of heat pumps is the COP relating the heat output to the total energy consumed in the system; the precise definition of COP has been the subject of considerable discussion among the various bodies associated with the industry and it is expected that an agreed standard will be issued

The main factors involved are:

(1) All the energy consumed in the system must be taken account of, i.e. any fans or pumps needed to circulate fluids as well as the main drive motor.
(2) Losses between the heat pump and the heat output point must be deducted from the total output energy.

(3) Because the COP falls away at lower source temperatures, the COP figure must be quoted at a particular input temperature.

(4) In making assessments of the economics of a heat pump system a 'seasonal' COP should be used. This figure should take count of the variation of COP with different external temperatures and different heating requirements.

(5) There remains some uncertainty as to whether the COP quoted is theoretical, measured under laboratory conditions, or arrived at as a result of long term field trials. The latter figure is, of course, the one of interest to the user, but as conditions vary from site to site it is clearly prudent to play safe by using a COP lower than the theoretical or empirical figure sometimes quoted.

To form an assessment of the advantages of the engine-driven system reference may be made to the diagrams at **B22.6** from which it can be seen that the ratio of output to fossil fuel input rises from 0.81 for an electric-driven unit to 1.33 for a gas engine driven unit; this ratio is referred to as the coefficient of fuel utilization.

B22.3 Heat pump installations

Some of the points to bear in mind in considering the suitability of heat pumps as an alternative to conventional heating systems are listed below:

(1) Because air or water to water units are constrained to deliver heat at lower temperature than conventional wet systems, checks must be made if connected to radiators to ensure that the heat output will be adequate. Underfloor systems, because of their relatively large surface area, can function well at these lower temperatures, but the pros and cons of underfloor systems need careful evaluation.

(2) When operating at low external temperatures there is certain to be a 'frosting' problem with condensers, and controls need to be built in for a defrost cycle during which no heat will be delivered; checks must be made to ensure that this cycle is working effectively. The cycle should be of the order of 3 to 4 minutes duration, and means to dispose of condensate must be provided.

(3) When operating for long periods at low temperature the COP will be low and auxiliary heat will be needed. Ideally this should come from a conventional boiler connected to the same system, as sustained electrical boosting can be very expensive.

(4) With air to air units care must be taken in siting the discharge from the air intake as this discharge can be very cold and may affect comfort conditions if wrongly directed.

(5) Because heat pumps have many more moving parts than conventional boilers care must be taken that skilled attention is readily available in the event of failure; unless they are used in conjunction with a boiler it could be unwise to depend on a single heat pump to maintain comfort conditions.

(6) Heat pumps are inherently more expensive than conventional boilers and this necessitates a very careful investment appraisal if they are being considered solely for heating; if cooling is also involved the heat pump is much easier to justify.

B22.4 Calculation procedure

The charts at **B22.7** show clearly the effects of low source temperature and high output temperature on performance. In order to carry out an economic appraisal of a simple heat pump installation the following data may be assumed:

Electricity costs: Day units   5.6p/kWhr
Electricity costs: Night units   1.9p/kWkr
(The day charge includes an adjustment for standing charge and fuel surcharge)
Hence cost per GJ = £15.6 (day) and £5.3 (night)
Natural Gas at 33.5p/therm and 75% conversion efficiency costs £4.3 per GJ
Class D Oil at 17p/litre and 80% conversion efficiency costs £5.6 per GJ

B22.5 Application of heat pumps to timber drying

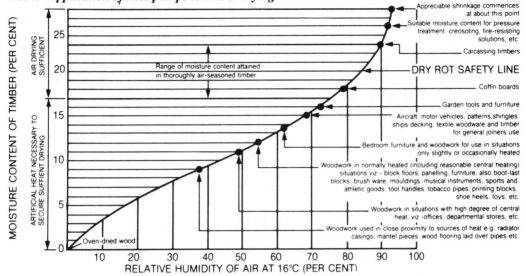

Figure B22.5(a) *Chart showing the moisture content of timber for various purposes*

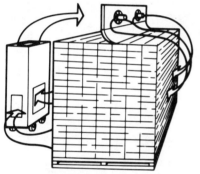

Figure B22.5(b) *Diagram showing typical airflow in a timber kiln*

B22.5 Typical drying schedule suitable for drying 25 mm thick beech wood (from G. W. Pratt, *Timber Drying Manual*, HMSO)

Moisture content (%)	Temperature (°C)	Relative humidity (%)
Green	40	85
60	40	80
40	40	70
35	45	60
30	45	50
25	50	40
20	60	30
15	65	30

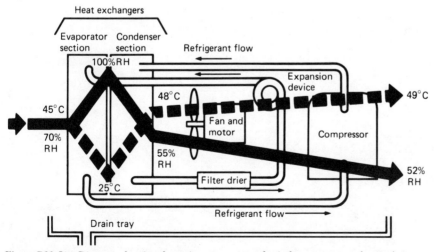

Figure B22.5(c) *Diagram showing the main components of a timber seasoner and typical air flow conditions. The figures used are for illustration*

From the charts at **B22.7** assuming 30 °C return temperature from the heating system and an outside temperature of 5 °C the COP could be up to 3.0, but for −5 °C it may fall to 1.7. It is probably reasonable to take a seasonal average COP at 2.3.

If we further assume that the night tariff can be used to provide 40% of the total output required from the heat pump the average cost of electricity will be

$$0.4 \times £5.3 + 0.6 \times £15.6 = £11.5 \text{ per GJ}$$

With a COP of 2.3 this will equate with

$$\frac{11.5}{2.3} = £5.0 \text{ per GJ}$$

This figure indicates that for heating alone the heat pump cannot compete with gas, but shows some advantage over oil.

For a convincing case to be made for the heat pump as an alternative to a gas fired boiler the night usage would need to be nearer 60%; this would be very difficult to achieve without provision for auxiliary heat from another source in daytime.

The above assessment has been based on a typical tariff for medium supplies; if a maximum demand tariff applies account should be taken of the influence of the heat pump load during peak demand periods.

The annual running cost should be calculated from the structure loss (**B11**) and the air change loss (**B23**); a balance should then be drawn between the reduced running cost compared with oil fuel and the higher capital cost of the heat pump installation.

B22.6 Sankey diagrams for electric and gas engine-driven heat pumps

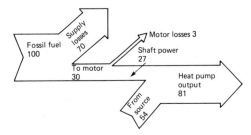

Figure B22.6.1 *Sankey diagram for a heat pump driven by an electric motor*

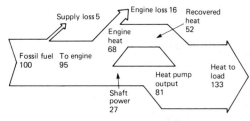

Figure B22.6.2 *Sankey diagram for a heat pump driven by a gas engine*

B22.7 The effects of temperature conditions on the output from heat pumps

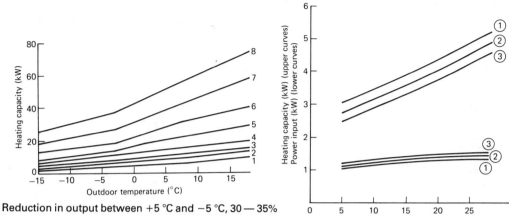

Reduction in output between +5 °C and −5 °C, 30 — 35%
Probable corresponding reduction in COP, 25 — 30%

Figure B22.7.1 *The effects of temperature conditions on the output of heat pumps: variation of output with supply temperature for a typical range of air to air heat pumps. Numbers are heat pump frame size*

Figure B22.7.2 *The effects of temperature conditions on the output of heat pumps: characteristics of 3-5 kW heat pump for hot water supply*
1 Curves for 30 °C output
2 Curves for 40 °C output
3 Curves for 50 °C output
COP at 20 °C, 3.0
COP at 5 °C, 2.5

(B22.7.1 and B22.7.2 are adapted from catalogue information supplied by Thermecon Ltd.)

B23 Mass air flow and energy losses from ventilation

Two case studies illustrate the simplified calculation procedures.

B23.1 Oil fired warm air heater supplying a plenum system

From the data in **Table B23.1**, the density of air at 55°C is

$$1.293 \times \frac{273}{273 + 55} = 1.076 \, \text{kg/m}^3$$

The volume flow (compression caused by fan and duct resistance being ignored) is

$$4.7 \, \text{m/s} \times 0.46 \, \text{m}^2 = 2.162 \, \text{m}^3/\text{s}$$

and the mass flow is

$$2.162 \, \text{m}^3/\text{s} \times 1.076 \, \text{kg/m}^3 = 2.326 \, \text{kg/s}$$

The heat capacity of the air at 55 °C is

$$2.326 \, \text{kg/s} \times 0.997 \, \text{kJ/kg} \times 55 \, °\text{C} = 128 \, \text{kJ/s} = 128 \, \text{kW}$$
(This figure reconciles with the rated output of 132 kW)

Annual flow of usuable energy:

$$\frac{128 \, \text{kW} \times 3600 \, \text{seconds} \times 8 \, \text{hours} \times 5\frac{1}{2} \, \text{days} \times 30 \, \text{weeks} \times 75\%}{10^6 \, (\text{converts from kJ to GJ})} = 455 \, \text{GJ}$$

B23.1 Data for oil-fired warm air heater supplying a plenum system

Output of heater	132 kW
Conversion efficiency	73%
Fuel	Class D oil
Fuel calorific value	38.0 MJ/l
Fan motor output	2.2 kW
Delivery duct	0.46 m²
Average velocity of warm air delivered	4.7 m/s
Temperature of air leaving heater	55 °C
Minimum external temperature	0 °C
Operating time	8 hr/day
	5½ day/week
Heating season	30 weeks
Utilization factor of heater	75%
Density of air at 0 °C (**A12**)	1.293 kg/m³
Specific heat capacity of air at 55 °C	0.997 kJ/kg K

$$\text{Input to heater} = \frac{455}{0.73 \text{ (efficiency)}} = 623 \text{ GJ per year}$$

$$\text{Fuel requirement} = \frac{623\,000 \text{ MJ}}{38 \text{ (MJ/litre)}} = 16\,400 \text{ litres}$$

Annual cost at 20p per litre = £3280

B23.2 Self contained production shop with three air changes per hour

From the data in **Table B23.2** (and **B23.1**), the density of air at 16 °C is

$$\frac{1.293 \times 273}{273 + 16} = 1.221 \text{ kg/m}^3$$

B23.2 Data for self-contained production shop heating

Floor area	521 m²
Average ceiling height	5 m
Annual hours of heating	2000 hr
External ambient temperature	0 °C
Nominal temperature of air	16 °C
Cost of natural gas	32p/therm

The heat capacity of air in the building is

$$\frac{521 \text{ m}^2 \times 5 \text{ m} \times 1.221 \text{ kg/m}^3 \times 0.988 \text{ (sp.ht.cap)} \times 16 \text{ °C}}{1000 \text{ (converts from kJ to MJ)}} = 50.28 \text{ MJ}$$

The heat loss per second is

$$\frac{3 \text{ (air changes)} \times 50.28 \text{ MJ} \times 1000 \text{ (MJ to kJ)}}{3600 \text{ (hr to s)}} = 41.9 \text{ kJ/s}$$

Hence, output of heater required is 41.9 kW.
The annual energy loss is

$$\frac{41.9 \times 3600 \text{ s} \times 2000 \text{ hr}}{10^6 \text{ (kJ to GJ)}} = 302 \text{ GJ}$$

With natural gas as fuel and a conversion efficiency 78%, the annual fuel requirement is

$$\frac{302}{0.1055 \text{ (GJ to therm)} \times 0.78} = 3670 \text{ therm}$$

Thus the annual cost at 32p per therm is £1170.

Note that this calculation covers air change loss only (*see* **B12**), and that no allowance is made for the humid content (*see* **B25**).

B23.3 Energy loss from air leaks

At 15 °C The specific heat capacity of air is

$0.988 \times 1.221 = 1.206 \text{ kJ/m}^3\text{K}$

Under average United Kingdom winter conditions with an external temperature of say 7 °C the loss per m³ of air leakage is

$1.206 \times (19 - 7) = 14.5 \text{ kJ}$

Three possibilities may be examined:

(1) Factory door 5 × 3 m, air speed 5 m/s
Loss per hour:

$$\frac{5 \times 3 \times 5 \times 3600 \times 14.5}{1\ 000\ 000} = 3.90 \text{ GJ}$$

From **B1** the cost of usable energy may be taken at £3.50 for gas and £11 for electricity; thus the cost of the open door per hour is £14 for gas or £43 for electricity.

(2) Pedestrian door left open for 2 hours each day of a 5 day week.
Door size 3 m × 1 m, air speed 5 m/s.
Loss per week

$$\frac{3 \times 1 \times 5 \times 3600 \times 2 \times 5 \times 14.5}{1\ 000\ 000} = 7.83 \text{ GJ}$$

Cost £27 or £86 depending on fuel.

(3) 10 mm gap at the bottom of the 5 m × 3 m factory door throughout the heating season (assumed at 1400 hrs)
Loss per heating season

$$\frac{0.01 \times 3 \times 5 \times 3600 \times 1400 \times 14.5}{1\ 000\ 000} = 10.96 \text{ GJ}$$

Cost £38 or £120 depending on fuel.

The use of air curtains is a well established method of reducing losses through doors but in order to ensure their effectiveness a careful assessment of site conditions is essential, and some of the points for consideration are:

(1) Heating of the air stream is normally essential and the cost of doing this together with the cost of fan power must be related to potential savings.

(2) A closed air circuit is desirable and depending on space available for fans and ducting the flow may be either vertical or horizontal.

(3) Extract ventilation within the building must be checked to ensure that the ventilation system does not create a negative pressure such that external air is being drawn in through the opening under study.

(4) Air curtains are not a substitute for an air-lock where wind speed is high (4 m/s) leading to a positive air flow into the building.

The drive towards reducing air filtration and preventing draughts in the interests of energy savings can lead to secondary problems with air pollution and condensation. Occupied buildings should not be hermetically sealed; some ventilation is essential but it needs to be controlled.

B24 An outline of heat transfer theory

The calculations associated with heat transfer can become very complex and in the ultimate much of the technology rests on empirical and experimental data; the main burden of making these calculations must rest with specialist manufacturers who design and manufacture the various types of heat exchanger. The application engineer is interested mainly in obtaining from the manufacturer sufficient data to ensure that the equipment is correctly applied; nevertheless an understanding of the principles and terminology of heat transfer theory is essential for a proper appraisal of the merits of products offered. The three mechanisms by means of which heat is transferred are conduction, convection and radiation, and these are reviewed in the sections which follow; in practical situations at least two of these mechanisms apply and in some cases all three.

B24.1 Conductive heat transfer

Conduction is the only way in which heat can be transferred in solids and results from the energy exchanged in the molecules of the substance which in turn is influenced by the temperature difference between the ends or opposite faces of the conducting medium; this temperature change is referred to as temperature gradient or driving force. Conduction plays a limited role in heat transfer in liquids and has negligible influence in gases.

Section B11 defines the terminology and calculation procedures for conductive heat transfer in buildings; these procedures apply equally to metals and other materials used in heat exchangers and the same terms are used, i.e. Thermal Conductivity, Thermal Resistance and Thermal Transmittance. There are two ways of expressing the rate of heat transfer, Q (watt) i.e.

$$Q = \frac{\text{thermal conductivity (W/m K)} \times \text{area (m}^2) \times \text{temperature difference (K)}}{\text{Thickness (m)}}$$

or

$$Q = \frac{\text{Temperature difference (K)} \times \text{area (m}^2)}{\text{Thermal resistance (m}^2 \text{ K/W)}}$$

A special case with particular relevance to heat exchangers is conductive flow through the walls of a cylindrical pipe; in this case the area per unit length of pipe across which the heat is flowing increases continuously from the inside to the outside of the pipe and for this reason must be expressed as a logarithmic mean for calculation purposes. The logarithmic mean radius may be determined from the expression

$$\frac{r_o - r_i}{\log_e r_o/r_i}$$

where
r_o is outside radius
r_i is inside radius

The logarithm to the base e may be taken from the chart at **B24.7** which shows the relationship between y and x where $y = e^x$ or $x = \log_e y$.

This is the exponential function and can be arrived at by evaluating the series

$$y = 1 + x + \frac{x^2}{2} + \frac{x^3}{3 \times 2} + \frac{x^4}{4 \times 3 \times 2} \text{ etc.}$$

The ratio (r_o/r_i) (y) is unlikely in practice to exceed the maximum value shown on the chart. Values of $\log_e x$ or $\ln x$ can be obtained from any scientific calculator. The logarithmic mean area of the pipe per metre of pipe length will be 2π times the logarithmic mean radius.

B24.2 Convective heat transfer

Convection occurs because hotter and therefore lighter molecules tend to rise and set up circulation; this tendency is described as buoyancy in situations where free convection takes place; forced convection is that created by fans. The mechanism for heat transfer between a hot and a cold fluid on either side of a solid plate is by means of the so called stagnant film adjacent to the inside and outside surfaces.

The temperature gradient is steep across this film on both the hot and the cold sides of the plate and the bulk of the energy transfer takes place within the two stagnant films.

The rate of heat transfer Q (watt) may be expressed by the relationship:

Q = overall heat transfer coefficient (W/m^{2}K) × area of surface (M^2) × temperature difference (K)

The overall heat transfer coefficient may be treated in exactly the same way as the 'U' value for building structures which is defined in **B11**. The next step is to segregate the overall heat transfer coefficient into its three elements and this is best done (as with building structures) by using reciprocals and regarding the reciprocals as resistances to heat flow thus:

$$\frac{1}{\text{Overall heat transfer coefficient}} = \frac{1}{\text{inside film coefficient}} + \frac{\text{thickness of plate}}{\text{conductivity of plate}} + \frac{1}{\text{outside film coefficient}}$$

(If the transfer instead of being through a flat plate is through a pipe then the inside and outside areas are different and the overall coefficient must be based on one or the other.)

The resistance of the plate is defined as above by its thickness and conductivity so the remaining problem is to define the film heat transfer coefficients; it is also necessary to allow for the increase in resistance to heat transfer occasioned by contamination of the surfaces and this is done in practice by adding an arbitrary fouling factor.

It is around the evaluation of film heat transfer coefficients that much of the complexity of heat transfer theory arises and the central problem concerns the number of variables involved in analysing and drawing conclusions from experimental work. The variables are:

(1) The size and shape of the surface which for a pipe is the diameter and for other surfaces is expressed in terms of the diameter (m).
(2) The temperature difference or driving force (K)
(3) The properties of the fluid, i.e.

Velocity (m/s)
Kinematic viscosity (m^2/s)
Dynamic viscosity (kg/m s)
Thermal conductivity (W/mK)
Density (kg/m^3)
Thermal expansivity (1/K)
Specific heat capacity at constant pressure (J/kg K)

So as to avoid manipulating all these variables either when conducting experiments or when predicting rates of heat transfer, three dimensionless numbers are used each owing its origin to experimental work by pioneers in fluid flow and each bearing the name of that pioneer.

B24.2.1 The Reynolds number (Re)

Reynolds' work in this context was concerned with laminar (streamline) flow in pipes and the velocity at which laminar flow breaks down and turbulent flow begins to take over.

$$Re = \frac{\text{Diameter (m)} \times \text{velocity (m/s)} \times \text{density (kg/m}^3)}{\text{Dynamic viscosity (kg/m s)}}$$

[Since (m $\times$ m/s $\times$ kg/m^3) $\div$ kg/m s $= 1$ the number is dimensionless]
Re may also be expressed as

$$\frac{\text{Diameter (m)} \times \text{velocity (m/s)}}{\text{Kinematic viscosity (m}^2\text{/s)}}$$

In very general terms it may be stated that for numbers below 2000 the flow will be laminar and above 10 000 it will be turbulent. The transition from one to the other depends on the roughness of the pipe and other sources of interference such as vibration.

The flow of fluids in pipes at the velocities occurring in normal engineering practice is likely to be turbulent except for high viscosity fluids such as oils. In order to ensure satisfactory performance of heat exchangers turbulent flow is essential, as transfer of heat in laminar flow depends on conduction and conductivity of fluids is low.

B24.2.2 The Prandtl number (Pr)

Prandtl is best know as 'the Father of Aerodynamics', but he played a major role in the establishment of a sound theoretical basis for fluid mechanics.

$$Pr = \frac{\text{Specific heat capacity (J/kg K)} \times \text{dynamic viscosity (kg/m s)}}{\text{thermal conductivity (W/m K)}}$$

(The number is again dimensionless and is concerned solely with the properties of the fluid.)

The range of practical values of Pr is:

Liquid Metals	0.004 to 0.04
Gases	0.6 to 0.9
Low viscosity liquids	1 to 10
High viscosity liquids	50 to 1000

B24.2.3 The Grashof number (Gr)

$Gr = $ [Diameter (m)]$^3 \times$ [density (kg/m^3)]$^2 \times$ gravitational acceleration (m/s^2) $\times$
thermal expansivity (1/K) $\times$ temperature difference (K)

$$\times \frac{1}{[\text{viscosity (m}^2\text{/s)]}^2}$$

(Once again dimensionless)

The three dimensionless numbers are combined into the NUSSELT GROUP (Nu)

$Nu = C$(an experimental constant) $\times Re^x \times Pr^y \times Gr^z$

It has been established experimentally that in forced convection Gr is negligible in comparison with Re whereas in free convection the opposite is true, thus:

For forced convection $\quad Nu = C_1 \times Re^x \times Pr^y$
For free convection $\qquad Nu = C_2 \times Pr^y \times Gr^x$

The value of Gr at which free convection will occur is in excess of 10^8 and turbulence becomes evident if Gr reaches 10^{10}.

There are numerous derivitatives of these numbers and in this volume it is not possible to go into the ramifications of the various equations which have been postulated but there are a few simplified equations which may be applied to solving practical problems.

Dittus–Boelter has stated

(1) $Nu = 0.23 \times Re^{0.8} \times Pr^{0.3}$
(2) $Nu = 0.23 \times Re^{0.8} \times Pr^{0.4}$

Both equations apply to flow inside a circular pipe; the first covers calculations when the fluid inside is being cooled and the second when it is being heated.

Many practical problems are related to free convection in air and because, over limited temperature ranges, the properties of air are reasonably constant, some simplification of the Nusselt number for such problems can be made. The starting point for such simplification is

$Nu = C(Gr \times Pr)^x$
where for laminar flow $C = 0.13$ and $x = 0.25$
and for turbulent flow $C = 0.59$ and $x = 0.33$

By substituting values for air (at say 15 °C) of density, viscosity, specific heat capacity, thermal conductivity, expansivity and the acceleration due to gravity, the values of film heat transfer coefficient for horizontal pipes with natural convection may be approximated to

$$k \ (\mathrm{W/m^2\,K}) = 1.35 \times \left(\frac{\Delta t}{\mathrm{diam}}\right)^{0.25}$$

for laminar flow, where Δt is temperature difference, or

$$k \ (\mathrm{W/m^2\,K}) = 1.2 \times (\Delta t)^{0.33}$$

for turbulent flow, and for surfaces not less than 0.7 m in height

$$k = 1.7 \times (\Delta t)^{0.33}$$

for turbulent flow.

The Heilman equation offers an alternative approach for calculations involving free convection of air. The quantity of heat transferred per second across unit area, Q, is

$$Q \ (\mathrm{W/m^2}) = \frac{C \times (\Delta t)^{1.266}}{(\mathrm{diam})^{0.2} \times (\mathrm{mean\ temp.})^{0.181}}$$

where
C is a constant: 2.9 for horizontal pipes, 3.5 for vertical pipes, 2.5 for plates facing down, 5.1 for plates facing up and 4.0 for vertical plates.

The Langmuir equation can be used for forced convection:

$$Q \ (\mathrm{W/m^2}) = 1.95 \times (\Delta t)^{1.25} \times \left(\frac{\mathrm{air\ speed\ (m/s)} + 0.35}{0.35}\right)^{0.5}$$

The reader interested in pursuing this theory should consult some of the many references on the subject as this review is intended only to enable a grasp of the principles to be gained.

B24.3 Heat transfer by radiation

Radiation takes place from any source whose temperature is above absolute zero, but in heating parlance radiation will commence to be effective at around 50 °C and increase rapidly in intensity as the temperature rises. Electromagnetic radiation covers the complete spectrum of frequency (wavelength) as illustrated in **Chart A40**, but we are concerned primarily with radiation in the visible and infrared ranges from 8 to 900 THz (38 m to 330 nm). The hypothetical 'Black Body' has been postulated as a body which will absorb all the radiation falling upon it; in practice part is absorbed, part reflected and part transmitted. These fractions are defined as absorptivity, reflectivity and transmissivity with the sum of all three being unity.

A 'grey' body which is again hypothetical is one where emissivity is less than unity, but which otherwise behaves much as a 'black' body. All real surfaces are said to be 'coloured' which implies that their response to incident radiation varies according to the frequency (or colour) of that radiation. Heat transfer by radiation is governed by the Stephan–Boltzmann law and Planck's law. The Stephan–Boltzmann equation is

Emissive power (W/m^2) $= \sigma T^4$

where
σ is the Stephan–Boltzmann constant 5.67×10^{-8} W/m^{2}K^4)
T is absolute temperature

For practical calculations, this can be written

$$Q \text{ (kW)} = \frac{56.7 \times A \times T^4}{1000}$$

where
A is area of surface (m^2)
Q is rate of heat transfer
T is absolute temperature

The equation can be adapted to assess the radiant heat transferred from one surface to another:

$$Q = 56.7 \times A \times f \times e \left\{ \left(\frac{T_1}{1000} \right)^4 - \left(\frac{T_2}{1000} \right)^4 \right\}$$

where
f is a factor to allow for the heat absorbed by the intervening air (usually 0.9)
e is absorbtivity of the receiving surface
T_1 and T_2 are absolute temperatures of radiating and receiving surface respectively.

Yet a further development of the equation is concerned with transfer between surfaces and utilizes the Huttel factor which takes count of:

(1)   The fraction of radiation leaving surface A_1 which is intercepted by surface A_2 [the view factor(F) which is concerned with the relative attitudes of the two surfaces].
(2)   The relative areas A_1 and A_2.
(3)   The relative emissivities ε_1 and ε_2.

$$\text{Hottel factor} = \frac{1}{\left(\dfrac{1}{\varepsilon_1} - 1 \right) + \dfrac{1}{F} + \dfrac{A_1}{A_2} \left(\dfrac{1}{\varepsilon_2} - 1 \right)}$$

Finally an overall heat transfer coefficient may be evaluated for a given situation enabling the rate of heat transfer to be expressed in terms of surface area and temperature difference exactly as for conduction and convection.

Planck's law enables monochromatic emissive power to be calculated according to wavelength and absolute temperature. For practical purposes it may be written:

Emissive power (W/m^2) at wavelength λ (μm)

$$= \frac{C_1/(\lambda)^5}{\exp{(C_2/\lambda)} - 1}$$

where
λ is wavelength in micrometre
T is absolute temperature
C_1 has the value 3.743×10^8 numerically
C_2 has the value 1.4387×10^4 numerically

The charts at **B24.3.1** have been drawn using the Planck equation to calculate the relationship between emmissive power and wavelength at three different temperatures, i.e. 373, 473 and 573 K. Note:

(1) The sharp increase in output with increasing temperature.
(2) The depression of the frequency for maximum emissive power as the temperature increases. This depression is in accordance with Wien's law which can be written in the form.

Wavelength (μm) $\times$ T = constant for maximum power
where the constant has the value $2898 \, m^{-6} \, K$

The chart also indicates the frequencies for maximum power at the higher temperatures of 1000 K (power $= 12\,700 \, W/m^2$) and 2500 K (power $= 46 \times 10^6 \, W/m^2$).

(3) The area under the curve is the total emissive power and it can be shown by integration that the value of power equates with the value obtained from the Stefan–Boltzmann equation.

All these relationships are on the 'black' body concept and it can be seen that the frequencies for maximum emission vary from $1 \, \mu$m to $8 \, \mu$m according to temperature.

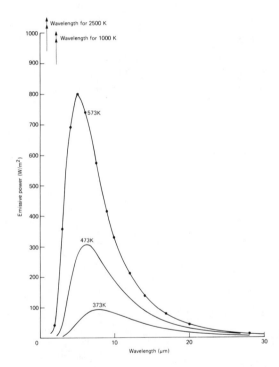

Figure B24.3.1 *Relationship between emissive power and wavelength for black bodies*

A 'grey' body may be loosely described as a less than perfect 'black' body, whose monochromatic emissivity is not strictly related to wavelength. 'Grey' bodies at the same temperature as the 'black' bodies shown on the chart will have lower peak emissive powers.

All real bodies are 'coloured' and they will exhibit variations in emissivity with variations in frequency (or colour) of the radiant energy. The spectral distribution charts for various light sources as depicted in **B27.6** are analagous to the behaviour of coloured bodies.

The terms short and long wave radiation are sometimes used in relation to radiant heat sources; short wave referring to the high temperature sources which are towards the visible end of the spectrum and long which are in the medium infrared band (*see* **Chart A40**). Short wave radiation is associated normally with incandescent electrical sources and long with black radiant gas heaters or hot water radiators. These classifications are by no means hard and fast and there are many overlaps. Reference is made at **B11.3** to emissivity in relation to building structures and the table at **A21.8** gives values for some common materials. Data on radiation from pipe surfaces is given in **Tables A32.1** to **A32.4**.

B24.4 Calculation procedures for conduction

B24.4.1 Through a steel sheet

The simple case is illustrated by a sheet of steel of thickness 10 mm and size 2 m $\times$ 3 m; the hot face is held at say 100 °C and as a result of forced fan cooling the cold face is held steady at say 95 °C. From **Table A20.1** thermal conductivity may be taken at 48 W/m K. The rate of heat transfer is thus

$$\frac{48 \,(\text{W/m K}) \times 2 \times 3 \,(\text{m}^2) \times (100 - 95)\,(\text{K})}{0.010 \,(\text{m})} = 144 \,\text{kW}$$

It is interesting to compare the rate of transfer assuming natural convection on the outside surface in which case the temperature difference is unlikely to exceed 0.20 °C; under these conditions the rate of transfer becomes 5.8 kW only. Calculations of conduction through composite, multilayer, materials may be made using identical procedures to those in **B11** and a practical example of furnace heat loss in given in **B14**.

B24.4.2 Through a pipe

The heat transfer through a pipe may be illustrated by reference to a schedule 40 standard pipe of nominal size 2 inches with outside diameter 60.3 mm and inside diameter 52.5 mm. Taking an inside temperature of 80 °C, an outside temperature of 79.9 °C and a standard length of one metre the calculation is as follows:

$$\text{Ratio of radii} = \frac{30.15}{26.25} = 1.15$$

$\log_e 1.15$ (from **Chart B24.6**) $= 0.13$

The logarithmic mean radius is

$$\frac{30.15 - 26.25}{0.13} = 30 \,\text{mm}$$

The logarithmic mean area per metre length is thus

$$\frac{2\pi \times 30}{1000} = 0.188 \,\text{m}^2$$

$$\text{Pipe thickness} = \frac{30.15 - 26.25}{1000} = 0.0039 \,\text{m}$$

Therefore the rate of heat transfer is

$$\frac{48 \,(\text{W/m K}) \times 0.188 \,(\text{m}^2) \times (80 - 79.9)\,\text{K}}{0.0039 \,(\text{m})} = 230 \,\text{W/m}$$

B24.4.3 Composite pipe

The same procedure can be used for a composite pipe (for example with external insulation) by calculating the resistance of each layer and using the relationship

$$\text{Rate of heat transfer} = \frac{\text{Temperature difference (K)}}{\text{Sum of resistance of layers (m}^2\text{K/W)}}$$

As an example assume that the pipe is insulated by 20 mm radial thickness of glass fibre having a thermal conductivity of 0.040 W/m K (**Table A20.1**) and the external temperature is 25 °C.

The logarithmic mean radius of insulation is

$$\frac{50.15 - 30.15}{\log_e\left(\frac{50.15}{40.15}\right)} = \frac{20}{\log_e 1.66} = \frac{20}{0.501} = 39.9 \, \text{mm}$$

and the logarithm mean area is, per metre length,

$$\frac{2\pi \times 39.9}{1000} = 0.25 \, \text{m}^2$$

The resistance of the insulation per metre length is

$$\frac{\text{Thickness}}{\text{Thermal conductivity} \times \text{log mean area}}$$

$$\frac{0.020}{0.040 \times 0.25} = 2.0 \, \text{m}^2 \, \text{K/W}$$

and the resistance of the pipe is

$$\frac{0.039}{48 \times 0.188} = 0.0043 \, \text{m}^2 \text{K/W}$$

The new rate of heat transfer is

$$\frac{80 - 25}{2.0 + 0.0043} = 27 \, \text{W/m}$$

The large reduction in heat transfer illustrates the effect of insulation.

Arising from the April, 1982 amendment to the Building Regulations various studies have been made to determine the economic thickness of pipe insulation. Both convection and radiation are involved and calculation procedures have been suggested relating energy loss to the cost of insulation.

From the example above it is clear that the major saving in heat loss comes from 20 mm radial thickness, as increasing this to 30 mm would reduce the loss by no more than 7 W/m compared with 200 W/m already achieved.

It is always obvious when a pipe is poorly insulated and an experienced eye coupled with the touch of a hand goes a long way towards ensuring that the job is well done!

B24.5 Calculation procedures for convection of air

As already indicated, calculation procedures for convective heat transfer through fluids other than air are beyond the scope of this manual; however, some straightforward examples involving air are of interest.

B24.5.1 Convection from a pipe in air

Pipe	Schedule 80, nominal 2½ in
Outside diameter	73 mm
Inside diameter	59 mm
Surface temperature	70 °C
Air temperature	20 °C

The film heat transfer coefficient (from **B24.2**) is

$$1.35 = \left(\frac{(70 - 20) \, °C}{0.073}\right)^{0.25} = 6.9 \, \text{W/m}^2\text{K}$$

The rate of heat transfer is

$$6.9 \times 50 = 345\,\text{W/m}^2$$

The surface area of the pipe is, per metre length,

$$\pi \times 0.073 = 0.229\,\text{m}^2$$

Therefore the heat transfer per metre length of pipe is

$$345 \times 0.229 = 79\,\text{W}$$

If now the pipe is insulated with 25 mm of felt having a thermal conductivity of 0.045 this will have a resistance to conduction based on the log mean area derived as in **B24.4**:
The logarithmic mean radius is

$$\frac{98 - 73}{\log_2 (98/73)} = 89.3\,\text{mm} = 0.0893\,\text{m}$$

Hence resistance per metre length is

$$\frac{0.025}{0.045 \times 0.0893 \times 2\pi} = 0.99\,\text{m}^2\text{K/W}$$

The resistance of the pipe itself will be negligible in comparison with that of the felt and may be disregarded. Hence new rate of heat transfer is

$$\frac{70 - 20}{0.99} = 51\ \text{W/m}^2$$

The new area per metre length will be

$$\frac{2\pi \times \left(\dfrac{73}{2} + 25 \right)}{1000} = 0.386\,\text{m}^2$$

and the new rate of heat transfer per metre length

$$0.386 \times 51 = 20\,\text{W}$$

B24.5.2 Use of the Heilmann equation

Using the Heilmann equation (**B24.2**) as an alternative for the same pipe calculation the rate of heat transfer may be arrived at:

$$\frac{2.9 \times (70 - 20)^{1.266}}{(0.073)^{0.2} \times 4.5^{0.181}} = \frac{2.9 \times 141.6}{0.5926 \times 1.992} = 348\,\text{W/m}^2$$

which shows a remarkably close correlation with Nusselt.

B24.5.3 Use of the Langmuir equation

Applying the Langmuir equation for forced convection and assuming a velocity of 3 m/s the rate of heat transfer becomes:

$$1.95 \times 50^{1.25} \times \left(\frac{3.35}{0.35} \right)^{0.5} = 1.95 \times 132.7 \times 3.094 = 800\,\text{W/m}^2$$

Similar procedures may be applied to flat surfaces using the simplified formula derived from Nusselt or the Heilmann equation with the appropriate constant.

B24.6 Calculation procedures for radiation

B24.6.1

A gas fired tubular type radiant heater, overall length 4 m, outside diameter of tube 60 mm. Surface temperatures ranging from a maximum of 500 °C near the burner to 150 °C at the exhaust end of the tube. Assume 30% of the length at maximum temperature, 20% at minimum and the balance mid-way.

The total area of the radiating surface is

$$\pi \times 0.06 \times 4 = 0.75\,\mathrm{m}^2$$

Of this total, $0.225\,\mathrm{m}^2$ is at 500 °C, $0.150\,\mathrm{m}^2$ is at 150 °C and $0.375\,\mathrm{m}^2$ is at 325 °C. The total output is

$$56.7 \times \left\{ \left[0.225 \times \left(\frac{775}{1000}\right)^4 \right] + \left[0.15 \times \left(\frac{423}{1000}\right)^4 \right] + \left[0.375 \times \left(\frac{578}{1000}\right)^4 \right] \right\} = 7.5\,\mathrm{kW}$$

In practice this outpuit will be reduced according to the extent to which the emissivity falls below unity.

A heater to this outline specification is likely to have a total output of 12–13 kW with the balance coming from convection.

B24.6.2

To assess the heat transferred by such a radiant heater to surfaces at 20 °C, assuming a factor of 0.9 for loss to surrounding air and incidental losses and also assuming an absorptivity factor of 0.85 for the receiving surfaces, we may use the second formula in **B24.3**.

Total heat transferred

$$= 56.7 \times 0.9 \times 0.85 \times \{[0.225 \times (0.773^4 - 0.293^4)] + [0.15 \times (0.423^4 - 0.293^4)] + [0.375 \times (0.598^4 - 0.293^4)]\} = 5.5\,\mathrm{kW}$$

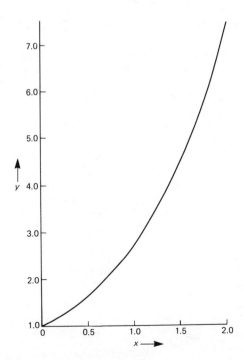

Figure B24.7 *The exponential curve:* $y = \mathrm{e}^x$, $y = \exp x$, *or* $x = \log_e y$

184

B25 Heat exchangers

B25.1 *The main categories of heat exchanger*

The function of heat exchangers is to extract heat from the primary (hot) fluid and transfer it to the secondary (cool) fluid; the fluids may be gas, air or liquid and are normally kept entirely separate from one another. One exception is the heat wheel where a very small carry-over from primary to secondary does occur; it goes without saying that this is permissible only if the risk of contamination is neglible. Counter-flow exchangers (those where the direction of flow of the two fluids is opposite) have inherent advantages in terms of effectiveness for heat transfer. Some common examples are:

(1) 'Gas to air' in which, as a typical example heat is extracted from the flue of a boiler, heater or furnace and supplied to working areas as warmed air.
(2) 'Air to air' in which warmed air from a process is used to supply space heating requirements.
(3) 'Gas (or air) to water' where water is the secondary fluid and is used for example in a low pressure hot water heating system.
(4) 'Run around coils' where a liquid flows in a closed circuit with a coil in one location exposed to the heat source and a coil in a second location exposed to the heat recovery medium.
(5) 'Heat pipes' which rely on the principle used in absorption type refrigerators. These have a closed circuit containing a refrigerant fluid which evapourates at one end of the circuit and recondenses at the other. The evaporation process takes the heat in and the condensing process releases it.
(6) 'Heat pumps' operate on the same principle as 'heat pipes' but are assisted by a mechanical refrigeration system and depend for their effectiveness on their ability to absorb low grade (low temperature) heat and deliver it at higher temperature.
(7) 'Thermal wheels' are based on a slowly rotating wheel designed to absorb and release heat with minimum loss and controlled cross contamination. The ducting carrying the primary and secondary air streams is split across a diameter of the wheel with one half carrying the primary flow and the other half the counter flow secondary.

Details of these appliances are best obtained from manufacturers but they are illustrated in diagramatic form at **B25.4**.

There are, of course, many other devices which receive heat from one source and supply it to another, but in general these are associated with heating equipment, i.e. recuperators and economizers, unit heaters, radiators and convectors.

B25.2 *Heat exchanger characteristics*

B25.2.1

Temperature efficiency is the generally accepted basis for measuring the effectiveness of a heat exchanger and this measurement really states how much of the total dry enthalpy available in the primary fluid can be transferred to the secondary.

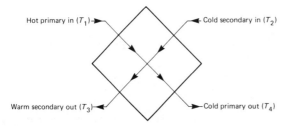

Figure B25.2.1 *Counter flow heat exchanger diagram*

The temperature efficiency will rise from its nominal level if the mass flow of secondary air exceeds that of the primary air; if the reverse applies the temperature efficiency will fall. It must be emphasized that this method of stating efficiency ignores the effect of the enthalpy released as a result of condensation of the moisture content of the primary air; in practice this source of heat energy can have a very significant effect on economics.

The diagram **B25.2.1** illustrates a counter flow air to air heat exchanger and the normal definition of temperature efficiency is:

$$\eta = \frac{(\text{temperature gain to secondary})}{(\text{initial temperature difference})} = \frac{T_3 - T_2}{T_1 - T_2}$$

B25.2.2

The rate of heat transfer depends partly on the characteristics of the exchanger and partly on the four temperatures T_1, T_2, T_3 and T_4. Because within the exchanger itself the temperature varies between inlet and outlet a logarithmic mean temperature is used, derived from the formula:

$$T_m = \frac{(T_1 - T_3) - (T_4 - T_2)}{\log_e \left(\dfrac{T_1 - T_3}{T_4 - T_2} \right)}$$

The rate of heat transfer is

$$Q \text{ (kW)} = U \times A \times T_m$$

where
U is overall heat transfer coefficient (kW/m²K)
A is surface area (m²)

B25.2.3

The rate of heat transfer may also be referred to either the primary or the secondary fluid.
  For primary

Q = mass flow rate (kg/s) × (specific energy In − specific energy Out)

  For secondary

Q = mass flow rate (kg/s) × (specific energy Out − Specific energy In)

B25.2.4

Account must be taken of the latent heat supplied with the primary air and the extent to which the moisture will be condensed in the exchanger. Availability of latent heat will raise the efficiency of the exchanger, but it will, of course, introduce the problem of disposing of the condensed moisture.

B25.2.5

Materials from which heat exchangers are manufactured impose a limit on the maximum operating temperature particularly in handling flue gases with a corrosive content; for high temperature operation stainless steel or even vitreous finish may be needed.

B25.2.6

A compromise must be found between the volume the exchangers will handle and the back pressure created by the resistance of the exchanger to fluid flow. In the case of gas or air a high resistance will increase the size and cost of fans. On the other hand a large design volume for the exchanger (with a correspondingly low back pressure) will increase the cost of the exchanger.

B25.2.7

The proximity of the source of heat to the area where it can be used influences selection. The ductwork associated with counter flow exchangers, heat wheels and heat pipes can cause problems if distances are appreciable. Run-around coils lessen these problems, but tend to be less efficient.

B25.2.8

The modular approach is frequently used and manufacturers will normally supply pressure drop and efficiency figures (related to volume flow) for standard modules. The chart at **B25.5** based on information supplied by Dantherm illustrates this approach. The single module will handle up to $0.8\,m^3/s$ with high pressure drop and lowest efficiency at maximum flow. The 'four module' and 'nine module' units have relatively higher efficiencies and are able to handle larger volumes. In any of these three configurations additional modules may be used 'in parallel' so as to share the flow without change in pressure drop–efficiency characteristics. Thus three 'nine module' units in parallel will handle up to $4.0\,m^3/s$.

B25.3 Heat exchanger calculations

A specific example of the application of air to air counter flow exchangers follows:

Primary air (from process)	$0.95\,m^3/s$ at 50 °C
Relative humidity, minimum	10%
Relative humidity, maximum	20%
Secondary air	$0.85\,m^3/s$ at 10 °C
Relative humidity	50%

Normal practice is to refer all calculations to 0 °C. Losses in the heat exchanger itself are ignored.

An efficiency figure is needed and this entails making a preliminary selection of heat exchanger. From chart **B25.5** it can be seen that the choice lies between a four module unit, which at $0.95\,m^3/s$ would give an efficiency at 68.5% with a pressure drop of 265 Pa, or two single modules in parallel giving 57.5% at 130 Pa. Assume that the higher efficiency can be justified.

The temperature of the warmed secondary air may now be arrived at, i.e.

$$0.685\;(\eta) = \frac{T_3 - T_2}{T_1 - T_2} = \frac{T_3 - 10}{50 - 10}$$

Hence $T^3 = 37.4$ °C (Allowing for intervening ductwork and assuming discharge above face level such a temperature is suitable for workshop heating) T_4 may be taken as $T_1 - T_3$, i.e. 12.6 °C.

The calculations which follow are presented in a rigorous manner in order to emphasize the procedures; checks are made at each stage against the psychrometric chart; in dealing with practical problems the chart would be used exclusively and many short cuts employed. Checks should be made with manufacturers to ensure that flow, pressure and efficiency are based on exchangers currently available.

B25.3.1 Enthalpy of incoming primary air

(a) Dry heat content
Ideal gas density of air at 15 °C = 1.225 (from **A12**). Hence at 50 °C.

$$1.225 \times \frac{288}{323} = 1.092\,kg/m^3$$

Hence mass per second =

0.95 (m³/s) × 1.092 = 1.037 kg/s

Specific heat capacity of air at 50 °C (from **A15.1**) = 0.997 kJ/kg K
Dry heat content = 1.037 × 0.997 × 50 = 51.69 kJ/s
(From the chart at **A36** a line through the intersection of the 50 °C dry bulb line with the zero moisture content line intersects the specific enthalpy line at 51 kJ/kg, i.e. 51 × 1.037 = 52.9 kJ/s)

(*b*) *Liquid phase water content at 10% (alternatively 20%) RH*
From the intersection of the 50 °C dry bulb line with the 10% and 20% humidity curves the moisture content can be read at 0.0078 and 0.0156 kg/kg respectively (these figures could, if needed, be established from first principles using partial pressures as described in **B15**). With the addition of moisture the total mass of moist air will rise to 1.045 and 1.053 kg/s respectively for 10% and 20% humidity.
From the chart the dew point may now be established by following the horizontal line from the intersection of the dry bulb line with the 10% and 20% RH curves giving 10.7 °C and 20.9 °C respectively.
From **Table A18** the specific heat capacity at constant pressure of liquid phase water is 4.18 kJ/kg K over a temperature range of 20 °C to 60 °C. Hence enthalpy of water content (from 0 °C to dew point)

At 10% RH   4.18 (kJ/kg K) × 0.0078 (kg/kg) × 1.045 (kg/s) × 10.7 (K) = 0.36 kJ/s
At 20% RH   4.18 × 0.0156 × 1.053 × 20.9 = 1.44 kJ/s

(*c*) *Latent heat content*
From the steam tables at **A32** the specific latent heat in kJ/kg is 2477 at 10.7 °C and 2452 at 20.9 °C and the masses will be the same as liquid phase water; hence latent heat
At 10% RH   2477 × 0.0078 × 1.045 = 20.19 kJ/s
At 20% RH   2452 × 0.0156 × 1.053 = 40.28 kJ/s

(*d*) *Water vapour content*
The specific heat capacity at constant pressure can be taken from the steam tables at **A32** and the value at 10 °C is 1.87 kJ/kg K and at 50 °C it is 1.91. Thus water vapour content is

At 10% RH   1.89 kJ/kg × 0.0078 × 1.045 × (50 − 10.7) = 0.60 kJ/s
At 20% RH   1.90 × 0.0156 × 1.053 × (50 − 20.9) = 0.91 kJ/s

The total enthalpy of the primary incoming air is thus

At 10% RH   51.69 + 0.36 + 20.19 + 0.60 = 72.84 kJ/s
At 20% RH   51.69 + 1.44 = 40.29 + 0.91 = 94.33 kJ/s

By reading from the intersection of the 50 °C dry bulb line with the 10% and 20% humidity curves on the psychrometric chart values of enthalpy can be read off at 71 kJ/kg and 91 kJ/kg i.e.

71 × 1.045 = 74.2 kJ/s   91 × 1.053 = 95.8 kJ/s

The incoming primary energy can thus be summarized at 51 kW for dry air, 73 kW with 10% humidity, and 94 kW with 20% humidity.

B25.3.2 Outgoing heat content of primary air

The next stage is to make an assessment of the outgoing heat content of the primary air using 12.6 °C as the provisional temperature. The total moisture content will be changed unless saturation point is reached and as the dew point at 10% RH is 10.7 °C there will be no condensation with an exit temperature of 12.6 °C. However, with 20% RH incoming air the dew point is 20.9 °C and there will be condensation in the exchanger. Based on a moisture content of $0.0078 \times 1.045 = 0.0082$ kg/kg and a dry bulb temperature of 12.6 °C the relative humidity is 87%, wet bulb 11.5 °C and specific volume 0.82 m³/kg. The specific heat capacity of air at 12.6 °C $= 0.984$ kJ/kg (**A15.1**). We can now assess the total enthalpy of the outgoing air, i.e.

Dry content	$1.037 \times 0.984 \times 12.6$	$= 12.86$ kJ/s
Liquid phase water	$4.18 \times 0.0082 \times 10.7$	$= 0.37$ kJ/s
Latent heat	2472×0.0082	$= 20.27$ kJ/s
Water vapour	$1.87 \times 0.0082 \times (12.6 - 10.7) =$	0.03 kJ/s
	Total	33.53 kJ/s

From **A36** 33 kJ/kg $\times 1.037 = 34.2$ kJ/s)
From the chart the moisture content for saturation at 12.6 °C dry bulb is 0.0091 kg/kg and under these conditions the new values of liquid phase and latent content will be

Liquid phase $\quad 4.18 \times 0.0091 \times 1.045 \times 12.6 = 0.50$ kJ/s
Latent heat $\quad 2472 \times 0.0091 \times 1.045 \quad\quad = 23.5$ kJ/s

There will be no water vapour content so the new total becomes

$12.86 + 0.50 + 23.51 = 36.87$ kJ/s

It is now possible to state the total energy available for transfer in the heat exchanger, i.e.

At 10% RH $\quad$ 72 kW $-$ 33.5 kW $=$ 38.5 kW
At 20% RH $\quad$ 94 kW $-$ 36.9 kW $=$ 57.1 kW

It is also possible to estimate the amount of condensate which must be collected from the exchanger with 20% RH primary supply.

Mass of moisture per second incoming $\quad 0.0156 \times 1.053$ kg/s
Mass of moisture per second outgoing $\quad 0.0091 \times 1.045$ kg/s

Quantity per hour for collection $(0.0164 - 0.0095) \times 3600 = 24.8$ kg.
The final stage is to check the incoming and outgoing enthalpy for secondary air.

(*a*) *Incoming secondary enthalpy*
Dry

$$0.85 \, (\text{m}^3/\text{s}) \times 1.225 \times \frac{288}{283} \, (\text{kg/m}^3) \times 0.984 \, (\text{kJ/kg K}) \times 10 \, (\text{K}) = 10.43 \, \text{kJ/s}$$

Liquid

4.18×0.0039 (kg/kg) (from Chart) $\times 1.247$ (kg/m³) $\times 0.85$ (m³/s) $\times 2$ (K) (from Chart the dew point is 2 °C) $= 0.003$ kJ/s

Latent

$2478 \times 0.0039 \times 1.247 \times 0.85 = 10.24$ kJ/s

Water vapour

$1.87 \times 0.0039 \times 1.247 \times 0.85 \times (10 - 2) = 0.06$ kJ/s

Total incoming enthalpy $= 20.76$ kJ/s

(*b*) *Outgoing secondary enthalpy*
Working from 37.4 °C and a moisture content of 0.0039 kg/kg, RH will be 10%, dew point 2 °C, wet bulb 17.5 °C and specific volume 0.887 m³/kg. Hence enthalpy out:

Dry

$$0.85 \times 1.225 \times \frac{288}{310} \times 0.996 \times 37.4 = 36.03 \,\text{kJ/s}$$

Liquid

$$4.18 \times 0.0039 \times 1.138 \times 0.85 \times 2 = 0.03 \,\text{kJ/s}$$

Latent

$$2412 \times 0.0039 \times 1.138 \times 0.85 = 9.10 \,\text{kJ/s}$$

Water vapour

$$1.90 \times 0.0039 \times 1.138 \times 0.85 \times (37.4 - 2) = 0.25 \,\text{kJ/s}$$

Total outgoing enthalpy $= 45.41 \,\text{kJ/s}$
Summary of secondary air enthalpy

Dry enthalpy In   10.8 kW   Dry enthalpy Out   36.0 kW
Total enthalpy In   21.1 kW   Total enthalpy Out   45.4 kW

Finally a check may be made against the 'temperature' efficiency based on primary and secondary enthalpy:
Dry energy to heat exchanger from primary flow

$$51.69 - 12.86 = 38.83 \,\text{kW}$$

Dry energy extracted by secondary flow

$$36.03 - 10.43 = 25.60 \,\text{kW}$$

Efficiency

$$\frac{25.60}{38.83} = 65.9\%$$

This figure shows reasonable correlation with the 'temperature' figure but is more accurate being based on the actual 'dry' enthalpies. Calculation of the apparent efficiency based on total enthalpies as calculated above yields unrealistic figures; this is because in practice the humid element will raise T_3 beyond 37.4 °C and this in turn will increase the enthalpy added to the secondary flow with a consequent higher efficiency.

Until saturation occurs the bulk of 'wet' enthalpy is carried through the heat exchanger by the primary flow; however, when saturation point is reached (as at 20% RH) it can be seen from the above calculations that an additional enthalpy of some 18 kJ/s is available for transfer.

If this is to be absorbed by the secondary flow the bulk of it must be added to the dry secondary flow 'out' giving $36 + 18 = 54 \,\text{kJ/s}$ and this in turn would increase the secondary air temperature 'out' to

$$\frac{54}{36} \times 37.4 = 56.1$$

This is of course *higher* than the primary 'in' temperature and cannot be achieved in practice.

If this additional energy is to be absorbed by the dry secondary air the volume of secondary flow must be increased to say, 1.25 m³/s. This increase will have repercussions on the efficiency of the heat exchanger and therefore necessitate a fresh calculation. Normal practice is to check at the outset for the contribution from the humid content of the primary air and to proceed with the calculation accordingly.

Some indication of the improved efficiencies which can result from the humid content can be seen from the charts at **B25.6** which are reproduced by courtesy of Heatex (represented in UK by A. Johnson and Co. Ltd.)

B25.4 Diagrammatic details of heat exchangers

Supply air coil   Exhaust air coils

4.4 kg/s 30% glycol at 8.6°C

5.1 m³/s −1°C   11.6°C

21°C D.B.
16°C W.B.
1.9 m³/s each coil

11.2°C D.B.
11.2°C W.B.

Pump

4.4 kg/s at 13.2°C

Figure B25.4.1 *Run around heat recovery system*

(Reproduced by courtesy of A.A.F. Ltd.)

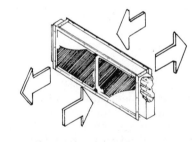

Heat pipe exchanger

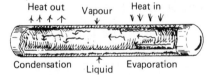

Heat out   Vapour   Heat in

Condensation   Liquid   Evaporation

Heat pipe detail

Figure B25.4.3 *Heat pipe exchanger and detail of heat pipe*

(Reproduced by courtesy of Ductwork Engingeering Systems)

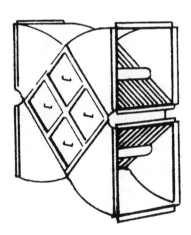

Figure B25.4.5 *Cross flow air to air heat exchanger*

(Reproduced by courtesy of Dantherm)

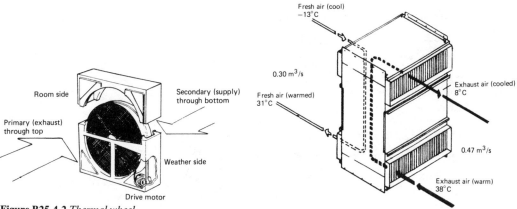

Figure B25.4.2 *Thermal wheel*

(Reproduced by courtesy of Acoustics and Environmetrics Ltd.)

Figure B25.4.4 *Two duct air to air heat exchanger*

(Reproduced by courtesy of S & P Coil Products Ltd.)

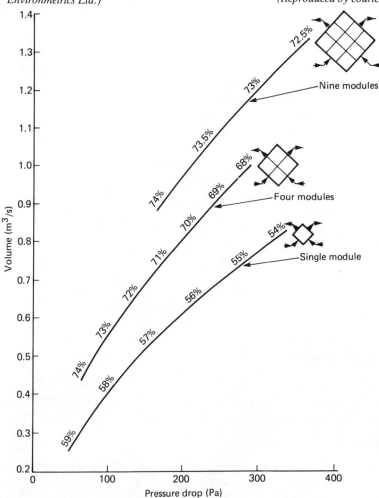

Figure B25.5 *Counter flow gas to air heat exchangers: relationship between volume flow and pressure drop. Reduction in efficiency with increasing volume is shown on the curves (adapted from Dartherm catalogue information).*

Parallel operation of two or more of any of the above configurations will enable two or more times the volume to be handled without increase in pressure drop.

B25.6 Air to air counter flow heat exchangers

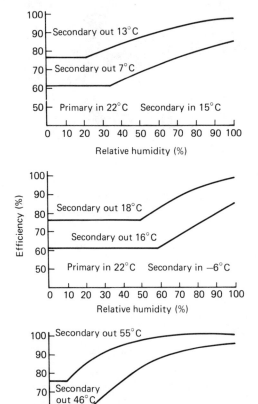

Figure **B25.6.1** *Air to air counter flow heat exchangers: effect of humidity on efficiency*
The temperatures shown on the graphs are secondary output temperatures, with primary input temperatures in parentheses

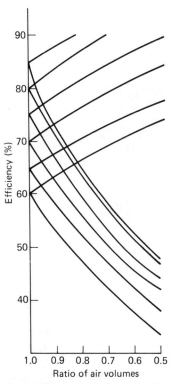

Figure **B25.6.2** *Air to air counter flow heat exchangers: relationship between air flow and efficiency*

When the secondary air volume is greater than the primary air volume the efficiency is low.
When the primary air volume is greater than the secondary air volume the efficiency is high.

(B25.6.1 and B25.6.2 are adapted from Heatex catalogue information supplied by A. Johnson and Co. Ltd.)

B26 Liquefied petroleum gas installations

The main characteristics of commercial propane and butane are given in **Table A6** but in order to make comparisons between these fuels, natural gas and fuel oils some conversions and calculations are necessary; these are best illustrated by case histories:

B26.1 Small scale propane installation

Tanks are supplied by LPG supplier and remain their property. Standard tanks are 380, 1200 and 2000 litres. The tank size designates the volume of *liquid* propane held [1960 litres per tonne (**A6**)].

At 15p per litre the cost per tonne is thus

$$\frac{15}{100} \times 1960 = £294$$

The calorific value is $95\,MJ/m^3$ (vaporized) and as there are $544\,m^3$ of gas per tonne (**A6**) the calorific value may be stated as

$$\frac{95 \times 544}{1000} = 51.7\,GJ/tonne$$

Hence cost per GJ supplied is

$$\frac{294}{51.7} = £5.69$$

B26.2 Industrial complex using 2000 tonnes per year

In this case bulk storage facilities are normally leased but can, of course, be owned by the consumer. A storage capacity of 100 tonnes would be needed with twice weekly deliveries during peak consumption periods. The capital element in the storage facilities must, of course, be quantified. Assuming a cost of £230 per tonne for bulk deliveries the price per GJ becomes

$$\frac{230}{51.7} = £4.45$$

B26.3 Bottle supplies

If used for operating fork lift trucks or for emergency heating it is normal for 15 kg bottles to be supplied at an inclusive charge which on a basis comparable with the above two examples could well be £7.70. Thus the cost per tonne would be

$$\frac{1000}{15} \times 7.7 = £513$$

and the cost per GJ

$$\frac{513}{51.7} = £9.90$$

In passing it may be observed that the 15 kg bottle represents

$$\frac{15}{1000} \times 51.7 = 0.78\,GJ \text{ or } 780\,MJ$$

By way of comparison with the above examples, natural gas at 33p per therm, represents

$$\frac{33}{0.1055 \times 100} = £3.13 \text{ per GJ}$$

and Class D oil at 18p per litre and $38\,MJ/l$ (**A7**)

$$\frac{0.18}{38} \quad £4.74 \text{ per GJ}$$

B26.3 Comparison of fuel costs

Fuel	Cost (£ per GJ)
Natural gas	3.13
Propane in tonnage supplies	4.45
Propane in 380–2000 l tanks	5.69
Class D oil	4.74
Propane in bottles	9.90

Electricity, depending on tariff, can vary from £10 to £17 per GJ. The comparisons are shown in **Table B26.3**. In each case adjustments must be made for conversion efficiencies and allowances made for convenience of use, particularly for portable or emergency heating equipment.

B27 Electric lighting

Lighting is a subject of great complexity and the design of lighting schemes requires specialized knowledge and consideration of the many factors which have a bearing on what is in the end a subjective reaction by people whether they be at work or at leisure; however, an appreciation of the terminology used, some knowledge of the major factors influencing design and up-to-date values of the performance of light sources must be the starting point for value judgements on the effective use of energy. **Table A39** gives some of the more important data on light sources and illumination levels; the notes below will perhaps help with utilization and interpretation of the data.

B27.1 Terms used

Sections A1.1 and **A1.4** define the basic and derived SI units associated with lighting and **A2.8** lists the major units and their definitions. Some other common terms are listed:

Glare Discomfort caused by excessive brightness of the light source, controlled by cutting off the light rays (as with louvred or recessed luminaires) so as to prevent the light source coming directly into the normal field of vision.

Colour The Munsell system classifies surface colours using scales of hue, value and 'chroma'.

Colour rendering For simplicity may be defined as the ability of a light source to produce colours similar to those seen in good daylight.

Spectral distribution The colour composition of light as measured by a series of bands of wavelengths (frequencies) over the visible range (*see* **Charts at B27.6***).

Visible range The human eye is sensitive to wavelengths from 300 to 760 nanometer (*see* **Chart A40**). These wavelengths start from the violet end of the spectrum and go through blue, green and yellow to red.

Luminaire The term which has succeeded 'lighting fitting'.

Light ratio There are a number of ratios which are broadly associated with the proportion of light falling directly onto the working plane and the extent to which the full output of the lamp is utilized or dissipated as a result of the design of the luminaire.

Room index A measure of the effect of room shape on a lighting installation.

Working plane and mounting height These terms speak for themselves and lead to spacing – height ratios.

Utilization factor The total lighting flux reaching the working plane divided by the total lamp flux.

Maintenance factor A measure of the extent to which light output from a lamp/luminaire will be lost in practice because of accumulation of dirt.

Light loss factor A measure of deterioration of lamp/luminaire with age.

Standard service illuminance This is the value measured in lux at the working plane of the recommended illuminance for a particular task; this value may be modified in the light of actual working conditions such as contrasts, duration of task, influence of windows and consequences of operator error. Some general levels are given in **Table A39** and comprehensive information is available in the CIBS code for interior lighting.

General lighting and task lighting These are self-defining.

*Reproduced by courtesy of Thorn Lighting

Correlated colour temperatures The temperature (K) of a full radiator which emits radiation having a chromaticity nearest to that of the light source being considered. The lowest values (around 2100 K for sodium discharge lamps) emphasize the yellow strongly and the highest (around 6500 K for colour matching fluorescent) come closest to north skylight.

Efficacy This is referred to in **A2.8**; in lighting parlance two values are used: *luminous efficacy* relates lumen output to watts consumed by the lamp and *circuit efficacy* takes count also of the watts consumed by control geat.

Lumen outputs It is usual to distinguish between *initial lumens*, i.e. output available from a new lamp up to 100 hours running and *lighting design lumens* which is based on expected output after 2000 hours and is the output used for design of lighting schemes.

Fluorescence The effectiveness of fluorescent tubes depends on the ability of the phosphors with which the interior of the tube is coated to intercept ultraviolet radiation (around 250 nm) and transform it to visible light. Triphosphors are now used to improve the efficacy of krypton filled lamps and to enable the same output to be achieved for colour temperatures from 3000 K to 4100 K.

B27.2 The major factors in lighting design

A very much simplified list of factors to be considered is as follows:

(1) Luminance required at working surface (determined by nature and complexity of task).

(2) Colour rendering necessary; in many situations this may be purely subjective but indices and colour rendering groups are available for detailed consideration.

(3) The need to avoid glare.

(4) Mounting height available; the greater the brightness of the source (or size of the lamp in terms of luminous output) the higher the mounting height should be if glare is to be avoided, and if full spread is to be delivered.

(5) Decorative condition and colour of surfaces in the area to be lit.

(6) Whether entirely general lighting, entirely task lighting or a combination of both.

(7) Daylight availability and its influence on the lighting scheme.

(8) From the first cost viewpoint and from the maintenance requirements the smaller the number of luminaires the better.

(9) A compromise must be found between either frequent cleaning and regular re-lamping (which will ensure that lighting level remains as designed) or a higher initial level of lighting (which will enable less frequent maintenance to be tolerated).

(10) Development of improved light sources is going on continuously with particular reference to energy saving; manufacturers should be consulted.

(11) Attention must be paid to power factor correction, most fluorescent luminaires have correction capacitors included, but this should be checked.

(12) Method of starting fluorescents should be checked; in general krypton filled tubes should be used with switch start circuits, but where starterless circuits are in use (as with replacements) argon filled tubes are needed.

(13) Make allowance for run-up time and re-strike time when planning mercury or sodium discharge lighting.

(14) Overlighting by unnecessarily generous allowances leads to a waste of energy.

(15) Circuit planning should be such that switching is related to proximity of daylight and wherever possible to variations in need of lighting during normal working times.

(16) Study carefully the claims of sophisticated lighting control systems and relate their cost to accurate assessments of potential energy savings.

B27.3 Light sources

(1) Tungsten filament lamps have such low efficacy compared with discharge lighting that their use can be justified only when low first cost is paramount and use is occasional or in situations where decor is the major factor.

(2) Tungsten–halogen and mercury blended lamps have efficacies only slightly higher than tungsten filament and are therefore best reserved for specialized applications.

(3) The efficacy of mercury fluorescent lamps is now among the lowest for discharge lighting but their colour properties make them attractive for certain applications.

(4) Tubular fluorescent lighting is now used in the complete spectrum of lighting applications and is surpassed only by sodium in efficacy. Krypton filling and the use of triphosphors for coating have boosted the efficacy of the tubes and these, apart from the 2400 mm, are available in 26 mm diameter.

(5) Metal halide offers high intensity light sources (up to 2000 watt) having colour and efficacy comparable with tubular fluorescent.

(6) High pressure sodium offers very big advantages in light output and enables substantial energy savings to be made where adequate mounting height can be achieved. The predominance of yellow colour can be a disadvantage but recent developments on the arc tube (SONDL) have produced a colour rendering which is much closer to white light.

(7) Low pressure sodium lamps offer low wattage units with the highest efficacy of all, but are suitable only where their predominantly yellow colour is acceptable.

B27.4 References for more detailed information

For closer study and a better appreciation of the implications of all the aspects of lighting design, reference may be made to:

The CIBS Code for Interior Lighting.
The Lighting Industry Federation Lamp Guide.
Thorn Lighting Technical Handbook and Comprehensive Catalogue.
Osram – G.E.C. Lighting Catalogue.

B27.5 Calculation procedures

B27.5.1 Alternative lighting layouts in medium engineering shops

The diagramatic plan at **B27.7** shows four different lighting layouts in the same factory taken from an actual case study and provides an interesting comparison. All the shops were 'high bay' and therefore without restriction on mounting height.

Analysis of lighting levels and electricity consumption yields the following:

Bay 4 (400 W SON)
Electricity consumption

$$\frac{26 \times 400 \,(\text{W})}{1000} = 10.4 \,\text{kW}$$

Design lighting level

$$\frac{26 \times 45\,000 \,(\text{lumen})}{61 \times 18.3 \,(\text{m}^2)} = 1048 \,\text{lux}$$

Allowing a maintenance factor of 80% and a utilization factor of 90%, the probable lighting level at the working plane becomes

$$1048 \times 0.8 \times 0.9 = 750 \,\text{lux}$$

Consumption per m^2

$$\frac{10.4 \times 100}{61 \times 18.3} \times 1000 = 9.3 \,\text{W}$$

Bay 3 North (700 W MBF)
Electricity consumption

$$\frac{12 \times 700}{1000} = 8.4 \, \text{kW}$$

Design lighting level

$$\frac{12 \times 38\,000}{30.5 \times 18.3} = 816 \, \text{lux}$$

Allowing the same maintenance and utilization levels, the lighting level at the working plane is

$$816 \times 0.8 \times 0.9 = 590 \, \text{lux}$$

Consumption per m^2

$$\frac{8.4 \times 1000}{30.5 \times 18.3} = 15 \, \text{W}$$

Bays 1 and 2 (fluorescent)
Electricity consumption

$$\frac{144 \times 138}{1000} = 19.9 \, \text{kW}$$

Design lighting level (8400 lumen for 3600 K tubes)

$$\frac{144 \times 8400}{41 \times 46} = 641 \, \text{lux}$$

As decorative condition of the shop was good and layout well planned, allow 95% utilization factor giving probable level at working plane

$$641 \times 0.8 \times 0.95 = 490 \, \text{lux}$$

Consumption per m^2

$$\frac{19.9 \times 1000}{41 \times 46} = 10.6 \, \text{W}$$

Bay 3 South (125 W MBF)
Electricity consumption

$$40 \times 125 = 5 \, \text{kW}$$

These fittings had deteriorated and decorative condition of the shop was poor, so allow 75% maintenance factor and 80% utilization factor, hence lighting level at working plane

$$\frac{40 \times 5800}{30.5 \times 18.3} \times 0.75 \times 0.80 = 250 \, \text{lux}$$

Consumption per m^2

$$\frac{5 \times 1000}{30.5 \times 18.3} = 9 \, \text{W}$$

The advantages of SON lighting is clearcut. The order of merit of the forms of lighting is shown in **Table B27.5.1 (a)**.

An interesting comparison can be made by substituting the characteristics of krypton filled triphosphor fluorescents in Bays 1 and 2 (in practice this would involve new gear and probably new luminaires):

Consumption

$$\frac{144 \times 112}{41 \times 46} = 16.1\,\text{kW}$$

Consumption per m^2

$$\frac{16.1 \times 1000}{41 \times 46} = 8.5\,\text{W}$$

The all-up cost of electricity including maximum demand change and fuel surcharge was 3.85p per kWhr. Assuming 2000 hr running per year the cost of lighting the different bays was as shown in **Table B27.5.1 (b)**.

B27.5.1(a) Order of merit of four frames of lighting

Lighting	Illuminance (lux)	Consumption (W/m^2)
SON	750	9.3
Fluorescent	490	10.6
MBF 700 W	590	15.0
MBF 125 W	250	9.0

B27.5.1(b) Lighting costs for the factory described

Bay	Calculation	Cost £/yr	Cost £/m^2 yr
4	$\dfrac{10.4 \times 2000 \times 3.85}{100}$	800	0.71
3N	$\dfrac{8.4 \times 2000 \times 3.85}{100}$	650	1.16
1 and 2	$\dfrac{19.9 \times 2000 \times 3.85}{100}$	1530	0.81
35	$\dfrac{5 \times 2000 \times 3.85}{100}$	385	0.69

In assessing the economics of alternative lighting schemes it must be remembered that a reduction in the kW lighting load is very likely to bring about a reduction in winter maximum demand.

B27.5.2 Lighting in a large office complex

As an example of the very high running cost of an extravagant lighting scheme some details of an actual case history are quoted in **Table B27.5.2**.

The consumption of electricity per m^2 per annum is

$$\frac{4140}{5000} = 0.83\,\text{GJ}$$

This figure excludes of course fuel for the heating system which consumed 2850 GJ per annum.

The lighting was based on a number of continuous lines of luminaires recessed into the ceiling and utilizing 1200 mm argon filled warm white tubes. The building could be broken down into nine metre square modules each containing three lines of fittings and each line being triple tube. There were thus nine tubes across the width of the module and approximately seven along the length.

The 1200 mm warm white tube has a 2000 hours output of 2700 lumens so that the potential illuminance from the lamps is

$$\frac{9 \times 7 \times 2700}{9 \times 9} = 2100 \text{ lux}$$

As the luminaires were recessed into the ceiling behind prismatic panels the lighting level at the working plane was around 1500 lux. Other reasons for the very high consumption of the lighting system were:

(1) The architecture of the building had provided for uniform continuous lines of lighting regardless of need, with the result that corridors and vacant areas were all lit to the same very high standard.
(2) Block switching had been installed so that there was no way of switching off unwanted lights.
(3) No provision had been made to switch off lines of lighting adjacent to windows so as to take advantage of daylight.

The annual energy bill for the building was £61 000, of which £49 000 was for electricity and £25 000 for lighting alone. It is of interest to consider the economy which could result from a more realistic general level of lighting, say 750 lux and the use of maximum efficacy

B27.5.2 Data for lighting large office complex

Total annual electricty consumption (lighting, air conditioning, canteen, general services)	1150 000 kWhr (4140 GJ)
Consumption for lighting alone	570 000 kWhr (2050 GJ)
Winter maximum demand	360 kVA
Maximum demand for lighting alone	150 kVA
Total floor area	5000 m^2

fluorescent fittings with glare control, but without the inevitable loss from recessed tubes behind prismatic panels. A similar style could be achieved with three lines of three single 2400 krypton filled triphosphor tubes warm white with louvred luminaires.

With an 85% utilization factor the 2000 hour output from these on the nine m^2 modules would be

$$\frac{3 \times 3 \times 8900}{81} \times 0.85 = 840 \text{ lux}$$

The consumption would be $9 \times 100 = 900$ W compared with $9 \times 7 \times 40 = 2520$ W for the recessed lighting based on 1200 mm tubes.

Some estimation of the annual saving can be made by relating the total for the module to the total annual consumption for lighting, i.e.

New consumption

$$\frac{900}{2500} \times 2050 = 730 \text{ GJ}$$

Annual saving

$$2050 - 730 = 1320 \text{ GT}$$

At £11.50 per GJ, the annual saving would be £15 000

Pleasing lighting has long been an architectural feature, but with today's energy costs there is a clear need for a hard practical outlook in designing lighting schemes for commercial premises.

B27.6 Spectral distribution charts

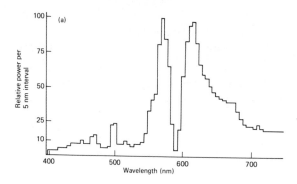

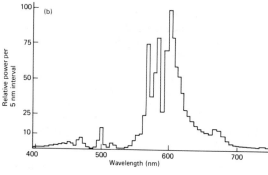

Figure B27.6.1 *Spectral distribution charts for high pressure sodium lamps*
(a) SON lamp with improved colour rendering
(b) Standard SON lamp

The upper chart (a) shows that the 570-620 nm bands are a relatively smaller proportion of the total power than with standard SON (b); this results in improved colour rendering.
Both charts show relative power, with the predominant 5 nm band taken as 100

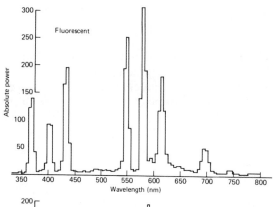

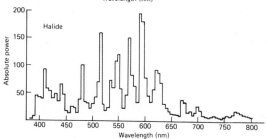

Figure B27.6.2 *Spectral distribution charts for mercury flourescent MBF lamps (upper) and mercury halide (MBIF) lamps (lower), showing the higher power of the fluorescent lamps and the better colour rendering of the halide lamps*
Absolute power is quoted in mW per 5 nm wavelength interval per 1000 lumen

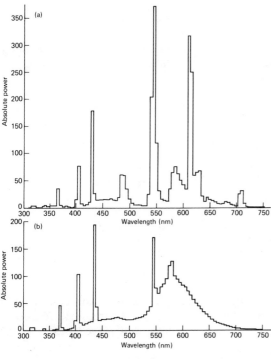

Figure B27.6.3 *Spectral distribution charts for krypton filled 1500mm 58W 26mm diameter fluorescent tubes, showing improved colour rendering and higher output of tubes using triphosphors*

(a) 3500 K with triphosphors
(b) 3500 K (standard)

The absolute power is quoted in mW per 5 nm wavelength interval per 1000 lumen and is the same for both charts

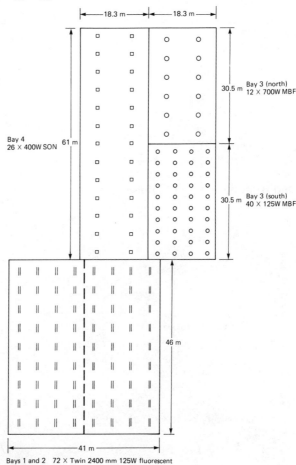

Figure B27.7 *Alternative lighting layouts in medium engineering factory*

Bays 1 and 2   72 × Twin 2400 mm 125W fluorescent

B28 Some basic concepts of electrical power systems

Avoiding deep involvement in electrical theory there are certain basic concepts which need to be understood in order to carry out straightforward calculations.

B28.1 The three phase system

This is best represented by the vector diagram showing a 'star' or 'wy' connection.

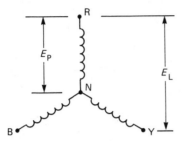

R, Y and B are the terminals of the red, yellow and blue phases of the system, N is the neutral point, E_L the line voltage and E_p the phase voltage. (The diagram can represent either the output from a supply transformer, or the windings of a three phase electric motor.)

The geometry of this diagram can be broken down as below:

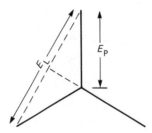

and from this a single right-angled triangle

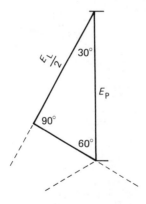

Since $\cos 30° = 0.866$ (or $\sqrt{3}/2$)

$$\frac{E_L/2}{E_p} = \sqrt{3} \quad \text{or} \quad E_L = \sqrt{3}E_p$$

The line voltage is thus $\sqrt{3} \times$ phase voltage. With star connection the current in the line will be the same as the current through the phase winding.

If we now consider a delta or D connection, it is clear from the diagram that the voltage across the line will be the same as the voltage across a phase winding:

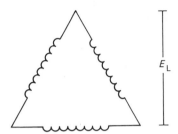

However, it can readily be shown from the geometry of the diagram that the current in the line will be greater than the current in the phase winding with the relationship

$$I_L = \sqrt{3} \times I_p$$

Thus for either star or delta connection the apparent power supplied is

Line volts $\times$ Line ampers $\times \sqrt{3}$ ($\sqrt{3}EI$)

B28.2 *Domestic and industrial supply systems*

There are still many different voltage and frequency systems in use for electricity supplies world wide, but these are gradually approaching three widely used standards, i.e.

UK   3-phase, 50 Hz, 415 volts
Much of Continental Europe   3-phase, 50 Hz, 380 volts
North America   3-phase, 60 Hz, 230–460 volts

With the UK and European systems the domestic voltage is phase to neutral on the 'star' connection, i.e.

$$\frac{415}{\sqrt{3}} = 240 \text{ or } \frac{380}{\sqrt{3}} = 220$$

In North America a series–parallel connection is used so that either 460 volt 3-phase or 230 volt 3-phase is available depending on the power demand on the system. The single phase 230 volt system is frequently again broken down by series–parallel connection to 115 volts.

Electric motors can often be used on either 380 volt 50 Hz 3-phase or 460 volt 60 Hz 3-phase, but on the 60 Hz supply they will run at a higher speed; to a lesser extent the same motor may be used on 415 volt 50 Hz as for 460 volt 60 Hz. Interchangeability between 415 volt 50 Hz and 380 volt 50 Hz involves some sacrifice in performance.

As far as single phase motors are concerned a compromise design can be available to cover from 220 volts 50 Hz to 230 volts 60 Hz.

For optimum performance and highest efficiency and power factor all electric motors should be designed for the system on which they will operate. For large indusial loads it is usual for the electricity authority to provide the supply at high voltage (11 000 volts is standard for UK). In these circumstances the consumer is responsible for transformers to provide distribution at 415 volts 3-phase, and there is invariably a special tariff for such 'bulk supply' loads.

B28.3 *The effects of inductive loads on power factor*

Active, Reactive and Apparent Power are defined in **A2.7**.

A resistive load is one where there is no self inductance in the circuit and where the power in watts is equal to volts × amperes ($W = EI$).

A straight length of resistance wire will show such a relationship. Self inductance results from a coil of wire and such circuits are said to be inductive. Electric motors are good examples where the windings are made up of coils and where the presence of iron or steel intensifies the inductance of the circuit. Inductive circuits have currents which lag behind the voltage, whereas capacitative circuits have leading currents.

95% of industrial loads will be inductive and therefore have a lagging power factor; power factor correction is dealt with in **B30** and at this stage it suffices to say that lagging power factor can be corrected by placing capacitors in the circuit.

The concept of current lagging or leading the voltage can conveniently be by-passed at this point and the diagram for power in **A2.7** used:

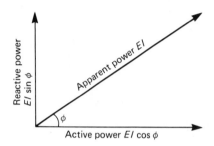

In this diagram we have a lagging power factor and this can be corrected by injecting leading reactive power from capacitors so as to reduce the angle and also reduce apparent power (*see* **B30**).

B29 Electricity maximum demand and its control

Electricity charges, their monitoring and control have been referred to in **B3** and **B4**; the table at **B29.1** is intended to give a perspective on the impact of maximum demand charges with particular reference to large industrial consumers. Electricity tariffs have become very complex, but in general their objectives are to discourage high demand in winter and at peak load periods but to encourage the use of electricity at night time; one of the major objectives of tariffs is to improve the load factor on electricity distribution networks.

In practice the tariffs tend rather to penalize the consumer by high prices at peak times as most industrial users have little choice but to use their electricity during normal working hours.

The impact of high winter demand charge is illustrated by **Table B29.1** which shows the extent to which the average cost per kWhr is boosted by MD charges. Industrial consumers normally consume more electricity during cold weather and it is likely that their peak demand will occur twice during the working day, once at around 9.30 hrs and again at between 15.00 and 15.30 hrs. Although substantial savings can be made by lopping these peaks the savings as a proportion of the total electricity bill are surprisingly small. From the figures at **B29.1** an estimate may be made of the savings which can be achieved by reducing maximum demand *without making any reduction in total consumption*.

If, for example the MD in January, February and December were held at 3000 kVA the annual savings would be:

Month by month (Jan to Dec)
  £67.5 × 12   =   £810

Jan and Feb winter surcharge
$2 \times 150 \times .£4.86$ = £1486
December winter surcharge
$100 \times £4.86$ = £486
Total   £2754

A worthwhile sum but only 0.8% of the total annual bill.

Holding demand down to 2850 kVA would produce savings of £5750 or 1.64% of the annual total.

For the industrial consumer the real reductions in charges must come from

(1) Changing major parts of the process to night shift so as to take advantage of the radically lower unit charge, remembering that 'night' does not start until midnight and usually finishes at 6.30 a.m.

B29.1 Electricity costs for a large industrial consumer

Mth	Day kWhr	Day p/kWhr	Day £	Night kWhr	Night p/kWhr	Night £	Maximum demand kVA (month)	Maximum demand kVA (year)	Monthly Charge (£)	Monthly Cost (£)	Winter Charge (£)	Winter Cost (£)	Fuel surcharge (£)	Total (£)
JAN	650 000	3.57	23 210	100 000	1.52	1520	3150	3150	see	1618	4.86	15 310	1800	43 458
FEB	650 000	3.57	23 210	100 000	1.52	1520	3150	3150	note	1618	4.86	15 310	1800	43 458
MAR	600 000	3.57	21 420	100 000	1.52	1520	3000	3150	(2)	1618	1.30	3900	1650	30 108
APR	550 000	3.57	19 640	80 000	1.52	1220	2900	3150		1618	—	—	1500	23 978
MAY	560 000	3.57	18 560	80 000	1.52	1220	2800	3150		1618	—	—	1400	22 798
JUN	520 000	3.57	18 560	80 000	1.52	1220	2700	3150		1618	—	—	1400	22 798
JUL	520 000	3.57	18 560	80 000	1.52	1220	2700	3150		1618	—	—	1400	22 798
AUG	450 000	3.57	16 070	60 000	1.52	910	2500	3150		1618	—	—	1250	19 848
SEP	520 000	3.57	18 560	80 000	1.52	1220	2800	3150		1618	—	—	1450	22 848
OCT	600 000	3.57	21 420	100 000	1.52	1520	2900	3150		1618	—	—	1700	26 258
NOV	650 000	3.57	23 210	100 000	1.52	1520	3000	3150		1618	1.30	3900	1850	32 098
DEC	550 000	3.57	19640	80 000	1.52	1220	3100	3150		1618	4.86	15 070	1650	39 198
Totals	6 780 000		£242 060	1 040 000		£15 850				£19 416		£53 490	£18 850	£349 646

Costs per kWhr and per GJ including maximum demand and fuel surcharges

	All-up p/kWhr	All-up £/GJ	Excluding night units p/kWhr	Excluding night units £/GJ
Average through year	4.47	12.4	4.92	13.7
Average April to October	3.80	10.6	4.15	11.5
Average March to November	4.29	11.9	4.73	13.1
Average January, February and December	6.11	17.0	6.83	19.0

Notes
(1) Consumption and demand figures derived from a case study but rounded and related to 1982/83 high voltage maximum demand tariff.
(2) Monthly demand charge: 1st 250 kVA, £0.68; 251–500 kVA, £0.54; 501–2000 kVA, £0.51; balance, £0.51. Once a peak demand is reached it applies for twelve months. Thus the January figure of 3150 kVA affects February to December.
(3) Fuel surcharge estimated for a year in which no change was made in tariff.

(2) Ruthlessly eliminating all use of electricity for heating whenever alternative fuels can be used. The heating load is almost certain to be in midwinter during peak demand periods and it therefore carries the extremely high marginal cost of up to £19 per GJ.

Changing to night shift is certainly not easy, involving as it does the problems of supervision, premium rates and integration of production requirements, but the eradication of expensive electrical heating must be a priority.

Reference is made to maximum demand control systems in **B33.4** and some of the specialist companies in this field are listed in the Manufacturers' Directory on page 229. The maximum demand meter as normally installed in industrial or commercial premises integrates the actual load over a series of short periods during each half hour and if, at the end of the half hour, the average load has reached a new peak then this peak will be recorded. Control instruments follow the same principle but anticipate the supply company's meter by signalling if after 15 minutes a new peak seems likely; this enables avoiding action to be taken either by manual load shedding or by automatic shut-down of selected processes. The need for extreme care in shutting down plant in an industrial complex is self-evident and the cost penalty of such shut-downs must be weighed against the saving which can be achieved.

B30 Calculation of power factor and the value of capacitors for power factor correction

Where a maximum demand tariff is used for medium or large supplies, it is normal for meters to be installed to read kWhr, kVAhr and kVARhr. Reference to **A2.7** enables these three measurements to be related to active, apparent and reactive energy supplied to the consumer. Both kWhr and kVARhr meters normally have rotating discs which can be timed to give an estimate of power factor at any particular time.

In a particular example the kVAhr meter was observed to complete nine revolutions in 220 seconds while the kWhr meter completed 18 revolutions in 223 seconds; on both meters each revolution represented 1 unit (kVAhr or kWhr). The active power is therefore:

$$\frac{3600 \text{ (seconds per hour)} \times 18}{223} = 291 \text{ kW}$$

and the reactive power:

$$\frac{3600 \times 9}{220} = 147 \text{ kVA}$$

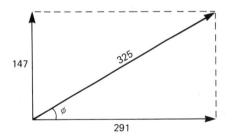

From the vector diagram apparent power is 325 kVA ($\sqrt{(291)^2 + (147)^2}$)

The power factor cos ϕ = $\dfrac{291}{325}$ = 0.90

Reading the meters in this way gives the actual load on the system and the power factor at the time the readings are taken.

A longer term average power factor may be obtained by taking readings of both meters on the digital or dial counters at intervals of say one hour during periods of maximum load; alternatively they may be taken daily or weekly to establish a pattern. Power factor can then be calculated by taking the differences of the two readings and applying the vector diagram as above. In these days of high maximum demand charges based on kVA, it is frequently economic to improve power factor to 0.96 or 0.97.

To calculate the value of capacitors in the above example first establish the value of apparent power at 0.97, i.e.

$$\text{kVA} = \frac{\text{kW}}{0.97}$$

i.e.

$$\frac{291}{0.97} = 300$$

The reactive power is thus

$$\sqrt{300^2 - 291^2} = 73\,\text{kVAr}$$

To correct from 0.90 to 0.97 the kVA of capacitors required is therefore

$$147 - 73 = 74\,\text{kVA}$$

and the revised vector diagram

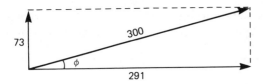

As an example of the economics of power factor correction the above figures may be related to an electricity tariff which charges £0.84 per month per kVA of maximum demand throughout the year *plus* £1.30 in addition for November and March and £4.86 in addition for December/January/February.

Assuming that power factor and maximum demand remain constant through the year the additional cost per kVA is

$$(0.84 \times 12) + (1.30 \times 2) + (4.86 \times 3) = £27.26$$

In the example above the annual saving is therefore

$$74 \times £27.26 = £2017$$

The annual saving must be set against the cost of installing capacitors and it will usually be found that improvement to 0.96 or 0.97 is fully justified.

B31 Applications of electric motors

As by far the majority of electric motor drives in industry are three-phase squirrel cage induction motors, an appreciation of the effect of these motors on the efficient use of energy is clearly essential. The performance details of a typical standard range of totally enclosed fan cooled motors are given at **A38**.

Some of the main factors are:

(1) Four pole speed (1420–1480 rev/min on 50 Hz, 1700–1770 rev/min on 60 Hz) is by far the most frequently used for motors up to 100 kW output.

(2) The bulk of standard production of such motors is geared to low first cost and as a consequence there is some sacrifice in optimum efficiency and a considerable sacrifice in optimum power factor. As can be seen from the performance figures the tail-off between full load and half load efficiency figures is quite sharp on the smaller sizes and the tail-off in power factor is even more noticable.

(3) Performance figures will vary considerably between one manufacturer and another, but the overall pattern will be similar when comparing standard ranges. Better figures can be achieved from special designs but at an inevitable penalty on first cost.

(4) Throughout industry and indeed in most electric motor applications there is a tendency to play safe on motor rating to avoid the risk of motor failure owing to overload; the result is that the proportion of motors developing their full output continuously is very small and consequently motors tend to operate below optimum efficiency and well below optimum power factor.

(5) On large motors (above say 100 kW) it is generally accepted practice to evaluate the advantages of high efficiency designs particularly where loading conditions are known with accuracy; however, for the bulk of smaller motors, life cycle cost is rarely considered.

(6) In most industrial situations the total installed power in electric motors is very much greater than the normal load on the supply system; this is, of course, due to intermittent loading and the margin of motor design output compared with the actual motor load. In a jobbing shop where many machine tools spend a high proportion of their life idle, the ratio of installed kW of motors to the kW drawn from the supply can be as high as ten to one.

All the above factors tend to create a situation where overall power factor and average efficiency are both unnecessarily low with consequent high electricity charges and waste of energy.

B28.3, B29 and **B30** deal with the problems of low power factor and maximum demand control but it is instructive to examine three specific examples:

B31.1

Actual power required	1.1 kW at 1440 r.p.m.
Cost of electricity (including standard, maximum demand and fuel surcharges)	4p per kWhr
Marginal cost per kVA per annum for additional maximum demand(**B30**)	£27

Three alternatives will be considered, namely:

(a) Special motor with full load P.F. 0.85 and full load efficiency 85%; output 1.1 kW
(b) Standard motor with full load P.F. 0.75 and full load efficiency 75%; output 1.1 kW
(c) Standard motor designed for 2.2 kW output, but operating at half load P.F. 0.56; efficiency 72%

Each of the above three is evaluated for continuous running (8760 hours per year) or alternatively day shift operation (2400 hr per year).

The operating costs of (a) will be

$$\frac{1.1 \text{ (kW output)} \times 4 \text{ (p per kWhr)}}{0.85 \text{ (efficiency)}} \times \frac{8760 \text{ (hr per year)}}{100 \text{ (p to £)}}$$

= £453 per year for unit costs

(£124 per year for 2400 hr)

This same motor will have a kVA demand of

$$\frac{1.1}{0.85 \text{ (efficiency)} \times 0.85 \text{ (P.F.)}} = 1.52\,\text{kVA}$$

For (b) the unit cost will be

$$\frac{1.1 \times 4 \times 87.6}{0.75} = £514$$

(£141 per year for 2400 hr)

and the kVA demand will be

$$\frac{1.1}{0.75 \times 0.85} = 1.73\,\text{kVA}$$

This motor will therefore add to the MD charge by

$$(1.73 - 1.52) \times 27 = £5.7$$

For (c) the unit cost will be

$$\frac{1.1 \times 4 \times 87.6}{0.72} = £535$$

(£147 per year for 2400 hr)

and the demand will be

$$\frac{1.1}{0.56 \times 0.72} = 2.73\,\text{kVA}$$

The addition to MD charge will be

$$(2.73 - 1.52) \times 27 = £32.7$$

For continuous operation the special 1.1 kW motor will save £67 per year but the underloaded 2.2 kW motor will add £68 to annual costs. The figures for 2400 hr are £23 saving and £31 addition to costs respectively.

A reasonable assumption with regard to first cost of the motors on a basis comparable with the electricity charges used would be

(a) or (c)   £100   (b)   £70

This example illustrates how motor annual running costs, particularly when operating continuously, can be far in excess of the first cost of the motor; it is also evident that the savings achieved by careful selection and application of motors can show payback periods of one year and sometimes very much less.

B31.2

Power required 7.5 kW; charges as **B31.1**.

Consider the alternatives of a fully loaded 7.5 kW motor and a half loaded 15 kW motor, both with performance details as **A38**.

(a)   7.5 kW   P.F. 0.82   efficiency 86%
Annual running cost (8760 hr)

$$\frac{7.5}{0.86} \times 4 \times 87.6 = £3,056$$

(£837 per year for 2400 hr)

Maximum demand

$$\frac{7.5}{0.86 \times 0.82} = 10.6\,kVA$$

(b) 15.0 kW operating at 7.5 kW   P.F. 0.76   efficiency 85%
Annual running costs (8760) hr

$$\frac{7.5}{0.85} \times 4 \times 87.6 = £3091$$

(£847 per year for 2400 hr)

Maximum demand

$$\frac{7.5}{0.85 \times 0.76} = 11.6\,kVA$$

Additional maximum demand charge

$$(11.6 - 10.6) \times 27 = £27$$

First cost of the two motors may be assumed at £175 and £340.
In this case correct application can save £165 on first cost and £63 per annum for 8760 hr or £37 per annum for 2400 hr.
There is also a case for evaluating a higher cost motor with better performance.

B31.3

110 kW motor operating continuously at 0.90 P.F. and 93% efficiency.
Annual unit cost

$$\frac{110}{0.93 \times 0.90} \times 87.6 \times 4 = £46\,050$$

Contribution to MD because P.F. is less than unity:

$$\frac{110}{0.93 \times 0.90} - \frac{110}{0.93} = 131 - 118 = 13\,kVA$$

Annual cost of low P.F.

$$13 \times 27 = £35$$

Running cost at 94% efficiency = £45 560
Annual saving = £490
In general, larger motors are less at a disadvantage in P.F. and efficiency when underloaded, but as shown in the last example, when operation is for 3000 or more hours per year, one percentage point can make a substantial difference to annual running costs.
By far the most serious wastage occurs when there are a large number of underloaded small motors operating for long periods. In the majority of engineering factories there is at least one compressor operating long hours, and the compressor motor is often a good starting point when checking for effective use of electric energy (*see* **B18**).
Some general guidelines for energy saving in electric motor application:

(1) Annual running costs of motors operating continuously are far in excess of first cost.
(2) Accurate matching of motor output to power required gives maximum economy.
(3) Special attention should be paid to motors running throughout shift as they are certain to affect maximum demand.
(4) Low overall power factor is inevitable when large numbers of small motors are underloaded; in such cases capacitors must be used to raise P.F. to around 0.97 (*see* **B30**).
(5) *See* **B33.4** regarding thyristor control systems

B32 Electric heating

Electrical heating by means of 'elements' is familiar to all and it consists of selecting a suitable resistance for the heating element which will cause a current to flow from a given supply voltage so that the required loading in watts is achieved. There are however, other methods of heating by electrical energy all of which are covered by the general term 'Electroheat' and in order to appreciate the means of applying these other methods some background of alternating current theory is needed. This section is therefore presented in three parts covering theory, method and application.

B32.1 Some simple concepts of alternating current theory

Section **B28** covers the essential practical aspects of AC systems at power frequencies (50 Hz or 60 Hz) but variation of frequency is an essential part of radio and communication circuits and frequencies other than 50 or 60 Hz are needed if introduction or dielectric heating is to be applied. Active and reactive power and the introduction of capacitors for power factor correction are covered by **B29** and **B30**; the notes which follow are intended to show how the effect on a circuit of both inductive and capacitative load *varies with frequency* and how resonance can be achieved by means of this variation.

By its very nature AC power comes from a rotating machine and the first step is the generation of a sine wave of voltage by rotation of the voltage vector, as shown in **B32.1(a)**.

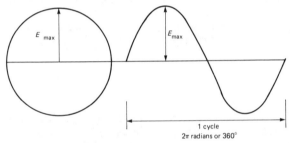

Figure B32.1(a) *Generation of a.c. sinusoidal waveform*

As the vector E_{max} rotates its height above or below the axis will determine the instantaneous voltage which is a maximum positive when the vector is vertical above the axis and maximum when vertical below. It can readily be seen that this will generate the sine curve on the right with complete rotation (2π radians or 360°) producing one cycle.

If the circuit consists of resistance only, the power factor will be unity and the current will be 'in phase' with the voltage.

Power in a resistance-only circuit is volts × amperes = watts and energy is produced in both the positive and negative parts of the cycle. This is because heating will take place

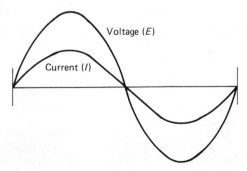

Figure B32.1(b) *Generation of 3-phase waveform*

regardless of the direction of current flow; it is also mathematically sound because if both voltage and current are negative and they are multiplied together the result will be positive. The power flow will thus be:

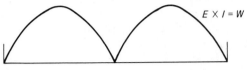

$E \times I = W$

Figure B32.1(c)

The average power in the circuit will be the average or mean height of the power curve and clearly it will be less than the peak power at the top of each wave.

From the definition of resistance $I = E/R$, and, power being $E \times I$, can be written as

Power $= I^2R$ or E^2/R

The instantaneous power thus varies as the square of the instantaneous voltage.

The effective average voltage as far as power in the circuit is concerned may therefore be stated as:

$$\sqrt{\text{Average of (instantaneous voltage)}^2}$$

or more simple Root–Mean–Square or RMS voltage.

The RMS voltage is that voltage which would produce the same heating effect with direct current instead of alternating current.

There is a straightforward mathematical method which establishes that the RMS value is equal to the peak value $\times$ sine 45° or $E_{max} \times 0.707$. Thus for a 240 volt single phase system

$$E_{max} = \frac{240}{0.707} = 340 \text{ approx.}$$

The vector and sine wave approach may be extended to a three phase system by considering three vectors 120° or $(2\pi/3)$ radians apart each generating a sine wave and separated in time by one third of the supply frequency, i.e. $\frac{1}{150}$ second for 50 Hz. finally the concept may be extended to inductive and capacitative circuits by introducing phase angle

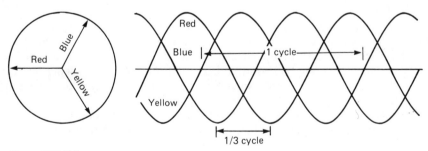

Figure B32.1(d)

or lagging/leading power factor as described in **B29** and **B30**. Without delving too deeply into electrical theory two basic relationships must be accepted which affect the behaviour of an inductive or capacitative load in a circuit of *varying* frequency, i.e.

Inductive reactance $(X_L) = \omega L$

Capacitative reactance $(X_C) = \dfrac{1}{\omega C}$

where L is the self inductance of a circuit (i.e. the characteristic causing lagging power factor) and is measured in henry. C is the capacitance of a circuit (i.e. the characteristic causing leading power factor) and is measured in farad. ω is angular frequency or the rate of rotation of the vector in radians per second (for 50 Hz, $\omega = 2\pi \times 50 = 314$ approx).

The impedance of a circuit (Z) is the nett effect of resistance, inductive reactance and capacitative reactance. The relationship between current and voltage in such a circuit is

Voltage = Current $\times$ Impedance

If we now consider all three elements in a single phase circuit we have:

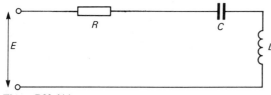

Figure B32.1(e)

As capacitative reactance causes lead and inductive reactance causes lag, the net effect on the total reactance of the circuit is

$$\omega L - \frac{1}{\omega C} \text{ or } \frac{1}{\omega C} - \omega L$$

If inductive reactance dominates we have

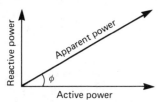

Figure B32.1(f)

with lagging power factor $\cos \phi$.

If capacitative reactance dominates we have

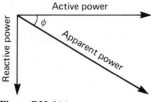

Figure B32.1(g)

with leading power factor $\cos \phi$.

Resonant circuits are those where $\omega L = 1/\omega C$ and where the power factor becomes unity. Resonance can be achieved in a given circuit by varying the frequency until the balance is achieved.

Very high voltages can exist across the capacitative and inductive parts of the circuit, but these voltages will cancel one another out at resonance and therefore not appear at the terminals.

The total impedance (Z) of the circuit is:

$$\sqrt{R^2 + \left(\omega L - \frac{1}{\omega C}\right)^2}$$

add the vector diagram:

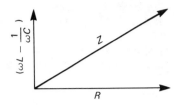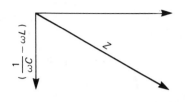

Figure B32.1(h)

and the full relationship between voltage, current and impedance becomes

$$E = \sqrt{R^2 + \left(\omega L - \frac{1}{\omega C}\right)^2} \times I \text{ or } E = Z \times I$$

B32.2 Methods of electrical heating

There are three basic methods of generating heat from electricity, i.e. Resistance, Inductive and Dielectric.

Resistance heating depends on passing a current through a resistance and applying the heat so generated by radiation, conduction, convection or a combination of these. Direct heating refers to passing the current through the material being heated and indirect heating applies to the use of electric heating elements. Resistance heating is normally confined to power frequencies (50 Hz or 60 Hz) and the practical effect of induction or capacity in the circuits is negligible.

Inductive heating is best appreciated by considering the heat which is generated in the laminated steel cores of electric motors or transformers. By using the workpiece or billet as the core in a specially designed winding, heat is produced directly where needed.

Dielectric heating. In a perfect inductance (unity power factor) there would be no heating effect; in the same way there would be no heating effect in a perfect capacitance (or condenser). In practice no capacitor produces a full 90° lead and there is in consequence some heating effect or dielectric loss. The principle is applied in dielectric heating and can be used to heat materials which are poor conductors of electricity.

In both inductive and dielectric heating a considerable range of frequencies is used and these are reviewed in the notes which follow. The spectrum of frequencies is illustrated in graphic form at **A40**. The complete electromagnetic spectrum divides fairly tidily into 'The Audible Range'*, 'Radio Frequencies' and 'Micro-Wave and Radar', 'From Infrared to Ultraviolet' and, for frequencies above around 10^{16} and up to 10^{32} Hz, 'X-rays, Gamma Rays and Cosmic Rays'.

The wavelength and frequency are related to one another by the velocity of light 299 793 km/s or for the sake of simplicity 300×10^6 m/s such that

$$\frac{300 \times 10^6}{\text{Wavelength (m)}} = \text{frequency (Hz)}$$

Thus 10 Hz has a wavelength of 30×10^6 m and
100 kHz has a wavelength of 3000 m

There are overlaps between the many bands in the complete spectrum and the divisions between the different classes of radiation are not hard and fast. Some relevant definitions:

Plasma This is of particular relevance to welding technology and is sometimes referred to as a fourth state of matter being neither solid, liquid nor gas. Plasma can be generated from solid or liquid, but is best described as the change of state of a gas resulting from the continuous injection of energy. It has been described as a 'soup' of ions, electrons,

---

*Caution. All electromagnetic waves travel in a vacuum at the same speed. We are talking here of audiofrequency *electromagnetic* waves, and not audiofrequency *sound* waves, which travel in air at the much lower speed of around 340 m/s, depending on conditions.

neutral atoms, molecules and electromagnetic radiation (or photons). As energy is injected into this 'soup' there is a continuous process of ionizing atoms, electrons recombining with ions to form neutral particles and photons being produced and absorbed.

Lasers The term laser is an acronym derived from 'Light Amplitude by Stimulated Emission of Radiation'; MASER (the subject of earlier work) was derived from 'Microwave (or Molecular) Amplitude by Stimulated Emission of Radiation'. Laser is concerned with the light spectrum and maser with the radio spectrum. The laser has by far the greater practical application and is in effect a very intense beam of coherent light of a very precise and accurately controlled frequency; this enables substantial power to be concentrated at the exact spot where needed. Apart from its implications for surgery the laser when based on carbon dioxide guns can produce power of the order of 30 kW for the precision cutting of metals.

Positive and negative ions Some comment is appropriate in view of recent work on the influence of ions in the atmosphere on the feeling of 'well-being' of a substantial proportion of people. In this context positive ions are the 'bad-boys' and are relatively persistent, being generated in particular by thunderstorms and being present in the less 'fresh' or heavy atmospheres. Negative ions, although short lived, have the ability to 'knock out' the positive ions and are generally found in country air, in the mountains and near waterfalls. Ionizers have been designed to generate negative ions in sufficient volume to suggest beneficial affects with a consumption of the order of 25–40 W. Positive ions have been linked with serotonin which is believed to influence migraine, high blood pressure and some other conditions.

Before reviewing the three methods of heating in more detail some comment on frequency bands is appropriate and reference may be made once again to **Chart A40**.

(1) All direct and indirect resistance heating is based on power frequencies, i.e. 50 or 60 Hz.
(2) Inductive (or induction) heating spreads from 2 Hz to 40 Hz.
(3) Radiofrequency heating uses the dielectric heating principle and overlaps the radio and TV bands starting from 70 kHz; in order to avoid radio interference two particular restricted bands are frequently used, i.e. 13.6 MHz and 27.2 MHz.
(4) Microwave heating (again using the dielectric principle) overlaps the RADAR spectrum with particular spot frequencies of 850 MHz and 1150 MHz.
(5) Infrared heating starts at 1000 GHz and goes on to overlap the visible spectrum (infrared heating can be generated from fossil fuels (*see* **B13.3**) and when applied electrically it is the result of high temperature electrical elements).

B32.2.1 Direct and indirect resistance heating

Indirect heating employs elements which in modern plant and domestic equipment are normally insulated; they are commonly made from nickel–chrome resistance wire held in the centre of metal tubes by compressed magnesium oxide. A variety of metal sheaths is available, i.e. copper, mild steel, monel, lead, nickel, inconel, incolloy, stainless steel and titanium.

Ratings vary from 2 to 8 W/cm^2 of surface area. Elements of this type can be made up into immersion heaters with ratings up to some hundreds of kW. They can also be used for air duct heating, and in fan assisted heaters. At appropriate temperatures and with suitable reflectors they are employed as infrared heaters. Any indirect heating system needs to have quite specific application advantages if it is to overcome the very high relative cost of electricity as a heating fuel. Cleanliness and ease of control can be factors in favour.

Direct heating uses the material to be heated as part of the electrical circuit and offers the advantages of very rapid heating (1.5 minutes to 1200 °C for 2 m steel bar 50 mm diameter is typical) and minimum scale formation. Very high currents can be involved necessitating special transformers so this method can be applied only to production processes after close study of the particular application. an interesting case is the use of current carrying steel cables encased in concrete road or runway surface to prevent icing in winter conditions. The same system has been applied to heating buildings, but it suffers from the inherent disadvantage of high relative running cost.

B32.2.2 Induction heating

In induction motors and transformers laminated sheet steel is used as the core, that is to say the flux path between the stator and rotor in the motor or between the primary and secondary windings of the transformer. This laminated steel is heated by alternating current because of eddy currents resulting from electromagnetic induction; the objective of the designer is to minimize this heat loss, but in induction heating the heat generated is put to useful purpose by using the work piece as part of the electromagnetic circuit. If a steel billet or workpiece is enclosed in a coil carrying a current, the workpiece will be heated and the interesting feature of this method is that there is a 'skin effect' which limits the penetration depth of the heating. This is extremely valuable in heat treatment where hardening to a limited depth of the workpiece is called for. The current density from the outer surface towards the centre decays according to an exponential law such that

$$I_x = I_0^{-(x/p)}$$

where
I_x is current density at distance x below surface
I_0 is current density at surface
p is penetration depth

The fundamental expression for penetration depth is

$$p = \sqrt{\frac{\rho}{\pi \mu_0 \mu_r f}}$$

where
ρ is resistivity (ohm m)
μ_0 is the magnetic permeability of a vacuum, which has the value $4\pi \times 10^7$ henry/metre $(1.26 \times 10^6 \, \text{H/m})$
μ_r is relative permeability of the conductor material
f is frequency

The expression can be written as

$$p = \frac{1}{2\pi} \sqrt{\frac{\rho}{\mu_r f}}$$

when resitivity is taken in units (ohm cm) $\times 10^9$, and the depth is now in *centimetre*, or

$$p = \frac{10}{2\pi} \sqrt{\frac{10^5 \rho}{\mu_r f}} = 503 \sqrt{\frac{\rho}{\mu_r f}}$$

when resistivity is in microhm m, and the depth is now in *millimetre*.

Nearly all applications of induction heating to magnetic materials involve temperatures beyond the CURIE point, so it is usual to take the ratio μ_r as 1.0. (The curie point is the temperature above which materials which are magnetic at low temperatures cease to be so; for most steels this is in excess of 700 °C).

In practical applications the penetration depth should not exceed half the radius of the workpiece; the minimum frequency for satisfactory penetration can therefore be obtained by substituting $r/2$ for p

$$\text{Minimum frequency} = 1.013 \times \frac{\rho}{r^2 \mu_r}$$

Values of resistivity are given in **Table A26** and taking as examples stainless steel at 0.72 microhm m and aluminium alloy (380 sc 84B) at 0.075 the minimum frequencies are as in **Table B32.2.2**. Note that, particularly for steel, ρ increases sharply with temperature thus increasing the penetration depth.

If induction heating is to be used for surface hardening where the penetration depth required is very much less the frequency required is correspondingly higher. The chart at **B32.5** gives the relationship between penetration depth and frequency for a range of values of resistivity (assuming $\mu_r = 1$). **Table A26.2** gives some values of the curie point.

Induction heating, by its very nature, leads to a low lagging power factor so that power factor correction capacitors are essential.

B32.2.2 Minimum frequencies for induction heating

Diameter (mm)	10	25	250
Stainless steel	29 kHz	4.7 kHz	47 Hz
Aluminium alloy	3 kHz	0.49 kHz	–

In summary, induction heating offers the advantages of cleanliness, rapid heating or melting, high efficiency, accurate temperature control, accurate location of heat, short heat treatment times, control of heat treatment depth and the facility if required to operate in a vacuum, but it does require specialized application, and is therefore mainly a volume production technique.

B32.2.3 Dielectric heating

Dielectric heating takes place in materials which are poor conductors of electricity, when these materials are place in high frequency electromagnetic fields. It depends on the capacitance effect of the electrodes connected to the high frequency supply and the amount of heat generated also depends on the characteristics of the material being heated, this is sometimes being referred to as the 'loss angle' of the material (or the extent to which the capacitance effect falls short of 90° lead).

The loss in materials subjected to high frequency heating is heavily dependent on moisture content and is also affected by the extent of compaction of the material.

The theory behind dielectric heating is quite complex, but the basic equation for rate of heating in watt per m^3 is

$$2\pi \times f \times \varepsilon_o \times \varepsilon_r'' \times E^2$$

where
f is frequency (Hz)
ε_o is permittivity of a vacuum (8.9×10^{-12} F/m)
ε_r'' is dielectric loss factor of the material
E is peak electric field strength (V/m)

To put an order of magnitude to this equation we may take 100 kHz as a starting point for frequency and 10 kV/m as a starting point for field strength. Moist paper would have a dielectric loss factor of say 1.5, hence

$$\text{Heating rate} = 2\pi \times 100\,000 \times 8.9 \times 10^{-12} \times (10\,000)^2$$
$$= 840 \text{ watts/m}^3$$

Dielectric heating is less prone to the 'skin effect' characteristic of induction heating and for radio frequency bands penetration depth is not a limitation; however, in the microwave bands it can be.

The expression for penetration depth in dielectric heating is, to say the least, awkward, but for the record it is given below:

$$\text{Penetration depth } p \text{ (m)} = \frac{\dfrac{\sqrt{2}c}{4\pi}}{f\sqrt{\varepsilon^1_r\left(\sqrt{1 + \left(\dfrac{\varepsilon''_r}{\varepsilon'_r}\right)^2} - 1\right)}}$$

where
c is speed of light
ε''_r is dielectric loss factor of material
ε'_r is relative permittivity (dielectric constant) of material
f is frequency

The value of $\sqrt{2}c/4\pi = 3.4 \times 10^7$ m/s. In the radiofrequency range, f is generally less than 3.4×10^7 Hz and, as the remainder of the denominator is certain to be less than unity, it can be seen that penetration depth is likely to be a limitation only in the microwave bands.

Frequencies used for dielectric heating are 13.6 MHz and 27.2 MHz in the radiofrequency spectrum and 986 MHz and 2450 MHz in the radar or microwave spectrum. It is perhaps worth stating that high frequency heating can be either dielectric or induction, noting that for certain induction heating applications 'high' frequencies can be called for in the radiofrequency bands. These frequencies are generated by inductive circuits supplied from a high frequency source and 'tuned' to give resonance.

B32.3 Applications of electrical heating

Applications can be sub-divided into six main categories.

B32.3.1 Metal melting and refining

Arc furnaces are dependent on the plasma effect and are largely used for melting and refining in the steel industry; they normally operate in air but can be supplied for vacuum melting for production of special alloys.

Induction furnaces are used for melting and holding either ferrous or non-ferrous metals and may be supplied direct from power frequency mains or through frequency converters. For special applications vacuum furnaces are available.

Resistance furnaces utilize electric heating elements and heat transfer is mainly by radiation to a crucible containing the metal; they are necessarily confined to lower temperatures (non-ferrous metsls) and to small sizes.

B32.3.2 Metal heating for hot working

Induction heating based on 'wrap around' coils for heating work pieces can also be used for melting.

Direct resistance heating applies to work pieces having a relatively high ratio of length to diameter and is based on direct connection of mains frequency supplies. This approach enables a very rapid heat-up to be achieved with the energy used confined very largely to the workpiece itself.

Indirect resistance heating based on the use of elements in an oven or furnace is difficult to justify because of the high cost of electrical energy but has advantages in cleanliness and relatively low capital cost.

B32.3.3 Heat treatment

Surface heat treatment by induction has substantial advantages because of its ability to control the depth of heat treatment and because of its suitability for production processes. The power source is a frequency converter either adjustable or designed to a specific application in the range up to 1 MHz.

Resistance furnaces are normally indirect, using elements, and as with hot working it is necessary to balance high energy cost against cleanliness and low capital cost.

B32.3.4 Welding, Brazing, glueing and joining

Electric arc welding There have been in recent years many advances in welding technology based on the application of plasma, such as MIG, and submerged arc; the advice of specialists in applying the technology to the process is essential.

Welding of plastics this is a dielectric process and involves frequencies in the range of 3 to 80 MHz.

Glueing Many adhesives can be heated and cured using the dielectric principle.

B32.3.5 Heating and drying process: non-metals

Indirect resistance heating may be used for a wide range of industrial applications, and of course electric ovens are used for commercial and domestic food preparation.

Radiofrequency dielectric heating is widely used in food production in the frequency range 10–2500 MHz. The principle is also applied to industrial drying of sheet materials, notable paper, textiles and wood products.

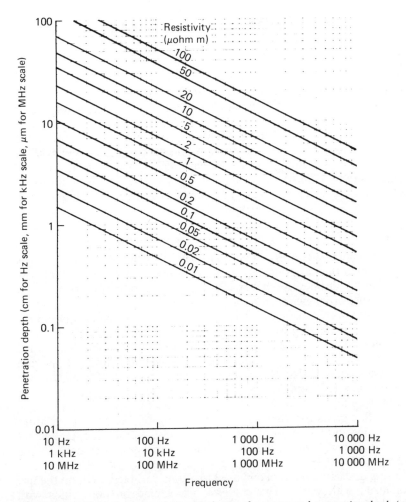

Figure B32.5 *Heat treatment: relationship betwen frequency and penetration depth ($\mu r = 1.0$)*

Infrared heating is applied to curing of surface coatings where advantages may be gained by matching the heat output to the curing requirements.

B32.3.6 Heating of liquids

Immersion heaters may be supplied in a wide range of outputs and sheathing to suit all types of industrial liquid heating requirements. Electrode boilers can be supplied using direct resistance as the means of heating for steam or hot water; heating is rapid and advantages are claimed where demand fluctuates; however, the high fuel cost must be borne in mind.

B32.4 Some notes on plasma technology

(1) The important aspect of the use of plasmas is that at temperatures in excess of 6000 K the plasma becomes highly ionized and is consequently a good conductor.
(2) Whereas oxyacetylene generates around $50 \, KW/cm^2$ plasma can reach $250 \, kW/cm^2$ giving it major advantages for metal cutting.
(3) It can be used for metal refining where it is necessary to melt metals in a controlled atmosphere.
(4) It has applications in melting refractories.
(5) Electrodeless plasma guns are available utilizing the induction principle and operating at 0.5 to 7 MHz with loadings of 20–1000 kW.
(6) Plasmas can operate in reducing, oxidizing or inert atmospheres.

B33 Energy control systems

The availability of the microprocessor has provided the spur for the development of a whole new industry in energy control; the subject is reviewed below under five headings, i.e. Combustion, Heating Systems, Lighting, Electrical Power and embracing all of these, Building Energy Management Systems.

The control systems now available open up entirely new opportunities for effective energy control, but it is well to remember that in most domestic and in many small commercial and industrial situations manual control still has much to commend it, offering as it does intelligent adaptibility in a constantly changing environment. The best approach lies somewhere between 'hands on switches and valves' at one extreme and sitting in front of a VDU at the other.

B33.1 Combustion controls

The fundamental control of boilers is concerned with ensuring that the steam pressure or water flow temperature is held at the design level and this is achieved by on/off or high/low control of the burner. **Section B7** outlines the theory of combustion and **Section B8** provides some notes on boiler and burner efficiencies. In the ideal situation the flue gas temperature should be the lowest consistent with satisfactory combustion, but at the same time above the acid dew point where sulphur bearing fuels are in use (180 °C).

The excess oxygen should be the lowest consistent with complete combustion whilst avoiding excessive stack emission with oil or solid fuels and keeping carbon monoxide content of flue gas within acceptable limits. No boiler can operate at peak efficiency if it is constantly subject to on/off control, and at low fire rates the oxygen requirement for optimum efficiency is relatively higher than when the boiler is on full load.

The chemical analysis of flue gases together with measurement of flue gas temperature has long been standard practice and based on the outcome of these measurements the best compromise of fuel and air delivery setting has been made with periodic checks to see whether 'drift' has taken place.

In recent years monitoring and control has been based to a considerable extent on the use of zirconium oxide stabilized with yttria. This material when clad on both sides with platinium operates as a fuel cell. The probe, which is in the form of a closed hollow cylinder, is exposed to a reference gas on the inside (instrument air) and when the outside is exposed to the flue gas a DC potential is generated related to the oxygen content of the flue gases by an inverse logarithmic law. The relationship is based on the Nerst equation:

Potential in millivolt $= 0.0496T \log_{10} (P_r/P_s)$

where
T is absolute temperature (K)
P_r and P_s are the partial pressures of oxygen in the reference and sample gases respectively.

The signal from the probe can be amplified and used to operate through suitable linkage on the damper controlling the air supply to the boiler. This enables the air supply to be varied according to feed-back from the oxygen content of the flue gas. These devices incorporate an in-built recalibration system to ensure that their accuracy is retained.

The economics of this approach can be assessed in very simple terms by relating the all-up cost of installing the simplest probe and its attended electronics, reference gas system and mechanical actuator at say £2500 to a 2½% saving in fuel. On this basis if a 2 year pay-back is required the minimum annual fuel bill would be £50 000.

The alternative approach for continuous monitoring of flue gas content is based on the measurement of carbon monoxide content using infrared absorption techniques. An infrared source is directed across the boiler flue and a receiver unit generates a signal based on CO content.

The arguments advanced in favour of the CO method as compared with O_2 control are:

(1) The CO control level is substantially constant with variation in boiler load whereas the required oxygen varies with firing rate.
(2) Errors can occur with the O_2 method due to infiltration of air.
(3) At low excess air levels the O_2 control becomes less sensitive.

The CO system is inherently more expensive than the zirconia approach, and is unlikely to be economic for boilers rated at less than 4000 kW.

Quite apart from combustion controls involving some form of feed-back the whole field of exhaust gas analysis has been transformed by the application of microprocessors to signals from probes which read temperature and oxygen content and compute efficiency based on NCV.

For routine checks on boiler efficiency the sampling and chemical analysis method is cheap and still widely used. A variety of analysis techniques is used including spectroscopic and chromatographic, but the advantages of direct reading instruments have been exploited by the application of microprocessors. When based on O_2 content these instruments use a zirconia probe and display efficiency based on NCV.

The so-called electronic flue-gas analyser uses a single probe to record temperature, CO^2 and CO content, but it is based on a sampling technique with the flue gases pumped electrically from the probe to the instrument. The relative ease with which flue gases can be analysed places a valuable tool in the hands of the engineer whose concern it is to ensure that boilers are operating as nearly as possible to peak efficiency.

B33.2 Heating and hot water system controls

The present state of the art is best reviewed by considering a system of heating and hot water supply based on low pressure hot water, the controls in approximate order of their significance being:

(1) Thermostatic control of primary flow.
(2) Simple time switch control preferably with four cams to give two 'off' and two 'on' switchings in each 24 hours.
(3) Simple thermostatic control of air temperature from a selected location judged to be typical of the air temperature needs of the heated zone.

(4)   Frost protection usually from and external thermostat which starts the boiler in the event of temperatures falling to a level where freezing or similar damage may occur.

Developing these basic control systems we have:

(1)   The facility for the primary flow control to maintain circulation in a pumped system for a limited time after boiler shut-down; this avoids heat stored in the boiler itself being lost entirely into the flue.

(2)   Simple time-switches can have day omission so that the on-off programme for week days is suspended in commercial premises on Saturday and Sunday, reliance being placed on the frost thermostat to avoid excessive drop in temperature. Where the programme varies from day to day a seven day switch is called for; these tend to be either physically larger than simple time switches or if made smaller, to have somewhat coarse time settings. The seven day switch is essential if building occupancy varies on a day to day basis.

(3)   To improve the effectiveness of thermostatic control zone thermostats may be used and in a LPHW system these must be associated with motorized valves so that the hot water flow in the various zones can be varied according to need. Beyond this individual thermostatic valves may be fitted to radiators so that room temperature can be controlled.

(4)   Compensator control uses the external thermostat to do more than just switch on at low external temperature. A signal from the external source can be used so that the heating system responds to variation in external temperatures. This has a clear potential for fuel saving when external temperatures rise unexpectedly and the normal controls may be slow to respond. The same system can be extended to provide for early shut-down on mild afternoons.

(5)   Optimum start control is designed to anticipate heating requirements according to overnight temperature conditions and can be particularly valuable for winter Monday morning early start in commercial and industrial premises. With this control the basic setting is for the earliest necessary overnight start which could be, say 1 a.m. Under the coldest conditions the boiler will start at 1 a.m. but start is delayed if external temperature is higher.

(6)   The so-called adaptive control is stated to utilize the system's ability (through a microprocessor) to adapt itself in the light of actual operations using compensator and optimum start controls so that the warm-up period is accurately tuned to prevailing weather conditions.

(7)   Where water heating and space heating are supplied from the same boiler system there is a clear need for independent control with suitable control valve, water thermostat and usually also a time switch. In larger installations there are good arguments for keeping the hot water system entirely independent of space heating.

(8)   Electronic controllers have been designed with operational variation according to the thermal inertia of the building, i.e. heavy, medium or light building.

(9)   Battery reserve is now a common feature enabling the system to retain its programme in the event of power failute.

(10) The sequencing of multiple boiler installations can be covered either by an independent control unit or it can be integrated into the total control system.

All these controls have been integrated in one form or another and a very wide range of control units is available from domestic units at a few hundred pounds to complete systems for major industrial installations.

B33.3 Lighting controls

B4.1 cites a particular example of the high cost of extravagant lighting installations with inadequate control over consumption; in the building in question 50% of electricity cost was for lighting.

The single fundamental problem in lighting control is that ease of installation leads to centralized switching positions and 'block' switching of large numbers of luminaires, whereas proper control of lighting requires local switching and in particular the facility to switch off lighting in those areas where daylight can at times render artificial lighting unnecessary.

The importance of ensuring that new installations are designed for energy economy cannot be over-emphasized as it is expensive and difficult to retrofit suitable switching arrangements.

There are three basic techniques in lighting control, i.e.

(1) The use of daylight sensors to switch off lights in areas adjacent to windows; this switch-off is frequently carried out in two stages so as to minimize disturbance to the people affected.

(2) Centralized control of all lighting to cover the conflicting requirements of working hours, lunch breaks, early starts, office cleaning, day-by-day occupational patterns, and the effects of daylight on the perimeters of buildings. Such control systems are based on supplementary low voltage wiring operating through relays to switch the appropriate circuits. Provision should also be made for selected local controls which can over-ride the central system when needed. The principle behind such a system is that lighting is automatically off unless the programme requires it to be on. The converse is true of conventional switching where lighting is always left on unless someone remembers to switch it off!

(3) The automatic reset switch is designed to give a degree of local control covering a single luminaire or a small group of luminaires. When the main programme turns all lights off the automatic reset switch remains in the off position until the local cord-pull is operated by someone requiring light in a particular area. This is best illustrated by the arrival of office cleaners who switch all lights on on a particular floor; *only those lights without automatic reset* will come on: the others can only be used by operating the local cord-pull. The advantage of the system is than when central control calls for 'off' all lighting except emergency lighting will be off and only local lighting can be restored by the reset switch. This arrangement clearly gives good opportunity to design the lighting programme for maximum economy.

Complicated and quite expensive hardware is required for a complete low voltage relay system, including such items as low voltage transformers, rectifiers, relays, master controllers for up to twenty-five circuits, time switches, photoelectric devices, control panels and multicore control cables.

Modifying existing lighting systems can clearly be expensive particularly if rewiring of 240 V circuits to regroup central switching arrangements is necessary. This need for re-wiring can be a particular problem with daylight switching as existing installations are rarely wired so that perimeter luminaires can be switched off in banks.

As with all control systems the microprocessor has had a major influence on the integration of switching programmes for large installations and lighting energy management must always be considered for projects involving large lighting loads.

B33.4 *Control of electrical power consumption*

Reference has been made to electricity maximum demand in **B29** and the penalty of high winter maximum demand charges is illusted by **Tables B3.2** and **B29.1**. The control of maximum demand by load shedding has been used in industry for approaching 50 years; what has happened recently is that increasing energy costs have provided the spur for the development of more sophisticated equipment and in particular have extended the awareness of the need for load shedding to commercial premises.

'Electrical Energy Management Systems' are now offered which are designed not just to deal with the impact of maximum demand charges but also to save energy by switching off plant and equipment when its operation is not essential. One is tempted to ask why it is running at all if its operation is not essential!!

Maximum demand control remains of paramount importance both to the company paying the electricity bill and to the electricity generating authority which has to have plant operational to meet peak loads. The balance in industry must be struck between the need to reduce MD charges and the necessity to keep the production process going. In commercial buildings the parameters can be different, i.e. shutting off fans or shutting down air conditioning equipment for short periods at peak demand times may well not be noticed at all by building occupants.

A second approach to electrical power economies seeks to exploit the problem of the underloaded induction motor which has been covered in **B31**. Starting from some work at NASA on small single phase motors the idea has been extended to three-phase machines. The approach is to reduce the terminal voltage of the motor when it is underloaded using thyristors and this has the effect of improving motor efficiency and power factor; where motors are very lightly loaded (20% or less) the economies can be substantial. The cost of these devices can be higher than the cost of the motor itself which prompts the suggestion that it might well be better to invest in a motor designed for higher power factor and higher efficiency! The paramount need as emphasized in **B31** is to match the output of the motor accurately to the load and to calculate life cycle costs with a view to justifying a more costly motor, but with better performance. It has been claimed that the additional first cost of high efficiency induction motors can be recovered in two years or less from reduced energy consumption.

B35.5 Building energy management systems

C8 lists some of the manufacturers in the control field and two-thirds of those listed offer building energy management systems; however, this list falls far short of the total tally of manufacturers who have moved into this field in recent years. The emphasis in most systems is towards large commercial buildings with their common needs of heating, cooling, hot water and lighting; the claims made for universal application of such systems must therefore be examined very carefully when considering industrial applications where each factory is likely to have its own special problems.

Once the input sensors have been installed it becomes possible to monitor, control and trim a range of data many times greater than can be handled by a manual system operated by the highest human intelligence. Some of the areas of control are listed:

External conditions Temperature on exposed and protected sides of the building, humidity, wind direction and speed, solar intensity and rainfall.

Internal conditions Temperatures according to zone and setting of thermostats, occupation pattern and working hours, lighting levels and daylight, domestic hot water temperature and usuage, heating system controls including pumps and motorized valves (all these controls can be tamper-proof).

Boiler conditions Flow and return temperature. High or low fire. Flue gas analysis and temperature and feed-back to boiler control.

Air-conditioning plant In traditional air-conditioning plant, heating and cooling go on simultaneously; by sensing the true enthalpy requirements and comparing these with the enthalpy in extract air and external air the system can also be controlled through actuators and motorized valves. This can lead to substantial energy economies.

Energy metering Gas, water, electricity, oil usage and rate of consumption.

Power demand and consumption MD control and load shedding, power consumption and monitoring.

Process heating and cooling An area which can be defined only in relation to the industrial process concerned.

Energy management systems can be designed to cover multiple sites by the use of stand-alone units linked by means of the public telephone network to central control. Such systems are particularly well adapted to the needs of banking groups and store chains. As a corollary to the design of energy management systems much effort has been devoted to software for the simulation of building energy requirements. Such systems can receive data on building construction, aspect and exposure, all seasons external conditions, heating, cooling, domestic hot water and lighting requirements and predict the effects of a range of changes aimed at the improvement of energy efficiency.

B34 Investment appraisal

In assessing and presenting capital investment projects for energy saving equipment by far the most important part of the study is to produce the most accurate possible audit of present energy use and cost; the study should also identify clearly where energy is being used extravagantly and where there is clear potential for saving. The Sankey diagram approach described in **B6** is a very effective way of doing this.

Having identified the areas for investment, i.e. new heating system or replacement boiler, a forecast of future energy consumption and cost must be made and budget proposals obtained from contractors for the new plant. A major point in favour of energy saving investments is that, so long as the business continues in existence and operates at near to its current level of activity, the savings will continue (within the life span of the new equipment). By contrast, investment in new production facilities or in warehousing and retail outlets must always be subject to the whim of the market place and the customer. On this standpoint the risk factor in investment in energy saving can be argued as minimal. However, the directors of any business are always faced with conflicting claims on financial resources and if new equipment or facilities are vital to the survival of the company energy saving investment will take second place!

From the early 1970s until recently, it has been possible to argue that energy prices have been rising more rapidly than inflation. However, with the pause in World energy prices (mid 1983) this argument carries less weight. In the long term World resources of fuels remain finite and the ultimate solution to the energy problems of the Western World is certainly not in sight; it therefore seems fair to argue that for the next decade or so the upward pressure of energy costs will continue.

This manual is devoted to the study and understanding of the ways in which energy is used and it strives to point out all those areas where there is potential for saving. Many of these savings can be achieved by better discipline with no capital expenditure at all; others involve only small expenditure where detailed presentation of a capital project is not called for. The writer's experience in recent years, particularly in dealing with smaller companies, is that we still have a long way to go before the simple, low-cost measures have been fully exploited.

An understanding of the basic accounting techniques by which capital projects are evaluated is essential if energy proposals are to get a fair hearing. Accounting methods used are sophisticated and can appear to be very complicated; this brief review will attempt to present them as simply as possible.

At the heart of all investment appraisal technique lies forecasting; it is a brave man indeed who stakes his reputation on a forecast made in today's World where economic activity, rates of inflation, interest rates, third world competition and political upheavals are all exerting their influence on the pattern of World trade. The forecast is, nevertheless, an essential criterion in assessing the relative merits of alternative projects, even if, in the final event it is the judgement or indeed the hunch of the entrepreneur which decides the path to follow.

Before a project is presented certain basic considerations need to be determined, i.e.

(1) Total capital cost and the approximate dates when funds will be needed; this is particularly important for projects which may be phased over a period of years. Without this information the accountant cannot plan cash-flow so that monies will be available when needed.

(2) A forecast of the savings which will result, when they will start to become effective, and how long it will be before the full savings are achieved.

(3) A much simplified statement of total investment cost and expected annual savings when fully operational.

(4) The best estimate possible of the true life span of the equipment; conflicting elements are involved here, i.e. experience of others, if available, with similar equipment; the likelihood of technical developments or changes rendering the equipment obsolete; the influence of long-term wear and tear, and maintenance costs on useful life. The life span from this consideration may well be different from the 'write off' period used for accounting purposes.

The techniques used by accountants to judge the viability of a capital project are outlined below.

B34.1 Return on capital employed (R.O.C.E.)

Investment	£100 000
Annual savings	£30 000
Life of equipment	5 years

On a 'straight line' write-off basis over five years £20 000 per year must be provided to cover depreciation charges. The nett annual return will therefore be £30 000 less £20 000 = £10 000. This represents an R.O.C.E. on the investment of

$$\frac{£10\ 000}{£100\ 000} \times 100\% = 10\%$$

An alternative approach, which will tend to show this investment in a more favourable light, is to argue that the depreciation charge *reduces* the value of the investment each year so that the *average* investment over the five years becomes

$$\frac{£100\ 000 + £80\ 000 + £60\ 000 + £40\ 000 + £20\ 000}{5} = £60\ 000$$

At £10 000 per year the R.O.C.E. now becomes

$$\frac{£10\ 000}{£60\ 000} \times 100\% = 18.3\%$$

B34.2 Pay-back period

This is very simple and enables an initial judgement to be made without detailed calculations.

£100 000 to secure savings of £30 000 per year will be recovered in

$$\frac{100\ 000}{30\ 000} = 3.33 \text{ years}$$

The popular answer when capital is not easy to come by is to say that unless a two year payback can be achieved the project is not on! This approach can rule out many worthwhile projects and persistence is needed if wise investments are to be made.

B34.3 Discounted cash flow (D.C.F.)

This is not really as complicated as it sounds; the principle behind the method is to translate the value of money in the future to equivalent values today. If we consider an annual inflation rate of 10%, £1 in a year's time will be worth

$$\frac{100\%}{110\%}$$ at *today's values*, i.e. £0.909

In two years it will be worth

$$\frac{100}{110} \times \frac{100}{110} \times £1,\ \text{i.e. £0.826}$$

and so on.

These values can be run off rapidly on a calculator with a simple memory for any annual percentage rate, but to save this labour tables as **B34.1** have long been used showing present values with depreciation rates up to 20%. It is important to appreciate the logic behind this approach; it is not to allow for inflation, but to allow for the fact that money available now can be used, i.e. to earn interest whereas money becoming available in the future cannot be put to work and therefore has a lower nett present value. Exactly the same approach has

B34.1 Present values of £1 discounted over ten years

Years	5	6	8	10	Discount rate (%) 12	14	15	16	18	20
1	0.952	0.943	0.926	0.909	0.893	0.877	0.870	0.862	0.847	0.833
2	0.907	0.890	0.857	0.826	0.797	0.769	0.756	0.743	0.718	0.694
3	0.864	0.840	0.794	0.751	0.712	0.675	0.658	0.641	0.609	0.579
4	0.823	0.792	0.735	0.683	0.636	0.592	0.572	0.552	0.516	0.482
5	0.784	0.747	0.681	0.621	0.567	0.519	0.487	0.476	0.437	0.402
6	0.746	0.705	0.630	0.564	0.507	0.456	0.432	0.410	0.370	0.335
7	0.711	0.665	0.583	0.513	0.452	0.400	0.376	0.354	0.314	0.279
8	0.677	0.627	0.540	0.467	0.404	0.351	0.327	0.305	0.266	0.233
9	0.645	0.592	0.500	0.424	0.361	0.308	0.284	0.263	0.255	0.194
10	0.614	0.558	0.463	0.386	0.322	0.270	0.247	0.227	0.191	0.162

been used for very many years by building societies who require payment of interest and capital from borrowers, but arrange the payments at a fixed annual or monthly rate, and this is best illustrated by **Table B34.2** which shows how £1000 per year paid to a building society will repay over five years a loan of £3790. This procedure is referred to in accounting parlance as discounting and merely means that a return of £5000 on a capital investment of £3790 spread evenly over five years will show a yield of 10% on the outstanding balance at any one time.

The relationship between the building society **Table B34.2** and the present value **Table B34.1** is immediately apparent by noting that the present value of £1 after five years at 10% is £0.621, i.e. £1000 being equivalent to £621. The capital repaid column in **B34.2** gives the

B34.2 Fixed mortgage repayments at interest rate 10% (£)

Years	Outstanding at start of period	Interest payable	Capital repaid	Total repayment	Dept at end of period
1	3790	379	621	1000	3170
2	3169	316	683	1000	2486
3	2486	249	751	1000	1735
4	1735	174	826	1000	909
5	909	91	909	1000	—

corresponding figures in **B34.1** in reverse over the five years, i.e. £621, £683, £751, £826, £909. The next step in applying D.C.F. is to sum up to present value year by year and the result of this is shown in **Table B34.3** (note that the 10% column indicates in reverse order the outstanding debt figures in **Table B34.2**).

Now to apply this approach to the £100 000 investment the simple pay-back figure is

$$\frac{£100\ 000}{£30\ 000} = 3.33 \text{ years}$$

On the five year line of **Table 34.3**, this indicates a D.C.F. rate of return of between 15% and 16% (between 3.352 and 3.274).

B34.3 Present values of £1 received annually

Years	5	6	8	10	Discount rate (%) 12	14	15	16	18	20
1	0.952	0.943	0.926	0.909	0.893	0.877	0.870	0.862	0.847	0.833
2	1.859	1.833	1.783	1.736	1.690	1.647	1.626	1.605	1.566	1.528
3	2.723	2.673	2.577	2.487	2.402	2.322	2.283	2.246	2.174	2.106
4	3.546	3.465	3.312	3.169	3.037	2.914	2.855	2.798	2.690	2.589
5	4.330	4.212	3.993	3.791	3.605	3.433	3.352	3.274	3.127	2.991
6	5.076	4.917	4.623	4.355	4.111	3.889	3.784	3.685	3.498	3.326
7	5.786	5.582	5.206	4.868	4.564	4.288	4.160	4.039	3.812	3.605
8	6.463	6.210	5.747	5.335	4.968	4.639	4.487	4.344	4.078	3.837
9	7.108	6.802	6.247	5.759	5.328	4.946	4.722	4.607	4.303	4.031
10	7.722	7.360	6.710	6.145	5.650	5.216	5.019	4.833	4.494	4.192

A further advantage of the D.C.F. approach is that it enables a minimum rate of return to be stipulated and then translated into present value over a given term. If, for example, the minimum rate is 14% over five years the present value is £3.433.

This means that for an investment of £100 000 the annual savings could still be acceptable, if as low as

$$\frac{£100\ 000}{3.433} = £29\ 000 \text{ approximately}$$

Yet another way of looking at this procedure is to establish the *present value of the investment* from the acceptable rate of return and the known annual savings. Thus for 14% the present value is £30 000 × 3.433 = £103 000 approximately, which, being greater than £100 000, makes the project acceptable.

If, however, the annual savings were £25 000 instead of £30 000, the present value becomes £25 000 × 3.433 = £86 000 approximately, which, being *lower* than £100 000, would render the project unacceptable.

Section C

Directory of products and services

The directory has been designed to present a brief coverage of the energy conservation industry; inevitably some names may have been omitted and the very few descriptive words used can do no more than indicate the scope of products offered. It is hoped, nevertheless, that the engineer with a specific problem to deal with will be able to use the Directory to make contact with organizations in the relevant field and thereby obtain comprehensive information and advice.

Those listed are mainly manufacturers or their agents but in appropriate instances companies with a service to offer have also been included as for example in the field of insulation.

Some devices have been used to avoid unnecessary repetition, i.e.

(1) Companies are not listed in every area in which they may have products or services to offer, but rather in those areas of their primary interest.
(2) Manufacturers of boilers are assumed to be able to offer oil or gas burners and a special note is made where they also offer solid fuel.
(3) There is a simple reference to the limit of size or output in the range and this is stated in SI units.
(4) Trade names are mentioned where they appear to be of significance.
(5) Such abbreviations as are used in describing products have their origins in the appropriate B section of the manual.
(6) C15 is devoted primarily to Heat Pumps, but it must be appreciated that many companies listed have their major activity in air conditioning.

A list of entries received too late for the main directory is given on page 250.

C1 Burners and boilers (including waste firing and incinerators)

Symbols: Bu = burner manufacturer, Bo = boiler manufacturer; oil or gas firing unless otherwise stated, SF = solid fuel firing, St = steam or high pressure hot water in addition to low pressure hot water, Wa = waste fired, Flu = fluidized bed firing. Maximum available output shown in kW

A. B. Generator (UK) Ltd, PO Box 31, Macclesfield, Cheshire, SK10 2EX	(0625) 613666	Bo, St, Flu	6000 kW	
A. P. Boilers Ltd, 295 Aylestone Rd, Leicester, LE2 7PB	(0533) 833581	Bo, St, Wa	12 000 kW	'Elboma'
Aeromatic Industrial Ltd, 47 Station Rd, Gerrards Cross, Bucks, SL9 8ES	(0753) 887249	Bu		
Allen Ygnis Tipton Ltd, 268–270 Vauxhall Bridge Rd, London SW1V 1BQ	(01) 834 2006	Bo, St	3500 kW	
Allied Boilers Ltd, Belgrave Mills, Honeywell Lane, Oldham, Lancs, OL8 2LY	(061) 633 1131	Bo, St, Flu	8000 kW	

Babcock Power Ltd, 65 Livery St, Birmingham, BS 1HA	(021) 236 7881	Bu, Bo, SF St, Flu	20 000 kW	Comprehensive range
D. Baldwin and Sons Ltd, Parkwood Boiler Works, Parkwood St, Keighley, Yorks, BD21 4NW	(0535) 65225	Bo, St	4400 kW	
B. & E. Boilers Ltd, Easthampstead Rd, Bracknell Berks, RG12 1NP	(0344) 21341	Bo, St	10 000 kW	
Beeston Boiler Co (Successors) Ltd, Beeston, Nottingham, NG9 2DN	(0602) 254271	Bo, St	900 kW	
Beverley Chemical Engineering Ltd, Billingshurst, West Sussex	(0403) 812091	Bo, Bu		'Comtro' incinerators
Bigwood Joshua, Ltd, PO Box 23, Wednesfield Rd, Wolverhampton, WV10 0DP	(0902) 54125	Bu, SF		
Boulter Bros Ltd, Meteor Close, Norwich Airport Industrial Estate, Norwich, NR6 6HG	(0603) 41187	Bo	1000 kW	
British Gas and Oil Burners Ltd, Burrell Way Thetford, Norfolk, IP24 3RA	(0842) 4444	Bu	900 kW	'Selectos'
Broag Ltd, Molly Millars Lane, Wokingham, Berks, RG11 2PY	(0734) 783434	Bu, Bo, SF	2300 kW	
Chaffoteaux Ltd, Concord House, Brighton Rd, Salfords, Redhill, Surrey, RH1 5DX	(0293) 3472744	Bo	1000 kW	'Econoflame'
Cleerburn Ltd, 1 High St, Slough, Bucks, SL1 1DY	(0753) 26764	Bo, Wa	300 kW	Incinerators
Clyde-Combustions Ltd, Cox Lane, Chessington, Surrey, KT9 1SL	(01) 397 5363	Bu, Bo	1100 kW	
Commercial and Industrial Boilers Ltd, 20A Jewry St, Winchester, Hants, SO23 8RZ	(0962) 56176	Bo	1200 kW	Chappee
Controlled Flame Boilers Ltd, Burymead Rd, Hitchin, Herts, SG5 1RT	(0462) 2378	Bo	1800 kW	
Cradley Boiler Co Ltd, Cradley Heath, Warley, Worchestershire, B64 7AN	(0384) 66003	Bo, St	6000 kW	
Danks of Netherton Ltd, 257 Halesowen Rd, Netherton, Dudley, DY2 9PG	(0384) 66417	Bo, St, Flu	20 000 kW	Comprehensive range
Deborah Fluidized Combustion Ltd, 6 Davydrive, N W Industrial Estate, Peterlee, Co Durham	(0783) 867411	Bo, St, Flu		Custom built
Dowson & Mason Ltd, Alma Works, Levensholme, Manchester, M19 2RB	(061) 432 0222	Bo, Wa, Flu		Incinerators
John Driscoll Ltd, 27 Albert St, Rugby, Warwicks, CV21 2SG	(0788) 73421	Bo	850 kW	
Dunphy Oil and Gas Burners Ltd, Queens Way, Rochdale, Lancs. OL11 2SL	(0706) 49217	Bu	7800 kW	
Electro-Oil International Ltd, London Rd, Woolmer Green, Knebworth, Herts	(0438) 813937	Bu	10 000 kW	
EMS Thermplant Ltd, Church Building, 321 Rookery Rd, Hansworth, Birmingham, B21 9PR	(021) 523 8070	Bo Flu	3000 kW	
The Energy Equipment Co Ltd, Energy House, Hockliffe St, Leighton Buzzard, Beds, LE7 8HE	(0525) 377600	Bo, St, Flu	17 000 kW	
Extracta Engineering Ltd, Holder Rd, Aldershot, Hants	(0252) 316661	Bo, Wa	400 kW	Timber/Waste
Frema Combustions Ltd, Queens Yard, Whitepost Lane, London, E9 5EN	(01) 985 8561	Bo, St, SF	1200 kW	
Fuelogic, Kirby Misperton, Malton, N Yorks	(0751) 74874	Bo, SF	200 kW	'HDG' safental high effy
General Engineering Radcliffe 1979 Ltd, Bury Rd, Radcliffe, Manchester, M269UR	(061) 723 3271	Bo Wa	6000 kW	'Consertherm'
Gibson Wells Ltd, 2 Town Gate, Calverley, Leeds, LS28 5NF	(0532) 550455	Bo, St, Flu		Custom built
G. I. S. Boilers, St. Albans House, Portland St, Leamington Spa, Warwickshire, CV32 5EZ	(0926) 36626	Bo, St	1800 kW	

G. P. Burners Ltd, Unit 3 EPH Industrial Estate, Faraday Rd, Dorcan, Swindon, Wilts, SN3 5HQ	(0793) 23484	Bu SF	1800 kW	Peabody Gordon-Piatt
E. Green & Son Ltd, Calder Vale Rd, Wakefield, W Yorks, WF1 5PF	(0924) 378211	Bo, Flu		
GWB Industrial Boilers Ltd, Burton Works, Dudley, W Midlands	(0384) 55455	Bo, SF, St, Flu	8500 kW	'Vekos'
Haat Sussex Ltd, 46 West St, Chichester, W Sussex, PO19 1RP	(0243) 786402	Bo Wa		Incinerators
Hamworthy Engineering Ltd, Fleets Corner, Poole, Dorset, BH17 7LA	(0202) 675123	Bo, Bu, SF	250 kW	Modules
Hartley & Sugden Atlas Boiler Works, Halifax, W Yorks, HX1 4DB	(0422) 55651	Bo, St, SF	3500 kW	
Heenan Environmental Systems Ltd, PO Box 14, Shrub Hill Rd, Worcester, WR4 9HA	(0905) 23461	Bo, Wa	5000 kW	Incinerators
Hilticroft Energy Conservation Ltd, The Science Park, Wynne St, Salford M6 6AD				
Hirt Combustion Engineers Ltd, Dane Works, Water St, Northwich Cheshire CW9 5HP	(0606) 47815	Bu		
Hotwork International Ltd, Low Rd, Earlsheaton, Dewsbury, W Yorks, WF12 8By	(0924) 463325	Bu	750 kW	Recuperative
Hoval-Farrar Ltd, Northgate, Newark, Notts, NG24 1JN	(0636) 72711	Bo, SF, Wa	500 kW	Boilers and pyrolitic incinerators
The Incinerator Co, St Neots, Cambridge	(0408) 213171	Bo, Wa		Incinerators
Robert Jenkins Systems Ltd, Wortley Rd, Rotherham, S61 1LT	(0709) 558701	Bo, WA		
Kayanson Engineers Ltd, PO Box 57, 1/3 Market Square, Chesham, Bucks, HP5 1BH	(0494) 782128	Bo, SF, St	4000 kW	
Laidlaw Drew and Co Ltd, Sight Hill Industrial Estate, Edinburgh, EH11 4HG	(031) 443 4422	Bu, Bo, Wa	16 000 kW	Burners and waste fired boilers
M. E. Boilers Ltd, M. E. House, Fengate, Peterborough, PE1 5BQ	(0733) 68471	Flu, Bo, SF, St	3000 kW	
N. E. I.-Thompson Cochrane Ltd, Newbie Works, Annan, Dumfrieshire, DG12 5QU	(046) 22111	Bu, Bo, SF, St, Flu	20 000 kW	Comprehensive range
Nordsea Gas Technology Ltd, Broadway, Globe Lane Indus. Estate, Dukinfield, Cheshire SK16 4UU	(061) 339 8689	Bu		
Nuway Heating Plants Ltd, Droitwich, WR9 8NA	(0905) 72331	Bu	50 000 kW	Volume prod and special
Peabody Holmes Ltd, Huddersfield, HD1 6RB	(0484) 22222	Bu, Bo, SF, Wa		Custom built
Perkins Boilers Ltd, Mansfield Rd, Derby, DE2 4BA	(0332) 48235	Bu, Bo, ST	2700 kW	
Perrymatics (SARB) Ltd, Fort Rd, Littlehampton, Sussex, BN17 7AF	(09064) 3991	Bu, Bo	1000 kW	'Riello' Italy modular
Potterton International Ltd, Portobello Works, Emscote Rd, Warwick, CV34 5QU	(0926) 493471	Bo, SF	2500 kW	Volume prodn domestic
Powermatic Ltd, Winterhay Lane, Ilminster, Somerset, TA19 9PQ	(04605) 3535	Bu, Bo,	600 kW	
Radiant Super-Jet Ltd, Clapgate Lane, Woodgate, Birmingham, B32 3BP	(021) 422 7221	Bu	3000 kW	
Robey of Lincoln Ltd, PO Box 23, Canwick Rd, Lincoln, LN5 8HV	(0522) 21381	Bo, St, Wa	17 000 kW	
Rossfor Associates Ltd, 37 Lowlands Rd, Harrow, Middlesex, HA1 3AW	(01) 423 0330	Bo, SF, Wa	5000 kW	Woodwaste burning
H. Saacke Ltd, Fitzherbert Rd, Farlington, Portsmouth, Hants, PO6 1RX	(0705) 383111	Bu, Bo, SF, Wa		Custom built
Scanfield Boilers Ltd, Hamstreet near Ashford, Kent, TN26 2EH	(0223) 732257	Bo, Wa	100 kW	Woodwaste and straw fired
Speedaire Supply Ltd, Cromford Rd, Langley Mill, Nottingham, NG16 4FL	(0773) 767412	Bu	350 kW	
Stelrad Group Ltd, PO Box 103, National Ave, Hull, N Humberside, HU5 4JN	(0482) 492251	Bo		Volume prodn domestic

Stone Boilers (Stone-Platt) PO Box 5, Gatwick Rd, Crawley, W Sussex, RH10 2RN	(0293) 27711	Bo, SF	1500 kW	Volume domestic
Stordy Combustion Engineering Ltd, Heathmill Rd, Wombourne, Staffs, WV5 8BD	(0902) 897654	Bu, SF	55 00 kW	Specialized oil/ gas burners
Strebel Ltd, Invincible Rd, Farnborough, Hants, GU14 7TA	(0252) 519846	Bo	1500 kW	
Talisman Heat Ltd, Lotherton Way, Aberford Rd Trading Estate, Garforth, Leeds	(0532) 861122	Bo, St, Wa	6000 kW	'Tomlinson' 'Tharm' denmark
T. I. Hartley and Sugden Ltd, Atlas Boiler Works, Halifax, Yorks, HX1 4DE	(0422) 55651	Bo, SF, St	3000 kW	
Trianco Industrial Boiler Division, Stewart House, Brookway, Kingston Rd, Leatherhead, Surrey	(0372) 376453	Bo	120 kW	
Trianco Redfyre Ltd, Stewart House, Brookway, Kingston Rd, Leatherhead, Surrey	(0372) 376453	Bo, SF	90 kW	Volume prod domestic
Tubeguard Boilers Ltd, Fiveaways House, Liverpool Rd, Neston, Wirral, Cheshire, L64 3TL	(051336) 3934	Bo	50 kW	
U. A. Engineering Ltd, Canal St, Sheffield, S4 7ZE	(0742) 21167	Bo, SF, Wa	50 kW	'Thermossi' also incinerators
Vaillant Ltd, Aerodrome Way, Heston Industrial Estate, Hounslow, TW5 9PU	(01) 897 6037	Bo	33 kW	Gas only
Wallsend Slipway Ltd, PO Box 8, Wallsend Tyne and Weir, NE28 6QH	(0632) 628961	Bo, St, Flu	3000 kW	
Weishaupt (UK) Ltd, Neachells Lane, Willenhall, W Midlands, WV13 3RG	(0902) 69841	Bu		
Wellman Incandescent Ltd, Cornwall Rd, Smethwick, Warley, W Midlands, B66 2LB	(021 558) 3151	Bu		'Basequip'
Wellman Selas Ltd, City Rd, E Manchester, M15 4PH	(061236) 2648	Bo	35 kW	
L. V. Williams, 34 High Lawn, Devizes, Wilts	(0380) 4577	Bo, St	400 kW	'Byworth'
Worcester Engineering Co Ltd, Diglis, Worcester, WR5 3DG	(0905) 356224	Bo, SF	30 kW	'Hoppamat'
T. C. Williams Burners Holme Manufacturing Co. Ltd., Bradshaw Road, Honley, Huddersfield HD7 2DT	(0484) 662185	Bu		

C2 Steam generators, calorifiers, water heaters and thermal fluid heaters

Symbols: SG = steam generators, C = calorifiers, HWB = hot water boilers, HWG = hot water generators, WSH = Water storage heaters, TFH = thermal fluid heaters, DFGH = direct gas fired immersion heaters. Maximum output in kW or maximum capacity in litre)

Advanced Services (Sales) Ltd, Ham Moor Lane, Weybridge Trading Estate, Weybridge, Surrey, KT15 2SD	(0932) 41124	HWB, WSH, DFGH	600 kW	A. O. Smith
Beaumont (UK) Ltd, Romsey Industrial Estate, Greatbridge Rd, Romsey, Hants, SO5 0HR	(0794) 516800	WSH	460 kW	
Bonair Engineering Ltd, Unit 2, Primrose Bank, Mill Friday St, Chorley, Lancs, PR6 0AH	(02572) 63691	DFGH	45 kW	
BSS Fleet House, Lee Circle, Leicester, LE1 3QQ	(0533) 23232	HWG		
Calomax (Engineers) Ltd, Lupton Ave, Leeds, Yorks, LS9 7DD	(0532) 496681	HWB	10 kW	
Clayton Thermal Products, 12 Rivington Court, Hardwick Grange, Woolston, Warrington, WA1 4RT	(0925) 823123	SG	5000 kW	
Crosse Engineering Ltd, Herriot House, North Place, Cheltenham, Glos, GL50 4DS	(0242) 34701	C	300 kW	
Direct Fired Calorifiers Ltd, 48 Cainston Lane, Dunchurch, Rugby, Warwicks	(0788) 814960	C, DGFH		
Gledhill Water Storage Ltd, Estate Squires, Gate Lane, Blackpool, Lancs, FY4 3RL	(0253) 401494	C, HWG, WSH	350 kW	Chaffoteaux
Grayhill Blackheat Ltd, 13 Cobham Rd, Ferndown Industrial Estate, Dorset, BH21 7PE	(0202) 896510	DGFH		

Hamworthy Engineering Ltd, Fleets Corner, Poole, Dorset, BH17 7LA	(02013) 5123	WSH, DGFH	120 kW	Modular
Heat Transfer Ltd, 3/4 Bath St, Cheltenham, Glos, GL50 1YE	(0242) 582777	C	2000 kW	
IMI-Range Ltd, PO Box 1, Stalybridge, Cheshire, SK15 1PQ	(061) 338 3353	C	5000 litres	
Johnson and Starley Ltd, Rhosili Rd, Brackmills, Northampton, NN4 0LZ	(0604) 62881	WSH	100 kW	'Janstor'
Lanemark Thermal Systems Ltd, D2 Greenwood Court, Veasey Close, Attleborough Fields Industrial Estate, Nuneaton, Warwicks, CV11 6RT	(0203) 341017	SG, TFH, DGFH	2000 kW	
Nordsea Gas Technology Ltd, Broadway, Globe Lane Industrial Estate, Dukinfield, Cheshire, SK16 4UU	(061) 339 8689	DGFH	600 kW	
State Boilers (UK) Ltd, Unit 4, Tower Industrial Estate, Eastleigh, Hants, SO5 5NZ	(0703) 641676	WSH	210 kW	
Stone Boilers, Gatwick Rd, Crawley, W Sussex, RH10 2RN	(0293) 27711	SG	1500 kW	
Twin Industries Agencies Ltd, Stoneyard Works, Part St, Camberley Surrey	(0276) 26152	SG	1200 kW	Fulton
Uttley Ingham (Copper Cylinders) Ltd, Windsor Works, Victoria Rd, Hebden Bridge, Yorks, HX7 8LN	(042) 284 2145	C	450 litres	
Wanson Co Ltd, 7 Elstree Way, Boreham Wood, Herts, WD6 1SA	(01) 953 7111	SG, TFH	2300 kW	'Vaporex'
Wellman Selas Ltd, City Rd, E Manchester, M15 4PJ	(061) 236 2648	DGFH	190 kW	

C3 Furnaces and industrial ovens

AEW Ltd, North Way, Walworth Industrial Estate, Andover, Hants, SP10 5AV	(0264) 61331
Born Heaters Ltd, Europa House, Goldstone Villas, Hove, E Sussex, BN3 3RZ	(0273) 722811
Boulton Industrial Furnaces Ltd, Newstead Trading Estate, Trentham, Stoke-on-Trent, ST4 8HZ	(0782) 658431
R. M. Catterson-Smith Ltd, Woodrolfe Rd, Tollesbury near Maldon, Essex, CM9 8SJ	(062) 1869342
Chatburn and Chantry Ltd, Millsborough Works, Langsett Rd, Sheffield, S6 2LS	(0742) 348068
Drayton Kiln Co Ltd, Newstead Trading Estate, Trentham, Stoke-on-Trent, Staffs, ST4 8HX	(0782) 657361
Drever (UK) Ltd, Shenstone House, Dudley Rd, Halesowen, W Midlands, B63 3XA	(021) 550 7361
David Etchells (Furnaces) Ltd, Stafford Rd, Darlaston, Wednesbury, W Midlands, WS10 8UA	(021) 526 3511
Ecotherm Ltd, Whitehall Rd, Great Bridge, Tipton, W Midlands, DY4 7JR	(021) 520 7561
Fairbank Brearley Ltd, Church St, Bingley, W Yorks, DB162DA	(0274) 560115
Furnace Construction Co, Newton Moor Industrial Estate, Hyde, Cheshire, SK14 4LS	(061) 368 8419
GKN Birwelco Ltd, Mucklow House, Mucklow Hill, Halesowen, W Midlands, B62 8DG	(021) 550 4777
Hedinair Ltd, 274 Whalebove Lane, S Dagenham, Essex, RM8 1BJ	(01) 593 7221
Industrial Combustion Systems, Unit 9, Thornleigh Trading Estate, Dudley, W Midlands, DY2 8UB	(0384) 235355
JLS Engineering Co Ltd, Brook St, Lakeside, Redditch, Worcs, B98 8NE	(0527) 67891

Kilns and Furnaces Ltd, Keele St, Tunstall, Stoke-on-Trent, ST6 5AS	(0782) 813621
Lingard Engineering Ltd, Westminster Rd, Wareham, Dorset, BH20 4SD	(09295) 6311
LTM Furnaces Ltd, Pinnox St, Tunstall, Stoke-on-Trent, ST6 6AH	(0782) 817419
A. N. Marr Ltd, Globe Rd, Leeds, LS11 5QL	(0532) 459144
Monometer Manufacturing Co Ltd, Rectory Grove, Leigh-on-Sea, Essex	(0702) 72201
Morganite Thermal Designs Ltd, PO Box 8, Steatite Works, Bewdeley Rd, Stourport-on-Severn, Worcs, DY13 8QS	(02993) 2271
Priest Furnaces Ltd, PO Box Southbank, 18 The Grange, Eston, Middlesborough, Cleveland, TS6 8DJ	(0642) 467171
Stanelco Ltd, 4 Elstree Way, Boreham Wood, Herts, WD6 1SE	(01) 953 4031
Stein Atkinson, Stordy Ltd, Midland House, Ounsdale Rd, Wombourne, W Midlands, WV5 8BY	(0902) 894171
Teisen Furnaces Ltd, Eckershall Rd, Kings Norton, Birmingham, B38 8SU	(021) 458 2284
Wellman Furnaces Ltd, Cornwall Rd, Smethwick, Warley, W Midlands, B66 2LB	(021) 558 3151
Wentgate Engineers (1976) Ltd, Industrial Estate St Ives, Huntingdon, Cambs PE17 4LU	(0480) 63984
Wild Barfield Ltd, Otterspool Way, Watford-by-pass, Watford, WD2 8HX	(0923) 26091

C4 Gas fired radiant heaters

Admiral Ltd, 3/4 Avon Riverside Estate, Portway, Bristol	(0272) 826612	Red Black	45 kW	
Ambi-Rad Ltd, PO Box 30, Cardale St, Rowley Regis, Warley, W Midlands, B65 0LZ	(021) 559 6411	Black	22 kW	
Bullfinch (Gas Equipment) Ltd, Diadem Works, Kings Rd, Tyseley, Birmingham, B11 2AJ	(021) 706 6301	Red	2½ kW	Also portable
Combined Radiant Tube Systems Ltd, 9 Mount Fort Close, Eynesbury, St Neots, Huntingdon, PE19 2NQ	(0480) 75264	Black	30 kW	
EMC Gas Appliances, Anson Rd, Martlesham Heath Industrial Estate, Ipswich, Suffolk, IP5 7BG	(0478) 625151	Red	11 kW	
Gas Fired Products (UK) Ltd, 4/5 Chapel Lane, Claydon, Ipswich, Suffolk	(0473) 830551	Black	51 kW	'Space-ray'
Gray Hill Blackheat Ltd, 52 Nuffield Estate, Poole, Dorset, BH17 7RS	(02013) 71828	Black	35 kW	
GWB Industrial Boilers, PO Box 4, Burton Works, Dudley, W Midlands	(0384) 55455	Red	21 kW	
Joule Manufacturing Ltd, Unit 1, Padgets Lane, Moons Moat South Industrial Estate, Redditch, Worcs, B98 0RA	(0527) 29517	Black	22 kW	
Hamworthy Engineering Ltd, Fleets Corner, Poole, Dorset, BH17 7LA	(0202) 135123	Black	30 kW	
ITT Reznor, Park Farm Rd, Folkestone, Kent	(0303) 59141	Red	30 kW	
Maywick (Hanningfield) Ltd, Rettendon Common, Chelmsford, Essex, CM3 5HY	(0245) 400637	Red	15 kW	
Mechanical Services (International) Ltd, Willment Way, Avonmouth, Bristol, BS11 8DJ	(0272) 821415	Black	40 kW	
Modern Industrial Heating, Station approach, Stoke, D'Abernon, Cobham, Surrey, KT11 3BN	(093) 26 5712	Black	18 kW	
Myson Industrial Space Heaters, PO Box 6, Marl Rd, Kirkby Industrial Estate, Liverpool, L33 7UJ	(051) 546 3541	Red	25 kW	
Pheonix Burners Ltd, 8 Prince Georges Rd, London, SW19 2PX	(01) 648 0964	Black	15 kW	Continuous flue systems
Pioneer Radiant Products Ltd, Kenmare, County Kerry, Ireland	(064) 41344	Red Black	45 kW 22 kW	'Re-verber-ray'

Radiant Superjet Ltd, Clapgate Lane, Woodgate, Birmingham, B32 3BP	(021) 422 7221	Red	16 kW	
Radiant Systems Technology Ltd, Roxby Place, London, SW6 1RS	(01) 381 4278	Black	23 kW	
SBM Radiant Heating Ltd, Southfield Rd, Nailsea, Bristol, BS19 1JE	(0272) 858202	Red	11 kW	
Schwank Ltd, 11a St Georges Rd, Wimbledon, London, SW19 4DR	(01) 946 9501	Red	28 kW	
Speedaire Supply Ltd, Cromford Rd, Langley Mill, Nottingham, NG16 4FL	(07737) 67412	Black	22 kW	'Raytube'

C5 Gas/oil fired warm air heaters

Symbols: IND = industrial heaters, available gas or oil fired except where indicated, DIR = direct fired, BAL = balanced flue gas fired. Input shown in kW

Balmforth Warm Air Ltd, Unit 11F, Cosgrove Way, Beds	(0582) 453555	BAL	9 kW	
Benson Heating Ltd, Ludlow Rd, Knighton, Powys, Wales, LD7 1LP	(0547) 528534	Ind	450 kW	
Bering Engineering Ltd, 49B Peach St, Wokingham, Berks	(0734) 790665	Ind	300 kW	
F. H. Biddle Ltd, Newtown Rd, Nuneaton, Warwicks, CV11 4HP	(0203) 384233	Ind	440 kW	'Waterbury'
Casaire Ltd, Raeburn House, Northolt Rd, Harrow, Middlesex, HA2 0DY	(01) 423 2323	Dir (gas)	3000 kW	
Colt International Ltd, New Lane, Havant, Hants, PO9 2LY	(0705) 451111	Ind	230 kW	
Combat Engineering Ltd, Oxford St, Bilston, W Midlands, WV14 7EG	(0902) 44425	Ind	290 kW	
Covrad Ltd, Sir Henry Parks Rd, Lanley, Coventry, CV5 6BN	(0203) 75544	Ind Dir (gas)	290 kW	'Dravo'
Dantherm, Hither Green, Clevedon, Avon, BS21 6XT	(0272) 876851	Ind	270 kW	
Drugasar Ltd, Deans Rd, Swinton, Manchester, M27 3JH	(061) 793 8700	Bal	15 kW	
G. R. Garbutt and Sons Ltd, Long Beck Trading Estate, Marske-by-Sea, Redcar, Cleveland, TS11 6HB	(0642) 48567	Ind	1000 kW	
ITT Reznor, Park Farm Rd, Folkestone, Kent	(0303) 59141	Gas only	120 kW	
Johnson and Starley Ltd, Rhosili Rd, Brackmills, Northampton, NN4 0LZ	(0604) 62881	Gas only	63 kW	
Mather and Platt Ltd, Fire Engineering Division, Park Works, Manchester, M10 6BA	(061) 205 2321	Gas only	110 kW	'Thermolier'
Wm May (Ashton) Ltd, Cavendish St, Aston-under-Lyne, Lancs, OL6 7BR	(061) 330 3838	Ind	300 kW	'Parmet'
Myson Industrial Space Heaters Ltd, PO Box 6, Marl Rd, Kirkby Industrial Estate, Liverpool, L33 7UJ	(051) 546 3541	Ind	300 kW	'Sunflame'
Niche, Unit 17, Broomhills Industrial Estate, Braintree, Essex, CM7 7RG	(0376) 23265	Dir (gas)	600 kW	'Mistrale'
Potterton International Ltd, Portobello Works, Emscote Rd, Warwick, CV34 5QU	(0926) 493471	BAL		
Powrmatic Ltd, Winter Hay Lane, Ilminster, Somerset, TA19 9PQ	(04605) 3535	Ind	440 kW	
Speedair Supply Ltd, Cromford Rd, Langley Mill, Nottingham, NG16 4FL	(07737) 67412	IND	350 kW	
Wanson Co. Ltd., 7 Elstree Way, Borehamwood, Herts WD6 1SA	(01953) 7111	IND	440 kW	

C6 Portable LPG/oil fired heaters

Andrews Industrial Equipment Ltd, Dudley Rd, Wolverhampton, WV2 3DB	(0902) 58111
Crown Gas Appliances (1981) Ltd, Unit 6, Trafalgar Trading Estate, Jeffreys Rd, Enfield, Middlesex, EN3 7UA	(01) 805 7700

Kongskilde (UK) Ltd, Holt, Norfolk	(026371) 3291		
Ritchie Bennie Ltd, Yeovil, Somerset	(0935) 6868		
Wysepower Ltd, Drove Rd, Everton near Gamlingay, Sandy, Beds, SG19 2HX	(0767) 50011		
W. C. Youngman Ltd, Manor Rd, Crawley, W Sussex, RH10 2QA	(0293) 23411		

C7 Heat recovery products and systems

Symbols: EX = heat exchangers and heater batteries, ECON = economizers, recuperators, flue gas heat recovery, ROT = rotary heat exchanges, HP = heat pipes, COOL = cooling equipment, STOR = thermal storage systems, WA HT = waste heat boilers, RAC = run around coils, CHP = combine heat and power, SYST = heat recovery systems.

AAF Ltd, Bassington Lane, Cramlington, Northumberland, NE23 8AF	(0670) 713477	ROT, RAC, SYST, EX	
Acoustics and Environmetrics Ltd, Winchester Rd, Walton-on-Thames, KT12 2RP	(01) 322 4764	ROT, EX, SYST, RAC	
Airaqua Heat Recovery Systems Ltd, PO Box 7, Wilmslow, Cheshire, SK9 2LG	(0625)615211	EX, ECON, STOR, WAHT	
APV-Hall International Ltd, Hythe St, Dartford, Kent, DA1 1EP	(0322) 27222	EX, COOL, RAC	
Auchard Development Co Ltd, Old Road, Southam, Leamington Spa, CV33 0HP	(092681) 2419	EX	
Auriema Ltd, 442 Bath Rd, Slough, Berks, SL1 6BB	(06286) 4353	EX	'Pyradyne'
Bahco Ventilation Ltd, Bahco House, Beaumont Rd, Banbury, Oxon, OX16 7TB	(0295) 57461	EX, RAC	
Boulten-Kanthal Stephen Newall Ltd, 67 James St, Helensburgh, Dunbar, G84 8XQ	(0436) 71111	SYST	'Kantherm'
Peter Brotherhood Ltd, Lincoln Rd, Peterborough, PE4 6AB	(0733) 71321	CHP SYST	
Burke Thermal Engineering Ltd, Mill Lane, Alton, Hants, GU34 2QG	(0420) 84159	EX, HP, RAC	
Carrier Ross Engineering Co Ltd, Dingwall Rd, Croydon, Surrey, CR9 2SH	(01) 686 0477	SYST	
Carter Industrial Products Ltd, Bedford Rd, Birmingham, B11 1AY	(021772) 3781	COOL	
Clayton Thermal Products Ltd, 12 Rivington Court, Hardwick Grange, Woolston, Warrington, WA1 4RT	(0925) 824279	WA HT	
Climate Equipment Ltd, Highlands Rd, Shirley, Solihull, W Midlands, B90 4NL	(021) 705 7601	EX, HP, RAC	
Corning Process Systems Ltd, Stone, Staffs, ST15 0BG	(0785) 812121	Ex (recovery from corrosive gases)	
Curwen and Newberry Ltd, Redhills Rd, Milton, Stoke-on-Trent, ST2 7ER	(0782) 537788	EX, ROT, HP, RAC, SYST	
Dantherm, Hither Green, Clevedon, Avon, BS21 6XT	(0272) 876851	EX, ECON	
Drytech Engineering Ltd, 78 Morland Rd, Burslem, Stoke-on-Trent, Staffs, ST6 1DY	(0782) 824777	EX, ECON, SYST	
Ellis Tylin, 118/120 Garratt Lane, London,SW18 4EF	(01) 874 0411	CHP	'Totem'
Encomech Engineering Services Ltd, 99/101 East St, Epsom, Surrey, KT17 1EA	(0372) 724344	ECON	Ceramics
Environmental and Thermal Engineering Ltd, Belgreen House, Fountain St, Macclesfield, Cheshire, SK10 1JN	(0625) 625057	EX, ECON	
Eurocoils Ltd, Unit 7A, Eurounit Industrial Estate, Sittingbourne, Kent, ME10 3RN	(0795) 75275	RAC	

Eurovent Environmental Control Ltd, Middlemore Lane, Aldridge, Walsall, W Midlands, WS9 8SP	(0922) 56336	EX, ECON, RAC	
Fabdec Ltd, Grange Rd, Ellesmere, Shropshire, SY12 9DG	(069171) 2811	EX (recovery 'Thermastor' from condensing units)	
Fercell Engineering Ltd, Unit 20, Swaislands Drive, Crayford Industrial Estate, Crayford, Kent, DA1 4HS	(0322) 53131	Briquetter for wood waste	
Flakt Products, Staines House, 158 High St, Staines, Middlesex, TW18 4AR	(0784) 57221	HP, RAC	
G. R. Garbutt Ltd, Longback Trading Estate, Marske-by-Sea, Redcar, Cleveland, TS11 6HB	(0642) 485367	ECON	
Greenbank-Darwen Engineering Ltd, Gate St, Blackburn, BB1 3AJ	(0254) 56401	Air, Radio frequency asisted dryers	
Greenlyon Ltd, Denton, Northampton, NN7 1Dl	(0604) 890144	Recovery from i.c. engines	
Hainault Thermal Services Ltd, 19/21 Fowler Rd, Hainault, Ilford, Essex	(01) 500 9981	COOL	
Heat Seeker Marketing Ltd, PO Box 72, Ipswich, Suffolk, IP4 5TW	(0473) 624168	ECON	
Heatsure Engineering, 32 New St, Ashford, Kent, TN24 8TS	(0233) 35224	SYST	
Heenan Marley Cooling Towers Ltd, PO Box 20, Pheasant St, Worcester, WR1 2DX	(0905) 26961	COOL	
Hiross Ltd, Totman Crescent, Weir Industrial Estate, Rayleigh, Essex, SS6 7UY	(0268) 781818	EX (recovery from condensing units)	
James Howden and Co Ltd, 195 Scotland St, Glasgow, G5 8PJ	(041) 429 2131	EX, ECON, ROT	
Hoval-Farrar, Northgate, Newark, Notts, NG24 1JN	(0636) 72711	EX	
Isoterix Ltd, Mill Works, Mill Crescent, Tonbridge, TN9 1PE	(0732) 358483	HP	
A. Johnson and Co Ltd, Alliance House, 9 Leopold St, Sheffield, S1 2GY	(0742) 737207	EX, HP	'Heatex' 'Furukawa'
Johnson-Hunt Ltd, Hall Works, Astley St, Dukinfield, Cheshire, SK16 4QT	(061) 330 0234	SYST	
Lamanco Heat Recovery Systems, Finway, Dallow Rd, Luton, Beds	(0582) 21697	EX	
MCM Manufacturing Ltd, 22 Arnside Rd, Waterlooville, Hants, PO7 7UP	(07014) 50611	EX (heat recovery ventilators)	
Metal Box Engineering, Chew Moor Lane, West Houghton, Bolton, Lancs, BL5 3JL	(0942) 815111	SYST	
Moducel Ltd, King St, Fenton, Stoke-on-Trent, Staffs, ST4 3ES	(0782) 321317	ROT, RAC	
Munters Rotaire Ltd, 2 Glebe Rd, Huntingdon, Cambs, PE18 7DU	(0480) 51201	EX, ROT	
NEI/International Combustion Ltd, Sinfin Lane, Derby, DE2 9GJ	(0332) 760223	CHP, SYST	
Nuway Ltd, Vines Lane, Droitwich, Worcs, WR9 8NA	(0905) 772331	ECON	
Pressure and Temperature Engineering Ltd, Glebe Rd, Urmston, Manchester, M31 1AL	(061) 747 5097	ECON	'Rondra ecomat'
S and P Coil Products Ltd, Evington Valley Rd, Leicester, LE5 5LU	(0533) 730771	EX, ECON, HP, RAC, SYST	
Serck Heat Transfer, PO Box 598B, Warwick Rd, Greet, Birmingham, B11 2QY	(021) 772 4353	Ex (Recovery from i.c. engines)	
S.O.S. Cooling Ltd, Unit 16, Saddington Rd, Industrial Estate, Churchill Way, Fleckney, Leics, L88 0UD	(0533) 403741	EX, COOL	
Specialist Heat Exchangers Ltd, Freeman Rd, North Hykeham, Lincoln, LN6 9AP	(0522) 683123	EX, COOL	
Stein Atkinson Stordy Ltd, Midland House, Ounsdale Rd, Wombourne, Wolverhampton, WV5 8BY	(0902) 894171	ECON (fume incinerators and high temp systems)	
Stone Boilers, PO Box 5, Gatwick Rd, Crawley, W Sussex, RH10 2RN	(0293) 27711	ECON (recovery from i.c. engines)	

Sulzer Bros (UK) Ltd, Farnborough, Hants, GU14 7LP	(0252) 544311	SYST	
Therm Tech Engineering Ltd, Vale Mill, Huddersfield Rd, Mossley, Ashton-under-Lyne, OL5 9JL	(04575) 5454	ECON, RAC	
Thermal Technology Ltd, Thermal House, 46 Hilporten Rd, Trowbridge, Wilts, BA14 7JH	(02214) 68111	EX, ROT, SYST (recovery from i.c. engines)	
Thermal Efficiency Ltd, Otterspool Way, Watford-by-pass, Watford, Herts, WD2 8HX	(0923) 26091	ECON	
Thermo Engineers Ltd, Chamberlain Rd, Aylesbury, Bucks, HP19 3BU	(0296) 87171	EX, ECON	
Thurley International Ltd, Ripon Rd, Harrogate, N Yorkshire, HG1 2BU	(0423) 61511	EX, SYST	
United Air Coil Ltd, Broadstairs, Kent, CT10 3JP	(0843) 63566	EX, RAC	
United Air Specialists (UK) Ltd, Cranford, Blackdown, Leamington Spa, CV32 6RG	(0926) 311621	EX	
Vent Air Ltd, Unit 4, Aston Rd, Aston Fields Industry Estate, Bromsgrove, Worcs, B60 3EX	(0527) 35566	EX (extract fans with heat recovery)	
Wanson Co Ltd, 7 Elstree Way, Boreham Wood, Herts, WD6 1SA	(01) 953 7111	ECON, RAC, WA HT	
Wilkins and Wilkins Ltd, Bridge approach Hamworthy, Poole, Dorset	(0202) 673174	ECON, WA HT	

C8 Heating system controls and building energy management systems

Symbols: EMS = total energy management systems, HSC = heating system controls, MD = maximum demand controls, TS = time switches, TRV = thermostatic radiator valves

AEG Telefunken (UK) Ltd, Eskdale Rd, Winnersh, Wokingham, Berks, RG11 5PF	(0734) 698330	EMS, TS	
Allen Martin Electronics Ltd, Marlborough Works, Thompson Ave, Wolverhampton, WV2 3NP	(0902) 58942	HSC	'People sensitive' controls
AMF-Venner Ltd, AMF House, Whitby Rd, Bristol, BS4 4AZ	(0272) 778383	HSC, TS	
Ancom Ltd, Devonshire St, Cheltenham, Glos, GL50 2LT	(0242) 513861	EMS, TS	
Atlantic Instruments, 1 Minto Ave, Altens Industrial Estate, Aberdeen, AB1 4JZ	(0224) 895024	EMS	
Building Systems (Marketing) 1 Crossland Rd, Redhill Surrey, RH1 4AN	(0737) 62507	EMS	
Burgess Energy Systems, Park St, Stafford, ST17 4AF	(0785) 47226	EMS, TS, HSC	
Chalmor Ltd, Unit 1, Albert Rd Industrial Estate, Luton, Beds, LU1 3QF	(0582) 429566	EMS	
Colt International Ltd, Havant, Hampshire, PO9 2LY	(0705) 451111	EMS	
Danfoss Ltd, Horsenden Lane, South Greenford, Middlesex, W36 7QE	(01) 998 5040	HSC, TRV	
Drayton Controls (Engineering) Ltd, Chantry Close, West Drayton, UB7 7SP	(0895) 444012	HSC, TRV	'Theta'
Electronic Control Systems (Southern) Ltd, 3 Wimbridge Close, New Wimpole, Royston, Herts, SG5 5QQ	(0223) 245191	MD	
Energy Management Ltd, Thames Tower, 99 Burleys Way, Leicester, LE1 3TT	(0533) 21312	EMS	
Energy Technology, 4 Mercia Way, Bells Close Industrial Estate, Newcastle-upon-Tyne, NE15 6UF	(0632) 644744	EMS	EM 2000
Faral Tropical Dor-to-Dor Depot, Gatwick Rd, Crawley, Sussex	(0293) 541705	TRV	
Ferranti Instrumentation Ltd, Moston, Manchester, M10 0BE	(061) 681 2071	EMS	'Cedrec'
Freeman Enercon Ltd, Science Park, Milton Rd, Cambridge, CB4 4BH	(0223) 315432	EMS	'Xyntax'
Gent Ltd, Temple Rd, Leicester, LE5 4JF	(0533) 730251	EMS, TS	

Company	Phone	Codes	Notes
Grasslin (UK) Ltd, Vale Rise, Tonbridge, Kent, TN9 1TB	(0732) 359888	TS	
Hamworthy Engineering Ltd, Fleets Corner, Poole, Dorset, BH17 7LA	(02013) 5123	HSC	
Holec Energy, Atlantic House, Jengers Mead, Billingshurst, Sussex, RH14 9PB	(04381) 4311	EMS	
Honeywell Control Systems Ltd, Charles Square, Bracknell, Berks, RG12 1EB	(0344) 24555	EMS, TS, TRV	
Horstmann Engineering Products, New Bridge Works, Bath, BA1 3EF	(0225) 21141	HSC, TS, MD	
Information Transmission Ltd (ITL), Unit 1, Bone Lane, Newbury, Berks, RG14 5PF	(0635) 35558	EMS	Building breathalyzer
Invicta Energy Management, Pegasus House, Elwick Rd, Ashford, Kent, TN3 1PD	(0233) 38344	EMS	Energy minda
ISS Clorius, Worton Drive, Reading, Berks RG2 0TG	(0734) 752000	EMS	
JEL Energy Conservation Services Ltd, Bramabl Moor Lane Industrial Park, Pepper Rd, Hazel Grove, Stockport SK7 5BW	(061) 487 1111		
Johnson Control Systems, PO Box 79, Stonehill Green, Westlea Down, Swindon, Wilts, SN5 7DD	(0793) 26141	EMS	
Landis and Gyr Ltd, Victoria Rd, North Action, London, W3 6XS	(01) 992 5311	HSC, TRV, MD	
Leading Edge Technologies Ltd, 58/66 Morley Rd, Tonbridge, Kent, TN9 1RD	(0732) 361177	EMS	
Micro Energy Management Systems, St Annes House, 45 Park St, Luton, Beds LU1 3JX	(0582) 421861	EMS	
Merlin Gerin, Stafford Park 4, Telford Shropshire, TF3 3BL	(0952) 618061	TS	
NEI Electronics Ltd Industrial Controls, Team Valley Trading Estate, Kingsway, Gateshead, Tyne and Wear, NE11 0QJ	(0632) 870811	EMS	
NEI Projects Ltd, Cuthbert House, All Saints Office Centre, Newcastle-upon-Tyne, NE1 2DP	(0632) 328880	EMS	
Potterton International Ltd, Portobello Works, Emscote Rd, Warwick, CU34 5QU	(0926) 493420	HSC	'EP2000'
Randall Electronics Ltd, Ampthill Rd, Bedford, MK42 9ER	(0234) 64621	HSC, TS	
Rapaway Ltd, 35 Park Ave, Solihull, W Midlands	(021) 705 3360	HSC, TS	
Robertshaw Controls Ltd, Raans Rd, Amersham, Bucks, HP6 6JT	(02403) 22311	EMS, TS	
H. Saacke Ltd, Fitzherbert Rd, Farlington, Portsmouth, Hants, PO6 1RX	(0705) 383111	EMS	
Salmon Electronics Ltd, PO Box 26, Croft, Darlington, DL2 2TN	(0325) 721368	HSC	
Sangamo Controls, Port Glasgow, Renfrewshire, PA14 5XG	(0475) 45131	EMS, TS, Fuel sensors	
Satchwell Control Systems Ltd, PO Box 57, Farnham Rd, Slough, Berks, SL1 4UH	(0753) 23961	HSC, TRV, EMS	
Satchwell Sunvic Ltd, Watling St, Motherwell, ML1 3SA, Scotland	(0698) 66277	HSC, TRV	
Sauter Automation Ltd, 165 Bath Rd, Slough, Berks, SL1 4AA	(0753) 39221	HSC, TS	
Sension Electronic Ltd, Denton Drive Industrial Estate, Northwich, Cheshire, CW9 7LU	(0606) 44321	MD Demand profile recorders	
Sensors and Systems Ltd, High St, Melbourne Derby	(03361) 2228	HSC	
SES Fuel Economy Systems Ltd, Sheep St, Stow-on-the-Wold, Glos	(0451) 30262		
Simplex Time Recorder Co (UK) Ltd, Homefield Industrial Estate, Holdsworth, Halifax, HX2 0RE	(0422) 246281	EMS, TS	
Smith Meters Ltd, Rowan Rd, Streatham Vale, London, SW16 5JE	(01) 764 5011	HSC, TS	
Square D Ltd, Cheney Manor, Swindon, Wilts, SN2 2QG	(0793) 36222	EMS	
Staefa Control System (UK) Ltd, Staefa House, 17 Staines Central Trading Estate, Staines, Middlesex	(0784) 62151	EMS, TS	
Superswitch Electric Appliances Ltd, 7 Station Trading Estate, Blackwater, Camberley, Surrey, GU17 9AH	(0276) 34556	TS	
Systemation Ltd, Systemation House, Pond Rd, Shoreham, W Sussex, BN4 5WU	(07917) 63840	EMS	

TA Controls Ltd, Lea Industrial Estate, Lower Luton Rd, Harpenden, Herts, AL5 5EQ	(0582) 767991	EMS
Tacotherm-Lamaco Ltd, 122/126 Kilburn High Rd, London, NW6 4HY	(01) 624 0448	EMS
Taylor Instruments Ltd, Gunnels Wood Rd, Stevenage, Herts, SG1 2EL	(0438) 312366	EMS, TS
Transmitton Ltd, Smisby Rd, Ashby-de-la-Zouch, Leics, LE6 5UG	(0530) 415941	EMS
Trend Control Systems Ltd, Foundry Lane, Horsham, W Sussex, RH13 5TS	(0403) 69612	EMS
Turnball Control Systems, Blackwater Trading Estate, Worthing, Sussex, BN14 8NW	(0903) 205277	HSC
Westinghouse Electric, Haden House, Argyle Way, Stevenage, Herts, SG1 2AH	(0438) 726177	EMS, MD

C9 Flue gas analysis, monitoring and control

Symbols: CHEM = chemical analysis, O_2 = oxygen measurement by zirconia probe. All manufacturers offer portable equipment and most are based on microprocessors.

Colwick Instruments Ltd, 9 New Vale Rd, Colwick, Nottingham, NG4 2EA	(0602) 249947	CHEM	
Dawe Instruments Ltd, Concord Rd, Western Avenue, London, W3 0SD	(01) 992 6751	ELEC	
Information Transmission Ltd, Unit 1, Bone Lane, Newbury, Berks, RG14 5PF	(0635) 35558	ELEC (also CO monitoring)	
Kane-May Ltd, Burrowfield, Welwyn-Garden City, Herts, AL7 4TU	(07073) 31051	O_2	
Land Combustion Ltd, Carrwood Rd, Chesterfield Trading Estate, Sheepbridge, Chesterfield, SA1 9QB	(0246) 453581	Infrared measurement	
Neotronics Ltd, Parsonage Rd, Takeley, Bishops Stortford, Herts, CM22 6PU	(0279) 870182	CHEM	
Shawcity Ltd, Unit 2, Pioneer Rd, Faringdon, Berks, SN7 7BU	(0367) 21675	CHEM	'Bacharach' (USA)
Taylor Instrument Ltd, Crowborough, Sussex, TN6 3DU	(0892) 62181	O_2	'Servomex'
Telegan Ltd, Legion House, Godstone Rd, Kenley, Surrey, CR2 5YS	(01) 668 8251	O_2	
Westinghouse Electric, Haden House, Argyle Way, Stevenage Herts, SG1 2AH	(0438) 726177	O_2	

C10 AC Induction motors: control of power factor and energy consumption

Symbols: PF = power factor correction capacitors, NASA = motor power input control by wave chopping as licensed by N.A.S.A.

AEG Telefunken (UK) Ltd, Eskdale Rd, Winnersh, Wokingham, Berks, RG11 5PF	(0734) 698330	PF
BICC Bryce Capacitors Ltd, Chester Rd, Helsby, Cheshire, WA6 0DJ	(09282) 2700	PF
Conder Electronics, Abbotts, Barton House, Winchester, Hants, SO23 7SJ	(0962) 63577	NASA
Denco Magna Controls Ltd, 14 St Martin, Hereford, HR2 7RE	(0432) 53434	NASA
GEC Measurements Ltd, St Leonards Works, Stafford, ST17 4LX	(0785) 3251	PF
Johnson and Phillips (Capacitors) Ltd, Hollands Road, Haverhill Suffolk, CB9 8PR	(0440) 3441	PF
Simon Relays (ERI) Ltd, 2 St Michaels Rd, Sandhurst, Surrey	(0252) 879383	NASA

C11 Instruments ·

Symbols: AIR = air flow, manometers, pitot tubes, GAS = gas flow, STEAM = steam flow, FLUID = fluid flow, TEMP = temperature, TACHO = revolution counters, ELEC = electrical instruments, REC = recording instruments, HUM = humidity, PROC = process instrumentation.

Air Flow Developments Ltd, Lancaster Rd, High Wycombe, Bucks, HP12 3QP	(0494) 25252	AIR, HUM	
Air Instrument Resources Ltd, Monument Industrial Park, Chalgrove, Oxford, OX9 7RW	(0865) 891190	AIR	
Analysis Automation Ltd, Southfield House, Eynsham, Oxford, OX8 1JD	(0865) 881888	PROC	
Arkon Instruments Ltd, Whaddon Works, Cheltenham, Glos, GL52 5EP	(0242) 27953	STEAM, GAS, FLUID, PROC	
Auriema Ltd, 442 Bath Rd, Slough, Bucks, SL1 6BB	(06286) 4252	TEMP	'Wahl'
AVO Ltd, Dover, Kent, CT17 9EN	(0304) 202620	ELEC	
Babcock Bristol Ltd, Oldington Vale Trading Estate, Stourport Rd, Kidderminster, Worcs	(0562) 743001	TEMP, REC	
Broyce-Marvid Ltd, Pool St, Wolverhampton, WV2 4HN	(0902) 773746	ELEC, TACH, TEMP	
Bruel and Kjaer (UK) Ltd, Cross Lances Rd, Hounslow, Middlesex, TW3 2AE	(01) 570 7774	ELEC, REC	Thermal comfort
Casella Ltd, Regent House, Britannia Walk, London, N1 7ND	(01) 253 8581	TEMP, HUM, REC	
James Clarke and Co Ltd, 111 Longfellow Rd, Dudley, W Midlands, DY3 3EF	(09073) 3923	AIR	Pitot static tubes
Comark Electronics Ltd, Rustington, Littlehampton, W Sussex, BN16 3QZ	(09062) 71911	TEMP	
Compact Instruments Ltd, Binary Works, Park Rd, Barnet, Herts, EN5 5SA	(01) 440 6663	ELEC, TACHO, TEMP	Thermo couples
Crompton Instruments Ltd, 50/52 Mayfair, Northampton, NN1 1NY	(0604) 30201	ELEC	
Dawe Instruments Ltd, Concord Rd, Western Avenue, London, W3 0SD	(01) 992 6751	GAS, STEAM, ELEC	
Fecon Ltd, Fecon House, Garth Rd, Morden, Surrey	(01) 330 2911	TEMP, HUM	'Novasina'
Gervase Instruments Ltd, Britannia Works, Cranleigh, Surrey, GU6 8ND	(0483) 275566	AIR, FLUID, STEAM	
Griffin and George, Ealing Rd, Wembley, HA0 1HS	(01) 997 3344	'Gipsi' portable modular multi-meter	
Gulton Ltd, Graphic Instruments Division, Maple Works, Old Shoreham Rd, Hove, Sussex, BN3 7EY	(0273) 77801	REC	
James Hugh Group Ltd, 150/152 West End Lane, London, NW6 1SD	(01) 328 3121	TEMP	
International Measurement and Control Systems, 72 Dinorben Ave, Fleet, Hants, GU13 9SH	(02514) 21759	GAS	'Quantometers'
ISS Clorius Ltd, Worton Drive, Reading, Berks, RG2 0TG	(0734) 752000	GAS, FLUID	Heat Metering
Jenway Ltd, Gransmore Green, Felsted, Dunmow, Essex, CM6 3LB	(0371) 820122	TEMP	
Kent Industrial Measurements Ltd, 4 Rosemary Lane, Coldhams Lane, Cambridge, CB1 3LQ	(0223) 249121	PROC	Infrared gas analysers
Land Pyrometers Ltd, Wreakes Lane, Dronfield, Sheffield, S18 6DJ	(0246) 417691	TEMP	Infrared
Moisture Controls and Measurement Ltd, Thorp Arch Trading Estate, Wetherby, Yorks, LS23 7BJ	(0937) 843927	HUM, TEMP	
MSA (Britain) Ltd, East Shawhead, Coatbridge, ML5 4TD	(0236) 24966	PROC, GAS	
Neotronics Ltd, Parsonage Rd, Takeley, Bishops Stortford, Herts, CM22 6PU	(0279) 870182	AIR	Micro manometer
Neptune Measurement Ltd, PO Box 2, Dobcross, Oldham, Lancs, OL3 5BD	(04577) 4822	PROC, FLUID	
Portec Instrumentation Ltd, 63 Castle St, Luton, Beds, LU1 3AG	(0582) 32613	TEMP	

Rayleigh Instruments, Brook Rd, Rayleigh, Essex, SS6 7XH	(0268) 747911	TEMP, TACHO, ELEC, REC	
The Record Electrical Co Ltd, PO Box 19, Atlantic St, Altrincham, Cheshire, WA14 5DB	(061) 928 6211	REC	
Response Co, Froxfield, Petersfield, Hampshire, GU32 1DX	(0730) 64645	ELEC	Elec consumption monitoring
Samson Controls (London) Ltd, 50 Holmethorpe Avenue, Redhill, Surrey, RH1 2NL	(0737) 66391	TEMP	Heat flow meters
Servis Recorders (Industrial) Ltd, 19 London Rd, Gloucester, GL1 3EZ	(0452) 24125	REC	
Solomat (UK) Ltd, Broadway, Woodbury, Exeter, Devon	(0395) 32199		
Tacotherm-Lamaco Ltd, 122/126 Kilburn High Rd, London, NW6 4HY	(01) 624 0448	TEMP	Heat flow meters
Terwin Instruments Ltd, Tollemache Rd, Spittlegate Level Industrial Estate, Grantham Lincs, NG31 7UN	(0476) 65797	TEMP, HUM	
Testoterm Ltd, Old Flour Mill, Queen St, Emsworth, Hants, PO10 7BT	(02434) 77222	AIR, TEMP, HUM	
Thorn EMI Instruments Ltd, Archcliffe Rd, Dover, Kent, CT17 9EN	(0304) 202620	ELEC	
Unity Power System Ltd, Legion House, Godstone Rd, Kenley, Surrey, CR2 5YS	(01) 668 8251	TEMP	Infrared
Vaisala (UK) Ltd, 11 Billing Rd, Northampton, NN1 5AW	(0604) 22415	TEMP, HUM	'Humicap' (Finnish)

C12 Heat pumps and heat pump drying

Symbols: AA = air to air, AW = air to water, WA = water to air, WW = water to water, DRYING = specialist in application to industrial drying

AAF Ltd, Bassington Lane, Cramlington, Northumberland, NE23 8AF	(0670) 713477	WA	Allis Chalmers (USA)
Accent Fano Technik (UK) Ltd, Riverside Industrial Estate, Sandy Lane, Stourport-on-Severn, Worcs	(02993) 78363	AW	
Airedale International Air Conditioning Ltd, Clayton Wood Rise, West Park, Leeds, LS16 6RF	(0532) 742011	AA, WW	
Anco Products (UK), Daish Way, Dodnor Industrial Estate, Newport, Isle of Wright	(0983) 521465	AW, WW	
Andrews Industrial Equipment, 36 Lewis Rd, Mitcham, Surrey, CR4 3XQ	(01) 648 6174	AA, AW, WA	
Applied Energy Systems, 1 Whippendell Rd, Watford, Herts, WD1 7LZ	(0923) 42222	AA, AW, WA, WW	
APV Hall International Ltd, Hythe St, Dartford, Kent, DA1 1EP	(0322) 27222	AA, AW, WA, WW	
Atlantic Heat Pumps Ltd, Service House, 156 North Rd, Cardiff, CF4 3BH	(0222) 621436	AW, WW	
F. H. Biddle Ltd, Newtown Rd, Nuneaton, Warwickshire, CV11 4HP	(0203) 384233	AA, AW, WW	
Carlyle Air Conditioning Co Ltd, Knightsbridge House, 197 Knightsbridge, London SW7 1RB	(01) 589 8111	AA, AW, WA, WW	
Conder M. and E. Products, Abbotts Barton House, Worthy Rd, Winchester, Hants, SO23 7SJ	(0962) 63577	AA, AW, WA, Conder, Hushon	
Coronet Heat Pumps, Unit 2, The Causeway, Maldon, Essex, CM9 7PU	(0621) 56611	AW, WW	
CRU Products, St Albans House, Portland St, Leamington Spa, Warwicks, CV32 5EZ	(0926) 36626	AW	
Dantherm, Hither Green, Clevedon, Avon, BS21 6XT	(0272) 876851	DRYING	
Delrac Ltd, 128 Malden Rd, New Malden, Surrey, KT3 6BR	(01) 942 2442	AA, AW	'Delchi (Italy)
Denco Air Ltd, PO Box 11, Holmer Rd, Hereford, HR4 9SJ	(0432) 277277	DRYING	
Drytech Ltd, 78 Moorland Rd, Burslem, Stoke-on-Trent, Staffs, ST6 1DY	(0782) 824777	DRYING	

Ductwork Engineering Systems Ltd, Airport Trading Estate, Biggin Hill, Kent, TN6 3BW	(09594) 71211	AA, AW, WA, WW	Carlyle
Eastwood Heating Developments, Portland Rd, Shirebrook, Mansfield, Nottingham, NG20 8TY	(062) 374 8484	AW, WW	
Environheat Ltd, Letts Rd, Far Cotton, Northampton	(0604) 66341	AA, AW	
IMI-Marstair, Unit 4, Armytage Road Industrial Estate, Wakefield Rd, Brighouse, W Yorkshire HD6 1PT	(0484) 714361	AA	
A. Johnson and Co (London) Energy Conservation Division, Aldwych House, Aldwych, London, WC2B 4EL	(01) 404 0755	AA	
Lennox Industries, PO Box 43, Lister Rd, Basingstoke, Hants, RG22 4AR	(0256) 61261	AA, AW, WW	
Myson Copperad Heat Pump Division, Old Wolverton Rd, Wolverton, Milton Keynes, Bucks, MK12 5PT	(0908) 312641	AA, AW	
Pioneer Air Conditioning, Beach House, Green Street, Green Road, Darenth, Kent, DA2 6PS	(0322) 25211	AA	
Qualitair (Air Conditioning), Castle Road Eurolink, Sittingbourne, Kent, ME10 3RH	(0795) 75461	AA	
RA (Air Conditioning) Company, Holland Rd, Haverhill, Suffolk, CB9 8PT	(0440) 702653	AA, AW, WA	
Searle Manufacturing, Newgate Lane, Fareham, Hants, PO14 1AR	(0329) 236151	AA	
Stal-Levin Ltd, River Pinn Works, Yiewsley High St, W Drayton, Middlesex, UB7 7TA	(08954) 46561	AA, AW, WA, WW	'Stal' Sweden
Stiebel Eltron Ltd, Brackmills, Northampton, NN4 0ED	(0604) 66421	AW, WW	W Germany
Sulzer Bros (UK), West Mead, Farnborough, Hants	(0252) 544311	AA, AW, WA, WW	W Germany, Switzerland
Temperature, Newport Rd, I.O.W. PO36 9PH	–	AA, WA	
Thermal Engineering Systems Ltd, Uffculme, Cullompton, Devon, EX15 8AJ	(0884) 40216	AW	
Thermecon Ltd, Baston Rd, Aston Fields Industrial Estate, Bromsgrove, Worcs	(0527) 78800	AA, AW	Electra Israel
Toshiba (UK) Ltd, Toshiba House, Frimley Rd, Camberley, Surrey, GU16 5JJ	(0276) 6222	AA	
Trace Heat Pumps Ltd, Eastways Industrial Park, Witham, Essex, CM8 3TQ	(0376) 515511	AA, AW, WA, WW	
Trane (UK) Ltd, Gastons Wood, Reading Rd, Basingstoke, Hants, RG24 0TW	(0256) 794731	AA, AW, WA, WW	
Trendpam Engineering, 62 High St, Ewell, Surrey, KT17 1RL	(01) 394 2555	AA, AW, WA, WW	
Westair Ltd, Thames Works, Central Ave, E Molesey, Surrey, KT8 0QZ	(01) 941 4184	DRYING	
Willison Controls Ltd, Dallas Rd, Bedford, MK42 9ES	(0234) 52286	DRYING	
York Division of Borg Warner, Gardiners Lane South, Basildon, Essex, SS14 3HE	(0268) 287671	AA, AW, WA, WW	

The above list does not attempt to give comprehensive cover of air conditioning even though this is the main interest of many of the companies listed.

C13 Fans and ventilation

Maximum fan outputs are quoted in m^3/s according to type. Symbols: PRO = propeller, AX = axial, CEN = centrifugal, BLO = blower, A.H. = air handling, F.V. = fire ventilation, N.V. = natural ventilation, R.U. = roof and wall units (powered), ACCESS = grills or diffusers, dampers and other accessories

Advanced Air (UK) Ltd, 3 Cavendish Rd, Bury St Edmunds, Suffolk, IP33 3TE	(0284) 701351	ACCESS
Airflow Developments Ltd, Lancaster Rd, High Wycombe, Bucks, HP12 3QP	(0494) 25252	AX 2.7 m^3/s, CEN 1.8 m^3/s
Airscrew Howden Ltd, Weybridge, Surrey, KT15 2QR	(0932) 45511	AX 95 m^3/s, PRO 9 m^3/s
Aldridge Air Control Ltd, Middlemore Lane, Aldridge, W Midlands, WS9 8SP	(0922) 56333	Ventilation systems

Applied Air Plant Ltd, Unit 4 Silver Road, White City Industrial Park, Wood Lane, London, W12 7SG	(01) 740 9293	A.H.
Bahco Ventilation Ltd, Bahco House, Beaumont Rd, Banbury, Oxon, OX16 7TB	(44295) 57461	A.H. to 40 m^3/s
B.O.B. Fans Ltd, Coleman St, Derby, DE2 8NN	(0332) 74112	CEN 55 m^3/s
Carter Industrial Products Ltd, Bedford Rd, Birmingham, B11 1AY	(021) 772 3781	Ventilation and dust collection systems
Clipper Air Handling Units Ltd, Raans Rd, Amersham, Bucks, HP6 6HY	(02403) 21212	A.H.
Colt International Ltd, Havant, Hampshire, PO9 2LY	(0705) 451111	N.V., F.V., R.U.
Dantherm Ltd, Hither Green, Clevedon, Avon, BS21 6XT	(0272) 876851	A.H. 6 m^3/s
Davidson and Co Ltd, Sirocco Engineering Works, Bridge End, Belfast, BT5 4AG	(0232) 57251	AX 44 m^3/s
Dougall Sutton Ventilation Ltd, 36 Wellingborough Rd, Northampton, NN1 4EP	(0604) 37331	A.C. Ventilation systems
Fan Systems Ltd, PO Box 8, Rochdale Rd, Greetland, Halifax, Yorks, HX4 8HB	(0422) 78131	CEN 95 m^3/s. Also special purpose axial
Fischbach Ventilation Ltd, 9 Jefferson Way Thame Industrial Estate, Thame, Oxon, OX9 3SY	(084421) 5611	CEN 5 m^3/s Variable speed
Fläkt Products, Staines House, 158 High St, Staines, Middlesex, TW18 4AR	(0784) 57221	A.H. and AX 100 m^3/s
GEC Xpelair Ltd, PO Box 220, Deykin Ave, Witton, Birmingham, B6 7JH	(021) 327 1984	PRO, R.U.
Gradwood Ltd, Edgeley Road Trading Estate, Edgeley Rd, Stockport, Cheshire, SK3 0TF	(061480) 9629	R.U. 25 m^3/s, F.V.
Greenwood Airvac Ventilation Ltd, PO Box 3, Brookside Industrial Estate, Rustington, Littlehampton, W Sussex, BN16 3LH	(09062) 71021	N.V., F.V.
I. and J. Hyman PLC, Adelaide Mill, Stampstone St off Shaw Rd, Oldham, Lancs, OL1 3LU	(061) 652 1311	PRO, AX, CEN 5 m^3/s
JJ Ventilation (International) Ltd, 13 Dowry Square, Bristol, BS8 4SL	(0272) 291295	A.H., R.U.
Keith Blackman, Tufnell Way, Colchester, Essex, CO4 5AN	(0206) 574230	CEN 100 m^3/s
Kiloheat Ltd, Vestry Estate, Sevenoaks, Kent, TN14 5EL	(0732) 459224	CEN 11 m^3/s R.U.
The London Fan Co Ltd, 75/81 Stirling Rd, London, W3 8DJ	(01 992) 6923	AX to 1000 mmm diam, PRO to 750 mm diam
McKenzie Martin Ltd, Eton Hill Rd, Radcliffe, Manchester, M26 9US	(061) 723 2234	N.V. F.V.
Mathews and Yates Ltd, Cyclone Work, Swinton, Manchester, M27 2AB	(061) 794 7311	AX 38 m^3/s CEN 83 m^3/s
Moducel Ltd, 165 King St, Fenton, Stoke-on-Trent, Staffs, ST4 3ES	(0782) 321317	R.U. 20 m^3/s, ACCESS
Myson Brooks Ltd, Vulcan Way, New Addington, Croydon, CR9 2AS	(0689) 41080	AX to 1900 mm diam PRO to 1000 mm diam
Norvent Ltd, Norris House, Crawhall Rd, Newcastle upon Tyne, NE1 2BZ	(0632) 322141	A.H.
Powrmatic Ltd, Winterhay Lane, Ilminster, Somerset, TA19 9PQ	(04605) 3535	N.V., F.V., R.U., 8 m^3/s
Radial and Axial Fans Ltd, Folly Lane, St Albans, Herts, AL3 5JU	(0727) 35181	AX to 75 m^3/s. CEN to 140 m^3/s
Redlaw (Environmental Equipment) Ltd, 21A Brighton Rd, South Croydon, Surrey, CR2 6EA	(01) 680 3440	ACCESS incl variable air volume terminals
H. H. Robertson (UK) Ltd, Cromwell Rd, Ellesmere Port, S Wirrall, Cheshire, L65 4DS	(051) 355 3622	N.V.
Roof Units Marketing Ltd, Peartree Rd, Dudley, W Midlands, DY2 0QU	(0384) 74062	R.U. to 14 m^3/s
Secomak Air Products Ltd, Honeypot Lane, Stanmore, Middlesex, HA7 1BE	(01) 952 5566	BLO
Smiths Industries Precision Fans, Witney, Oxon	(0993) 2929	CEN 16 m^3/s
Standard and Pochin Ltd, Evington Valley Rd, Leicester, LE5 5LS	(0533) 736114	CEN 940 m^3/s, AX 28 m^3/s, A.H., A.C.
Trane Air Conditioning, Sunbury-on-Thames, Middlesex	(09327) 80321	CEN 160 m^3/s, AX 20 m^3/s
Trox Bros Ltd, Caxton Way, Thetford, Norfolk, IP24 3SQ	(0842) 4545	AX 33 m^3/s, PRO 5 m^3/s
United Air Specialist (UK) Ltd, Cranford, Blackdown, Leamington Spa, CV32 6RG	(0926) 311621	Ventilation systems

Vent-Axia Ltd, Fleming Way, Crawley, W Sussex, RH10 2NN	(0293) 26002	R.U.
Ventilation Jones, Celcius House, Slough Lane, Saunderton, High Wycombe, Bucks, HB14 4HN	(024024) 4171	ACCESS
VES Andover Ltd, 2E West Way, Walworth Industrial Estate, Andover, Hants, SP10 5AR	(0264) 66325	AX 19 m³/s, CEN 7 m³/s, A.H.
Warm Air Components Ltd, Chase Park Industrial Estate, Ring Rd, Burntwood near Walsall, Staffs, WS7 8JQ	(05436) 73111	ACCESS
Woods of Colchester Ltd, Tufnell Way, Colchester, Essex, CO4 5AR	(0206) 44122	AX 200 m³/s, PRO 16 m³/s

C14 Filters and filtration systems

Symbols: MED = manufacturers of filtration media; others offer air cleaning or filtration units, i.e. ION = negative ionizers, ELEC = electrostatic precipitators, FILT = based on fabric or viscous filters, WELD = Weld fume control, SYSTEMS = supplies of industrial filtration plant

AAF Ltd, Bassington Lane, Cramlington, Northumberland, NE23 8AF	(0670) 713477	SYSTEMS
Air Filter Systems Ltd, Unit 14, Green Lane Trading Estate, 1st Ave, Smallheath, Birmingham, W Midlands	(021) 7720761	SYSTEMS
Aldridge Air Control Ltd, Middlemore Lane, Aldridge, W Midlands, WS9 8SP	(0922) 56333	SYSTEMS
Andee Electronics Ltd, 9 Chestnut Walk, Little Common, Bexhill-on-Sea, E Sussex, TN39 4PS	(04243) 3982	ION, ELEC
Auchard Development Co Ltd, Old Rd, Southam near Leamington Spa, CV33 0HP	(092) 681 2419	MED
Automet Filtration Ltd, Charles Lane, Mill Haslingden, Lancashire, BB4 5EQ	(0706) 229113	MED
BACC (Engineering) Ltd, Edgar Rd, Dover, Kent, CT17 0ES	(0304) 208422	ELC
Beltran Ltd, Sunderland St, Macclesfield, Cheshire, SK11 6JF	(0625) 615529	Fum Coalescers
Bondina Industrial Ltd, Greetland, Halifax, W Yorks, HX4 8NJ	(0422) 73528	MED
Carter Industrial Products Ltd, Bedford Rd, Birmingham, B11 1AY	(021) 772 3781	SYSTEMS
Colt International Ltd, Havant, Hampshire, PO9 2LY	(0705) 451111	FILT
Covec (Air) Ltd, 220 Ashley Rd, Parkstone, Poole, Dorset, BH14 9BY	(0202) 736584	SYSTEMS
C.T. (London) Ltd, Walnut Tree House, Woodbridge Park, Guildford, Surrey, GU1 1EL	(0483) 502020	ELEC
Davis Industrial (Filters) Ltd, Imperial Way, Croydon, CR9 3DR	(01) 686 7561	MED
Defuma Ltd, Stukeley Road Industrial Estate, Huntingdon, Cambs, PE18 6HZ	(0480) 56678	WELD
Donaldson Torit Ltd, 65 Market St, Hednesford, Staffs	(05438) 5515	SYSTEMS
Eaton-Williams Products Ltd, Station Rd, Edenbridge, Kent, TN8 6EG	(0732) 863447	ELEC, FILT
Electrostatic Ltd, 23/25 Lower St, Stanstead, Essex, CM24 8LN	(0279) 814993	ION, ELEC
Environco Ltd, Rampart House, Victoria St, Windsor, Berks, SL4 1EH	(07535) 67024	ELEC
E.P. Air Pollution Control Ltd, Brunel Rd, Earlstrees Industrial Estate, Corby, Northants, NN7 2JW	(0536) 200381	ELEC, SYSTEMS
Eurovent, Leighswood Grove, Aldridge, W Midlands, WS9 8SY	(0922) 5311	MED
The Filtermist Co, Faraday Drive, Stourbridge Rd, Bridgnorth, Shropshire, WV15 5BA	(07462) 5361	Centrifugal oil mist filtration
W. L. Gore and Associates (UK) Ltd, Queensferry Rd, Dunfermline, Fife	(0383) 26777	MED
Honeywell Control Systems Ltd, Charles Square, Bracknell, Berks, RG12 1EB	(0344) 24555	ELEC
Horizon Mechanical Services (International) Ltd, Wilment Way, Avonmouth Way, Avonmouth, Bristol, BS11 8DJ	(0272) 821415	ELEC
Hygeia Ltd, Priory House, Somerford Rd, Christchurch, Dorset, BH23 3PY	(0202) 476666	ION
Lake Style Ltd, 30 Thurnview Rd, Evington, Leicester, LE5 6HJ	(050981) 3624	SYSTEMS (grinding and fettling)
Medion Ltd, 4 Beadles Lane, Old Oxted, Surrey, RH8 9HQ	(08833) 2641	ION
Morley Bros (Huddersfield) Ltd, Lincoln St, St Andrews Rd, Huddersfield, HD1 6RU	(0484) 26223	MED
Myson Brooks Ltd, Electrostatic Filtration Division, Vulcan Way, New Addington, Croydon, CR9 2AS	(0689) 41080	ELEC

Nederman Ltd, 462 Walton Summit Centre, Bamberbridge, Preston, Lancs, PR5 8AX	(0772) 3472	WELD
Nilfisk Ltd, Newmarket Rd, Bury-St-Edmunds, Suffolk, IP33 3SR	(0284) 63163	WELD
J. Plymoth Ltd, 51 High St, Banbury, Oxon, OX16 8LA	(0295) 62503	WELD
PVH Engineering Ltd, Redhills Rd, Milton, Stoke-on-Trent, ST2 7ER	(0782) 534235	FILT
RDM Industrial Services Ltd, Parkfield Industrial Estate, Kemp St, Middleton, Manchester, M24 4AA	(061) 643 9333	Systems (spray painting)
Sangre Engineering Ltd, Heming Rd, Washford Industrial Estate, Redditch, Worcs	(0527) 24782	SYSTEMS
S. S. Stott Ltd, Haslingden, Rossendale, Lancs, BB4 6PE	(0706) 213666	SYSTEMS
Stubs Welding Ltd, Wilderspool Causeway, Warrington, Lancs, WA4 6QV	(0925) 50441	WELD
Trion Ltd, Brunel Gate, West Portway Industrial Estate, Andover, Hants, SP10 2SL	(0264) 64622	ELEC
Trox Bros Ltd, Caxton Way, Thetford, Norfolk	(0482) 4545	SYSTEMS
United Air Specialists (UK) Ltd, Cranford, Blackdown, Leamington Spa, CV32 6RG	(0926) 311621	ELEC
Visco Ltd, Stafford Rd, Croydon, Surrey, CR9 4DT	(01) 686 3861	MED
Vokes Air Filters Ltd, Barden Lane, Burnley, Lancs, BB12 0DU	(0282) 51121	MED

C15 De-stratication heat recovery

Colt International Ltd, Havant, Hants, PO9 2LY	(0705) 451111	Ceiling fans
Crompton Parkinson Ltd, Woodlands House, The Avenue, Northampton, NN1 5BS	(0604) 30201	Ceiling Fans
Gradwood Ltd, Edgeley Road Trading Estate, Stockport, Cheshire, SK3 0TF	(061) 480 9629	Units
J. J. Ventilation (International) Ltd, 13 Dowry Square, Bristol, BS8 4SL	(0272) 291295	Units
Martin Roberts Ltd, Sittingbourne, Kent, ME10 3JH	(0795) 76161	Units
Wm May (Ashton) Ltd, Cavendish St, Ashton-under-Lyne, Lancs, OL6 7BR	(061) 330 3838	Units 'Parmet'
Mechanical Services (International) Ltd, Wilment Way, Avonmouth, Bristol, BS11 8DJ	(0272) 821415	Units 'Horizon'
Millfield Engineering Ltd, 88 George Lane, South Woodford, London, E18 1JJ	(01) 989 0194	Units 'Genatherm.
The Package Heater Co Ltd, 331/335 Chiswick High Rd, London, W4 4HS	(01) 995 3752	Units 'Recoupak'
Redwood Industrial Products Ltd, Unit 14, Hither Green Industrial Estate, Clevedon, Avon, BS21 6XQ	(0272) 874591	Units
Telford Tools and Equipment Ltd, Unit A1, Haldane, Halesfield, Telford, Salop	(0952) 584882	'Strinex' heat syphon system
Thermoflo (Northern) Ltd, Higher Hillgate, Stockport, Cheshire, SK1 3QH	(061) 480 0317	Units 'Econotherm'
United Air Specialists (UK) Ltd, Cranford, Blackdown, Leamington SPA, CV32 6RG	(0926) 311621	Units 'Heat hog'

C16 Insulation and insulating materials

Symbols: BLDG = bats and panels for low 'U' value buildings, ROOF = panels and treatments for roof insulation, CAVITY = cavity filling, PIPE = pipe insulation, DRAUGHT = draught prevention seals, H. TEMP = furnace and kiln insulation, SYST = specialists in insulation application and systems, SOLAR = solar control films

Acousticabs Industrial Noise Control Ltd, Unit B1, Ebor Industrial Estate, Hallifield Rd, York, YO3 7XQ	(0904) 36441	SOLAR 'Isoflex'
ACR Thermal Insulation, Rollesby Rd, Kings Lynn, Norfolk, PE30 4LN	(0553) 63371	PIPE, 'Zippon'
B. R. Ainsworth and Co Ltd, Bracol House, 6 Thames Rd, Barking, Essex, IG11 0HZ	(01) 594 7277	SYST
Alcan Building Materials Ltd, Blackpole Trading Estate, Worcester, WR3 8TJ	(0905) 54030	BLDG Polyurethane panels

Armstrong World Products Ltd, 3 Chequers Square, Uxbridge, Middlesex, UB8 1NG	(0895) 51122	PIPE, BLDG
Baxenden Chemical Co Ltd, Paragon Works, Baxenden near Accrington, Lancs, BB5 2SL	(0254) 381631	Polyurethane and polyisocyanurate foams
Berkeley Invicta (UK) Ltd, Maidstone Rd, Matfield, Tonbridge, Kent	(0892) 722202	Urethane glass fibres
Bonwyke Ltd, Bonwyke House, 41 Redlands Rd, Fareham, Hants, PO14 1HL	(0329) 289621	SOLAR
R. N. Bradley, Brook Saw Mills, Bradshaw Brow, Bradshaw, Bolton, BL2 3EY	(0204) 51821	DRAUGHT
British Cork Mills Ltd, Vulcan St, Bootle, Merseyside, L20 4HL	(051) 922 1917	BLDG, PIPE
British Gypsum Ltd, Ruddington, Hall, Loughborough Rd, Ruddington, Notts, NG11 6LX	(0602) 844844	BLDG, PIPE, ROOF, SYST
Bulten-Kanthal Stephen Newall Ltd, 14B Shepcote Way, Tinsley Industrial Estate, Sheffield, S9 1TH	(0742) 446021	BLDG (bulk fibre)
Camfine Ltd, Barton under Needwood, Burton-upon-Trent, Staffs, DE13 8DG	(0283) 713638	ROOF (Retrofit)
Capricorn Industrial Services Ltd, 1 Sugar House Lane, Stratford, London E15 2QN	(01) 519 4933	
Coolag Ltd, Charlestown, Glossop, Derby	(04574) 61611	BLDG (rigid foam)
Craig Alford Ltd, 18 Tresham Rd, Orton, Southgate, Peterborough, PE2 0SG	(0733) 234750	BLDG (adhesive panels)
M. H. Detrick Co Ltd, 275/281 King St, Hammersmith, London, W6 9LZ	(01) 748 5056	H. TEMP
Dow Chemical Co Ltd, Swan Office Centre, 1508 Coventry Rd, Yardley, Birmingham, B258AD	(021) 707 2565	BLDG (polystyrene)
Duratherm Insulation, Reading Rd, Henley-on-Thames, Oxon, RG9 1EJ	(0491) 576275	Polycarbonate secondary glazing
Ecomax (UK) Ltd, Stone Circle Rd, Round Spinney, Northampton, NN3 4RA	(0604) 47143	BLDG, PIPE (Rockwool)
Euromatic Ltd, Maycrete House, Boston Manor Rd, Brentford, Middlesex, TW8 9JQ	(01) 560 6372	Ping-pong' balls for liquid surfaces
European Profiles Ltd, Llandybie, Ammanford, Dyfed, Wales SA18 3JQ	(0269) 850691	
Evode Roofing Ltd, Common Rd, Stafford, Staffs, ST16 3EH	(0785) 57755	ROOF (Retrofit)
Fibreglass Ltd, Insulation Division, St Helens, Merseyside, WA10 3TR	(0744) 24022	Fibreglass materials
Fibre Seal Ltd, Fernie Rd, Market Harborough, Leicestershire	(0858) 64414	Roof sealing
Stuart Forbes Insulation Ltd, 119/120 Maybury Road, Woking, Surrey, GU21 5JL	(04862) 70202	SYST
Foseco Steel Mills (Ltd), Drayton Manor, Tamworth, Staffs, B78 3TL	(0827) 289999	H. TEMP
Freeman Insulation Ltd, Willowcroft Works, Broad Lane, Cottenham, Cambridge, CB4 4SW	(0954) 50155	SYST
Imperial Chemical Industries Ltd, Mond Division, PO Box 8, The Heath, Runcorn, Cheshire	(09285) 73456	H. TEMP, 'SAFFIL'
Indenden Adhesives Ltd, 7 Brook Trading Estate, Deadbrook Lane, Aldershot, Hants, GU12 4XB	(0252) 311608	PIPE
Isodan (UK) Ltd, 12 Mount Ephraim Rd, Tunbridge Wells, Kent, TN1 1EE	(0892) 44822	CAVITY (granules)
Kay-Metzeler Ltd, 16/18 Robert Way, Wickford, Essex, SS11 8DG	(03744) 66301	DRAUGHT
Kitsons Insulation Products Ltd, 119 Cleppington Rd, Dundee, DD3 7NU	(0382) 455422	BLDG, PIPE, H. TEMP, SYST
Kleeneze Industrial Ltd, Martins Rd, Hanham, Bristol, BS15 3DY	(0272) 670861	DRAUGHT
Kooltherm Insulation Products Ltd, PO Box 3, Charlestown, Glossop, Derbyshire SK13 8LE	(04574) 61611	BLDG, PIPE
Micropore Insulation Ltd, 1 Arrowebrook Rd, Upton Wirral, Merseyside, L49 1SX	(051) 677 0131	H. TEMP
Morganite Ceramic Fibres Ltd, Tebay Rd, Bromborough, Wirral, Merseyside, L62 3PH	(051) 334 4030	H. TEMP
MPK Insulation Ltd, Hythe Works, Colchester, Essex, CO2 8JU	(0206) 573191	H. TEMP
Northlite Insulation Services Ltd, Webner Industrial Estate, Ettingshall Rd, Wolverhampton, WV2 2LD	(0902) 49400	ROOF (external pvc secondary glazing)
Pilkington Flat Glass Ltd, St Helens, Merseyside, WA10 3TT	(0744) 28882	Double glazing

Plasmore Ltd, PO Box 44, Womersley Rd, Knottingley, W Yorkshire, WF11 0DN	(0977) 83221	BLDG (blocks)
Protexulate Ltd, Curtis Rd, Dorking, Surrey, RH4 1XA	(0306) 886688	BLDG (Latex foam panels)
Rentokil Industrial Insulation Service, Felcourt, E Grinstead, Sussex, RH19 2JY	(0342) 833022	CAVITY
RMC Panel Products Ltd, Waldorf Way, Denby Dale Rd, Wakefield, Yorkshire, WF2 8HD	(0924) 362081	Window Shutters
H. H. Robertson (UK) Ltd, Cromwell Rd, Ellesmere Port, Cheshire, L65 4DS	(051) 355 3622	ROOF (Retrofit)
Robseal Ltd, 75/87 Eastcourt Ave, Earley, Reading, Berkshire	(0734) 661121	SYST
Rockwool Ltd, Pencoed, Bridgend, Mid Glamorgran, CF35 6NY	(0656) 862621	BLDG
Roehm Ltd, Makrolon Sheet Division, 18/19 Bermondsey Trading Estate, Rotherhithe New Rd, London, SE16 3LL	(01) 237 2236	BLDG (sheet material)
Ruberoid Contracts Ltd, Thomas St, Cirencester, Glos, GL7 2EW	(0285) 61281	ROOF
Scandura Ltd, PO Box 18, Cleckheaton, W Yorkshire, BD19 3UJ	(0274) 875711	H. TEMP
Schlegel Engineering (UK) Ltd, Henlow Industrial Estate, Henlow Camp, Beds, SG16 6DS	(0462) 815500	DRAUGHT
Sheffield Insulation Systems, Welland Close, Parkwood Industrial Estate, Rutland Rd, Sheffield, S3 9QY	(0742) 752881	BLDG
Suncell Ltd, Royal House, Datchet Rd, Windsor, Berks, SL4 1SL	(07535) 69661	SOLAR
TBA Industrial Products Ltd, PO Box 40, Rochdale, Lancs, OL12 7EQ	(0706) 47422	Glass fibre fabric
Thermalite Ltd, Station Rd, Coleshill, Birmingham, B46 1HP	(0675) 62081	BLDG (blocks)
Tremco Ltd, Key House, 199 Horton Rd, W Drayton, Middlesex, UB7 8HP	(08954) 40641	Weather proofing

C17 Lighting and lighting control

Symbols: LAMP = manufacturers of all main types of lamp, FTGS = lighting fittings and luminaires, PHO = photoelectric controls, SYST = control systems for energy management of lighting.

Allen-Martin Electronics Ltd, Marlborough Works, Thompson Ave, Wolverhampton, WV2 3NP	(0902) 58942	SYST
Beblec Ltd, 42 Station Rd, Taunton, Somerset, TA1 1NL	(0823) 54655	Impedance Dimming
Courtney Pope Lighting Ltd, Amhurst Park Works, Tottenham, N15 6RB	(01) 800 1270	Dimming controls
Crompton Parkinson Ltd, Woodlands House, The Avenue, Cliftonville, Northampton, NN1 5BS	(0604) 30201	LAMP, FTGS
Davis Engineering Ltd, 4/6 Rookwood Way, Haverhill, Suffolk, CB9 8PB	(0440) 4411	Thyractor converters mercury to H.P. sodium
Edison Halo Ltd, Eskdale Rd, Uxbridge Industrial Estate, Uxbridge, Middlesex, UB8 2RT	(0895) 56561	Task lighting specialists
Energy Conservation Systems Ltd, Gresham House, Twickenham Rd, Feltham, Middlesex, TW13 6HA	(01) 894 5511	SYST
Floorplan Electrical Ltd, Switchplan Division, Unit 7, Strawberry Lane, Willenhall, West Midlands, WV13 3RS	(0902) 61469	SYST
Gray-Campling Ltd, Magnalux House, Southcote Rd, Bournemouth, BH1 3SW	(0202) 291828	FTGS (vapour proof)
Harvey Hubbell Ltd, Ronald Close, Chantry Industrial Estate, Kempston, Bedford, MK42 7SH	(0234) 855444	FTGS
Holophane Europe Ltd, Bond Ave, Bletchley, Milton Keynes, MK1 1JG	(0908) 74661	FTGS
Horstmann Engineering Products Ltd, Newbridge Works, Bath, BA1 3EF	(0225) 21141	PHO
Londex Ltd, Oakfield Rd, London, SE20 8EW	(01) 659 2424	PHO
Marlin Lighting Ltd, Hansworth Trading Estate, Hampton Rd, West Feltham, Middlesex, TW13 6DR	(01) 894 5522	FTGS
Osram (GEC) Ltd, PO Box 17, East Lane, Wembley, Middlesex, HA9 7PG	(01) 904 4321	LAMP, FGTS

Philips Lighting Ltd, 420/430 London Rd, Croydon, CR9 3QR	(01) 689 2166	LAMP, FTGS
Royce Thompson Electric Ltd, 320 Cheapside, Birmingham, B5 6AX	(021) 622 7441	PHO
Simplex Lighting Ltd, Groveland Rd, Tipton, W Midlands, DY4 7XB	(021) 557 2828	FTGS
Superswitch Electric Applicances Ltd, 7 Station Trading Estate, Blackwater, Camberley, Surrey, GU17 9AH	(0276) 34556	SYST
Thorn-EMI Lighting Ltd, Commercial House, Lawrence Rd, London, N15 4EG	(01) 802 3151	LAMP, FTGS

C18 Industrial doors

Symbols: FOLD = folding, ROLL = roller, SLIDE = sliding, O'HEAD = 'up and over', AUTO = automatic operation, TRUCK = fork lift truck access, STRIP = strip curtains, A.C. = air curtains, REV = revolving, GEAR = automatic gear or thrusters, DOCK = dock shelters and loading bays.

Acme Doors Ltd, Great West Road, Brentford Middlesex, TW8 9AT	(01) 560 2233	FOLD, ROLL, STRIP
Air Curtain Engineering Ltd, 250/256 St Anns Rd, London, N15 5BN	(01) 802 1840	A.C.
Auto Doors International, Denton Island, Newhaven, Sussex, BN9 9BA	(07912) 3023	SLIDE, AUTO
Bahco Ventilation Ltd, Bahco House, Beaumont Rd, Banbury, Oxon, OX16 7TB	(44295) 57461	A.C.
Bell Industrial Services Ltd, 39 Clifton St, Lytham St Annes, Lancs, FY8 5ER	(0253) 738285	A.C.
Besam Ltd, Halesfield 19, Telford, Shropshire, TF7 4QT	(0952) 587001	GEAR
F. H. Biddle Ltd, Newton Rd, Nuneaton, Warwicks, CV11 4HP	(0203) 384233	A.C.
Bolton Gate, Turton St, Bolton, BL1 2SP	(0204) 32111	FOLD
Bostwick Doors (UK) Ltd, Mersey Industrial Estate, Stockport, Cheshire, SK4 3ED	(061) 442 7227	FOLD, ROLL, SLIDE
Cape Insulation Services Ltd, Rosanne House, Bridge Rd, Welwyn Garden City, AL8 6UE	(07073) 311155	SLIDE, AUTO, STRIP
Clark Door Ltd, Willow Holme, Carlisle, CAL 5RR	(0228) 22321	FOLD, SLIDE, AUTO, (also insulated)
Countrywide Aircurtains Ltd, Unit 2, Thorpe Way, Banbury, Oxon	(0295) 51638	A.C.
Covrad Ltd, Sir Henry Parkes Road, Canley, Coventry, CV5 6BN	(0203) 75544	A.C.
Crawford Door Ltd, Milton Rd, Milton, Stoke-on-Trent, Staffs	(078253) 5922	FOLD, ROLL, O'HEAD (also insulated)
Crisvale Ltd, 20 Orchard St, Crawley, Sussex	(0293) 511207	STRIP
Easilift Materials Handling Ltd, Spring Gove, Penistone Rd, Kirkburton, Huddersfield, HD8 0PL	(0484) 605235	DOCK
Ellard Sliding Door Gears Ltd, Works Road, Letchworth, Herts, SG6 1NN	(04626) 78421	STRIP, GEAR
Flexible Reinforcements Ltd, Queensway House, Clitheroe, Lancs, BB7 1AU	(0200) 25241	STRIP
Gradwood Ltd, Edgeley Road Trading Estate, Stockport, Cheshire, SK3 0TF	(061) 480 9629	A.C.
The Harefield Rubber Co Ltd, Bell Works, Harefield, Middlesex, UB9 6HG	(08958) 22551	TRUCK, STRIP, GEAR
Henderson Doors Ltd, Romford, Essex, RM3 8UL	(04023) 45555	ROLL, SLIDE, AUTO
Hillaldam Coburn Ltd, Red Lion Rd, Surbiton, Surrey, KT6 7RE	(01) 397 5151	FOLD, SLIDE, O'HEAD, STRIP
Mandor Engineering Ltd, Turner St, Ashton-under-Lyne, Lancs, OL6 8LU	(061) 330 6837	TRUCK, STRIP
Marley Buildings Ltd, Guildford, Surrey, GU3 1LS	(0483) 69922	FOLD, O'HEAD, GEAR
Martin Roberts Ltd, Sittingbourne Industrial Park, Sittingbourne, Kent, ME10 3JH	(07957) 6161	FOLD, ROLL, SLIDE, STRIP

250

Myson Engineering Co Ltd, 16 Radford Crescent, Billericay, Essex, CM12 0DG	(02774) 52177	A.C.
Neville Watts and Co Ltd, Quality House, 8/10 Fitzwilliam St, Sheffield, S1 3DU	(0742) 78831	TRUCK, STRIP
Newman-Tonks Engineering Ltd, Hospital St, Birmingham, B19 2YG	(021) 359 3221	GEAR
P. J. Air Curtains Ltd, 4A Hillingdon Parade, Uxbridge Rd, Hillingdon, Middlesex, UB10 0PE	(0895) 39250	A.C.
Pollards Industrial Doors Ltd, Tower Lane, Warmley, Bristol, BS15 2YT	(0272) 614141	FOLD, ROLL, STRIP, (also fire protection)
Ress International (Europe) Ltd, 10 Gordon St, Luton, Beds, LU1 2QP	(0582) 412354	REV
Shearflow Phoenix, Shearflow Works, 61 Markfield Rd, Tottenham, London, N15 4RE	(01) 808 0571	A.C.
Stokvis Industrial Doors Ltd, Poole Rd, East Molesey, Surrey, KT8 0HN	(01) 941 1212	FOLD, ROLL, SLIDE, O'HEAD, AUTO, TRUCK, STRIP, DOCK
Strip Doors Ltd, PO Box 36, Church St, Wellington, Telford, Salop, TF1 1DJ	(0952) 52134	STRIP
Thomas-Davenport Doors Ltd, Unit 11, Gainsborough Trading Estate, Rufford Rd, Stourbridge, W Midlands, DY9 7ND	(02993) 75318	TRUCK
Thermoscreens Ltd, Chandlersford Industrial Estate, Eastleigh, Hants, SO5 3DQ	(04215) 4731	A.C.

Late entries

C5

Vulcana Gas Appliances Ltd. Spindle Way, Crawley, W. Sussex RH1D 1HP	(0293) 24129	BAL

C7

Absolute Energy Systems and Engineering Products Ltd, 'Longfield', Birdham Rd, Nr W. Wittering, Chichester, Sussex, PO20 8QA	(0243) 512268	SYST
Aldridge Air Control Ltd, Middlemore Lane, Aldridge, W. Midlands W59 8SP	(0922) 56333	EX SYST
The APV Co., PO Box 4, Crawley, W. Sussex RH10 2QB	(0293) 27777	SYST
Bayliss Kenton Installations Ltd, Harwood St, Blackburn BB1 3DW	(0254) 60011	SPRAY RECUPERATION
Endless Energy Ltd, Century House, 56 Endless St, Salisbury, Wilts SP1 3UQ	(0722) 332153	SYST
Eurovent Environmental Contral Ltd, Middlemore Lane, Aldridge, W. Midlands WS9 8Sp	(0922) 56336	EX
Hayden Carrier-Ross Ltd, Carolyn House, Dingwall Rd, Croydon CR9 2SH	(01) 686 0477	EX SYST
Hubbard Heat Recovery, 26 Perivale Industrial Park, Horsenden Lane, South Perivale, Greenford, Middlesex UB6 7RJ	(01) 991 1881	SYST
Recuperator Ltd, 224 Station Rd, Kings Heath, Birmingham B14 7TE	(021) 444 7909	EX
Silkeborg Ltd, 415 Oakshott Place, Walton Summitt, Bamber Bridge, Preston PR5 8AT	(0772) 38116	'Pasilac' (Denmark)

C8

Atmospheric Control Engineering Ltd, St Anne's House, North St, Radcliffe, Manchester M26 9RN	(061) 724 9511	EMS
Delta Technical Services Ltd, Asser House, Airport Services Rd, Portsmouth, Hants PO3 5RA	(0705) 697321	EMS
Envirosystems Ltd, Hampsfell Rd, Grange-over-Sands, Cumbria LA11 6BE	(044 84) 4233	EMS

C9

Energy Technology and Control Ltd, 25 North St, Lewes, E. Sussex BN7 2PE	(07916) 6101	CHEM

C11

Ancom Ltd, Devonshire St, Cheltenham GL50 3LT	(0242) 513861	TEMP, HVM
Crest Energy Ltd, Hollins Lane, Marple, Stockport, Cheshire SK6 6AW	(061) 449 9096	ELEC, REC

Euro Electronics Ltd, Troyman House, 31 Camden Road, London NW1 1TE	(01) 267 5416	ELEC
C14		
Courtaulds Engineering Ltd, PO Box 11, Foleshill Rd, Coventry CV6 5AB	(0203) 88771	CATALYTIC OXIDATION
C16		
Alsto Film Ltd, Unit E2, Kingsditch Lane, Cheltenham	(0242) 521170	SOLAR
Compriband Ltd, Townfield Rd, Shoeburyness, Essex SS3 9QQ	(03708) 2341	DRAUGHT
Vencil Resil Ltd, Arndale House, Arndale Centre, Dartford, Kent DA1 2HT	(0322) 27299	CAVITY
C17		
Concord Controls Ltd, Unit 3, Dawson Rd, Mount Farm, Milton Keynes MK1 1LH	(0908) 644366	SYST
Electronic Control Systems (Southern) Ltd, Elcon House, Cambridge Rd, Orwell SG8 5QD	(0223) 208078	SYST
Endless Energy Ltd, Century House, 54 Endless St, Salisbury, Wilts SP1 3UQ	(0722) 332153	SYST
Moorlite Electrical Ltd, Burlington St, Ashton-under-Lyne, Lancs, OL7 0AX	(061) 330 6811	FTGS, SYST
C18		
Avon Industrial Doors, PO Box 1, Wooton-under-Edge, Glos GL12 8BG	(0454) 24762	FOLD, ROLL, O'HEAD
Pollastrip Ltd, Bath Rd, Bilton, Bristol BS15 6HZ	(0275) 884354	STRIP

Section D

Bibliography and Sources

D1 Sources of data

'Handbook of chemistry and physics', Ed. Robert C. Weast, CRC Press Inc., Florida, USA
'Tables of physical and chemical constants', G. W. C. Kaye and T. H. Laby, Longmans, London
'Technical data on fuel', Rose and Cooper, British National Committee, World Energy Conference
'Industrial oil fuels and LPG: properties and useful data', (MOR 817), Shell International Petroleum Co Ltd
'Handbook of useful data', Wanson Co Ltd
'Flow of fluids through valves, fittings and pipes', Publication 410M, Crane Ltd
'Systems balancing, heating and chilled water circuits', Publication 806A, Crane Ltd
'Chemistry data book (SI edition)', J. G. Stark and H. G. Wallace, John Murray, London (1975)

D2 Chartered Institution of Building Services Publications Guides

A3 Thermal properties of building structures
A4 Air infiltration
A5 Thermal response of buildings
B2 Ventilation and air conditioning – requirements
B3 Ventilation and air conditioning – systems and equipment
B13 Combustion systems
B14 Refrigeration and heat rejection
C1/2 Properties of humid air, water and steam
C3 Heat transfer
C4 Flow of fluids in pipes and ducts
C5 Fuels and combustion
Psychrometric charts: $-10\,°C$ to $60\,°C$; $10\,°C$ to $120\,°C$
Code for interior lighting

D3 Suggested further reading

'Heating services design', McLaughlin, McLean and Bonthron, Butterworth, London
'Refrigeration and air conditioning', A. R. Trott, McGraw-Hill, New York
'Applications of thermodynamics', Bernard D. Wood, Addison Wesley, New York

'Heat transfer', I. P. Holman, McGraw-Hill, New York
'Fundamentals of heat transfer', Lindon C. Thomas, Prentice-Hall, New York
'Applied fluid mechanics', Robert L. Mott, Charles C. Merril
'Management guide to modern industrial lighting', Lyons, Butterworth, London
'Woods practical guide to fan engineering', B. B. Daly, Woods of Colchester
'Industrial drying', A. Williams-Gardner, L. Brooks
'Thermal comfort analysis and applications in environmental engineering', P. O. Fanger, McGraw-Hill, New York
'The energy manager's handbook', G. A. Payne, Butterworth, London
'Energy management', W. R. Murphy and G. McKay, Butterworth, London
'The efficient use of energy', I. G. C. Dryden, Butterworth, London
'Physicochemical quantities and units: the grammar and spelling of physical chemistry', M. L. McGlashan, Royal Institute of Chemistry, London

D4 Manufacturers who have contributed data and information from their literature

AAF Ltd
Acoustics and Environmetrics Ltd
Airflow Developments Ltd
British Gypsum (White Book)
Bonwyke Ltd
Calor Gas Ltd
Crosse Engineering Ltd
Dantherm Ltd
Dick de Leeuw Co
Ductwork Engineering Systems
Duratherm Insulation
Eurisol-UK (The Association of British Manufacturers of Mineral Fibres)
Fishbach Ventilation Ltd
General Engineering Radcliffe (1979) Ltd
Glynwed Integrated Services Ld
Heat Transfer Ltd
Hydrovane Ltd
Imperial Chemical Industries PLC
Information Transmission Ltd
A. Johnson and Co Ltd
Morganite Ceramic Fibres Ltd
Morgan Refractories Ltd
Myson Brooks Ltd
Osram–G.E.C.
Radial and Axial Fans Ltd
Roof Units Group
S and P Coil Products Ltd
Sekomak Ltd Air Products
Spirax Sarco Ltd
Stelrad Group Ltd
Thermecon
Thermocell Ltd
Thorn Lighting
Thorn–E.M.I. Industrial Boilers
Westair
Woods of Colchester Ltd

Index